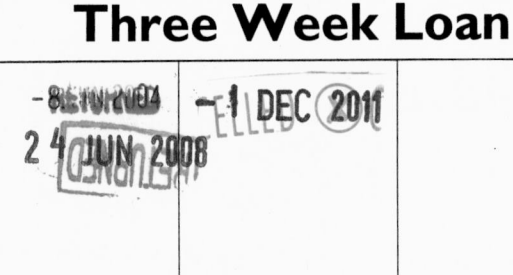

Technological Change in Agriculture

Also by Dominic Hogg

THE SAP IN THE FOREST: The Social and Environmental Impacts of Structural Adjustment Programmes in the Philippines, Ghana and Guyana

Technological Change in Agriculture

Locking in to Genetic Uniformity

Dominic Hogg
Senior Consultant
ECOTEC Research and Consultancy
Birmingham

palgrave

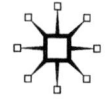

Published by PALGRAVE
Houndmills, Basingstoke, Hampshire RG21 6XS and
175 Fifth Avenue, New York, N. Y. 10010
Companies and representatives throughout the world

PALGRAVE is the new global academic imprint of St. Martin's Press LLC Scholarly and Reference Division and Palgrave Publishers Ltd (formerly Macmillan Press Ltd).

Outside North America
ISBN 0–333–75139–6

In North America
ISBN 0–312–22751–5

This book is printed on paper suitable for recycling and made from fully managed and sustained forest sources.

A catalogue record for this book is available from the British Library.

Library of Congress Catalog Card Number: 99–36207

10 9 8 7 6 5 4 3
08 07 06 05 04 03 02 01

Printed and bound in Great Britain by
Antony Rowe Ltd, Eastbourne

Contents

List of Tables and Figures

Tables

Figures

Preface and Postscript

This book arose out of concern for the way in which the food we consume is produced. The starting point was concern for the impact of modern agriculture on environment and health, documentation of which begins with the foundations of what has been termed industrial agriculture. This body of literature forces us to confront the question, why have such methods been employed for so long? This question has taken on added significance in the context of debates concerning 'sustainable development', and the emergence or, more properly, re-emergence of a variety of schools of 'sustainable agriculture'.

Despite interest in these 'alternatives', agriculture has stuck steadfastly to a path of increasing industrialisation and, arguably, an unsustainable exhaustion of soil, genetic resources, potash and nitrates. In many countries, it is a major consumer of increasingly scarce water resources. The efficiency of conversion of energy is low in industrial agricultural systems, and emissions of greenhouse gases are significant. Why, when the intellectual arguments in favour of sustainability appear to have been won, is agriculture still practised in what seems to be a manifestly unsustainable manner? The main reason cited in the literature is concern for global food supplies. Indeed, in such discussions, industrial agriculture has been shrouded in what Rosenberg (1982, 28), borrowing a phrase coined by Fogel, calls 'the axiom of indispensability'. Setting aside for one moment the thorny question of distribution (which no thorough investigation should), it is common to read of the need for more fertiliser and more pesticides in agriculture if the world's growing population is to be fed. But is the choice really so stark that we have no option but to fight, in Beck's words, 'the Devil of hunger with the Beelzebub of multiplying risks?' And if the answer might be 'no', why has this way of doing agriculture been able to fend off the strident criticisms of environmentalists for so long?

Questions of such enormity demand some degree of focus. The key to why modern crop agriculture produces so many negative externalities (to use the economists' language) lies not with one or other individual input, but in the pattern of their integration, which is to say, the particular technique employed. Increasingly pivotal in this integration is the genetic make-up of crops, in particular, their uniformity. Without such uniformity, the synchronisation of tasks and their mechanisation over large areas would not be possible. With it, the use of pesticides becomes more or less inevitable.

There are other reasons for asking whether things could be other than they are whilst still meeting the necessary production criterion to ensure the world's growing population is fed. The first is that as uniform crops replace diverse ones, important genetic resources, perhaps crops also, are lost which could otherwise be of critical importance in the future, particularly (and again somewhat ironically) if current breeding practices are maintained. Secondly, and somewhat more alarmingly, genetic uniformity implies vulnerability to crop losses. Historically, although there are many examples of crop losses, incidents leading to major loss of life are relatively rare, the most notable exception being the case of the potato famine in Ireland. However, crop losses in the future may range over greater geographical areas precisely *because of* increasing genetic uniformity. It is, therefore, *possible* that losses may be greater in magnitude, and, notwithstanding the ease with which goods now flow from one country to another, in impact, particularly if world food stocks fall to low levels (which they are likely to do as population grows). The hype surrounding the possibilities for biotechnology to improve matters in respect of global food production must be set in the context of these increasingly global developments in the population genetics of agricultural crops.

The point of departure for this book is that things *could* be other than they are, and that there are 'ways of doing and organising' agriculture that are more sustainable, and in particular, less genetically uniform. An implied claim is that the growing population of the world *could* be fed through agricultural techniques which are more sustainable. Evidently, this is a somewhat speculative position which cannot be grounded in hard evidence. Furthermore, industrial agriculture has received in excess of fifty years head start (in terms of continuous research support) that makes any comparison between it and any alternative somewhat unfair. One might argue, on a more moderate platform, that just as the lack of diversity within agricultural systems may prove problematic, so the lack of diversity in approaches to developing agricultural techniques limits the options available to cultivators, and may continue to do so in the future.

The plausibility of things being other they are is supported by recognition of the fact that the course of history is less well determined than accounts benefiting from hindsight tend to suggest. In particular, the history of technology is often presented as a history of technologies that become widely used without so much as a sideways glance at those technologies, the alternatives, from which the chosen one eventually emerges. The history of unsuccessful technologies, and of the reasons for their lack of success, remains largely untold. The same can be said of the avenues explored by research organisations which underpin the production of new technologies – counterfactual paths are never accorded the significance that they merit. Choices, once made, cannot simply be made differently with the benefit of hindsight. Each choice, once made, alters the course of history and shapes the menu of choices from which a new future will be moulded. These two facets of history, the element of choice, and the element of irreversibility implied by choice, make it necessary to employ the language of path-dependence when examining the history of research and technology. This work seeks to develop an understanding of technological change in agriculture as a path-dependent phenomenon.

The pre-eminent orthodox theory of technical change in agriculture, that of induced innovation, has long frustrated critics on the basis of its being too heavily determined by markets. I would say that its principal failing is that it fails to account for the importance of history in shaping technical change. The framework employed in this book helps answer the question posed at the beginning of this Preface, that of why technologies which are, arguably, damaging have been used in agriculture for so long (a question which, I argue, inducement theories could never even pose, let alone answer). The pursuit of one path in preference to other competing possibilities can lead to a situation where alternative paths effectively become locked out by factors that are institutional and systemic in nature. With this perspective in mind, it becomes easier to understand why it is that many environmental problems are so difficult to resolve, even when it is agreed that they are problems, and even when, narrowly defined, technical alternatives are, for all intents and purposes, available. Support is given to this framework through three case studies, two of which are historical, and one of which is current. These case studies seek to show how choices were, or are being made that effectively led, or are leading, to approaches to crop development based on genetic uniformity being chosen in preference to alternatives which implied, or imply, less genetically uniform agricultural systems.

At a somewhat different level, the book seeks to plug a gap by bringing an ecologically informed view to evolutionary economics. It surprises me that a discipline that relies so heavily on biological metaphors should show such a lack of concern for the very subject from which the metaphor was borrowed. Similarly, it can also be argued that ecological economists have failed in the main to inject evolutionary perspectives into their work. Ecological perspectives in economics are not new, and nor are evolutionary ones. Both are witnessing a revival and it would seem tragic that they should not in some way learn from each other, particularly since both have the merit of encouraging interdisciplinary approaches to the problems they seek to understand and resolve. It seems appropriate, therefore, that this book should seek some unity in these fields through a study of what one might call evolutionary ecological economics.

The book has a number of shortcomings. Distributional issues are largely ignored in the discussion in order to maintain the principal focus of the study. On the one hand, industrial agriculture may enable incomes in agriculture to keep pace with those elsewhere in the economy (hence the significance of 'livelihood' perspectives), whilst on the other, it may introduce new inequalities within agriculture. A separate piece of work of much broader scope is also needed to re-assess the linkages between agriculture and other sectors of the economy. Orthodox theories of the role played by agriculture in development, mostly fashioned in an era of full employment, seem to be increasingly irrelevant in a world where unemployment is of global concern, and where changes in agriculture appear to be creating factories in the field.

The book also suffers in its attempt to apply one theory to a broad range of technologies, and over a wide geographical range. The justification for these generalisations rests upon the view that agriculture is becoming more homogeneous across the globe. The techniques of production are founded upon the use of a combination of inputs which vary little in their function and in their pattern of use across a wide range of crops and environments. This process, however, though difficult to reverse, may have encouraged the emergence of alternatives, partly in reaction to the dominant, homogenising tendency. The issue, therefore, is less one of whether agriculture is becoming more standardised – it is – but more one of degree, and the extent to which alternative approaches to that under examination are making headway. Some argue passionately on behalf of the alternatives. I do so too, but I argue even more strongly that the composition of the world in which we live is such that however good our arguments, and however unpalatable the extrapolation of what passes for business-as-usual, we will not be listened to. This is, if you will, the sensory dimension of the phenomenon of lock-in I am seeking to explore.

I have not explored in depth the question of whether modern agriculture causes environmental damage. Opinions vary as to the extent to which this is the case. My own view is that it does, and this view underlies the question from which this work arose. However difficult the problems encountered by economists who would seek to reduce the complex consequences of modern agriculture to a single monetary numeraire, I maintain that one only needs to sit down in a room with five sensible people to arrive at agreement on this matter. We surely cannot formulate policy on the basis of a numbers game (awaiting for economists to determine our fate like anxious Saturday night TV addicts waiting for a lottery result) so we have no choice but to exercise our judgement in this matter. In any case, the omission does not unduly affect the rest of the argument which seeks to understand how things have come to be as they are. The emphasis is on how uniformity has been encouraged and how alternative strategies have been marginalised. In such a situation, even in the presence of economically viable alternatives which are more benign environmentally, modern high-external input agriculture, with uniformity its keystone, will be resilient to change.

Chapter 1 explores the significance of uniformity and its disappearance from agricultural systems. It argues, admittedly on the basis of selective citation, that the argument that the world can only be fed through industrial agriculture may be false. It thus raises questions as to why one way of doing agriculture has spread so widely. Chapter 2 turns to orthodox theories of technical change in agriculture and seeks to highlight some of their shortcomings. Chapter 3 seeks to construct a framework for understanding technical change in agriculture through the concepts of technological trajectories, path-dependence and lock-in. Some of the ways in which agriculture could become locked in to a particular way of doing agriculture are explored in Chapter 4.

Three case studies follow in Chapters 5, 6 and 7. These look, respectively, at the emergence of hybrid maize in the US in the early decades of this century, the early days of the Green Revolution in Mexico, and the changes being set in motion by biotechniques, and in particular, changes in intellectual property rights legislation. The book ends with a Conclusion that seeks to interpret some of the evidence from the case studies in the light of the framework developed in earlier chapters.

* * *

This book is essentially a PhD thesis with minor revisions. Through all but the first year of my thesis, I benefited from the guidance of Dr (now Professor) Peter Nolan. Not only has he enabled me to see things from many different perspectives, but his own willingness to approach all issues with an open mind has been an exemplary model to which any student would aspire. Learning itself being a path-dependent process, Peter has helped to make the path easier and more interesting to explore. Our discussions have always been fruitful ones, helping to sharpen my often raw and flawed ideas. I hope that through reflecting upon these (to my great benefit), he has learned at least a fraction of what I have learned from him.

I am especially grateful to my fellow students, and other participants in the Global Security Programme at Cambridge University, especially Dr Don Hubert, whose critical comments and friendship have been of great importance to me. Dr Jennifer Clapp and Dr Jim Whitman provided invaluable guidance along the way, and Chris Roberts produced the elegant Figures. Thanks are also due to Dave Hoisington of CIMMYT for arranging such an interesting programme for me in Mexico. I am indebted to the MacArthur Foundation for the PhD scholarship I received through the Global Security Programme. My wife Rita has had to put up with an attic room full of books and debris which accompany my attempts at scholarship. She deserves special thanks for her patience and understanding.

DOMINIC HOGG

Postscript

Because the subject of the final chapter of this book has been evolving since this work was drafted, some statements now appear incorrect. In particular a European Directive (98/44/EC) on the legal protection of biotechnological innovations was passed in July 1998. My view is that the arguments put forward are still valid despite, or because of, ongoing changes.

D. H.

List of Acronyms and Abbreviations

ABN	Agricultural Biotechnology Network
AD1	Agricultural Development (first edition)
AD2	Agricultural Development (second edition)
ADAS	Agricultural Development and Advisory Service
AR	Agricultural research
ASTA	American Seed Trade Association
BCM	Breeding, chemical and mechanisation mode
BML	Biotechniques-mechanisation-legislative mode
BPI	Bureau of Plant Industry
CADI	Centre for Agricultural Development Initiatives
CBD	Convention on Biological Diversity
CGIAR	Consultative Group for International Agricultural Research
CIAT	Centro Internacional de Agricultura Tropical
CIMMYT	Centro Internacional de Mejoramiento de Maiz y Trigo
CONASUR	Cooperación Agricola de los Paises del Area Sur (Agricultural Co-operation of the Countries of the Southern area of Latin America)
DNA	Dioxyribonucleic acid
DUS	Distinct, uniform and stable
ENDS	Environmental Data Services
EU	European Union
FAO	Food and Agriculture Organisation of the United Nations
FARO	Formal agricultural research organisation
FSR	Farming systems research
GATT	General Agreement on Tariffs and Trade
GSP	Generalised System of Preferences
GRAIN	Genetic Resources Action International (Barcelona)
H&R	Yujiro Hayami and Vernon Ruttan
HDRA	Henry Doubleday Research Association
HEIA	High external input agriculture
IBP	International Biological Programme
IBPGR	Internatonal Board for Plant Genetic Resources
ICPGR	International Committee for Plant Genetic Resources
IFOAM	International Federation of Organic Agriculture Movements

IFS	Integrated Farming System
IIA	Instituto de Investigaciones Agricolas
INIA	Instituto Nacional de Investigaciones Agricolas
INM	Integrated nutrient management
IPC	Innovation possibility curve
IPM	Integrated pest management
IPR	Intellectual property rights
IRRI	International Rice Research Institute
IUPGR	International Undertaking on Plant Genetic Resources
LEIA	Low external input agriculture
LEISA	Low external input sustainable agriculture
LER	Land equivalent ratio
MAFF	Ministry of Agriculture, Fisheries and Food (UK)
MAP	Mexican Agricultural Programme
NGO	Non-government organisation
NIAB	National Institute of Agricultural Botany
NIH	National Institutes of Health (US)
NRC	National Research Council
OECD	Organisation for Economic Co-operation and Development
OEE	Officina de Estudios Especiales (Mexico)
OES	Office of Experiment Stations
OTA	Office for Technology Assessment (US Congress)
PBR	Plant Breeders Rights
PPA	Plant Patent Act
PTO	Patent and Trademark Office
PVPA	Plant Variety Protection Act
QPM	Quality protein maize
RAFI	Rural Advancement Foundation International
R&D	Research and development
RF	Rockefeller Foundation
RYT	Relative yield total
SEASAN	Southeast Asian Sustainable Agriculture Network
TRIPs	Trade Related Intellectual Property Rights
UK	United Kingdom
UN	United Nations
UNCTC	United Nations Centre on Transnational Corporations
UNDP	United Nations Development Programme
UNEP	United Nations Environment Programme
UPOV	Union for the Protection of New Varieties and Plants
US	United States
USDA	United States Department of Agriculture

VCU	Value for cultivation and use
WSAA	World Sustainable Agriculture Association
WTO	World Trade Organisation

1
Genetic Diversity in Agriculture: Its Rise, Fall and Significance

The rise and fall of genetic diversity in agriculture

The development of diversity

For more than ten thousand years, human beings have sought to transform their environment to ensure that their basic food requirements are met. Through agriculture, societies have directed the evolutionary process of animals and crops.[1] The criteria for selection of crop and animal varieties for agricultural purposes substantially changes the selection pressures to which these organisms are exposed. Few domesticated crops would survive in the absence of human husbandry or cultivation.[2] They have been adapting to, and have been selected for their suitability in, agricultural systems which have changed over time in their rationale, geographical extent and location. Agricultural crops (and livestock) co-evolve with humans.[3]

It is worth considering what conditions are necessary in order for crops to adapt to changed socio-economic and environmental conditions. Natural selection in 'the wild' requires that there be diversity on which selection pressures can act. The process of selection imparts to adaptation a genetic, and therefore heritable, base.[4] Those combinations that are selected will constitute the best part of the genetic make-up of subsequent generations, resulting in the development of ecotypes adapted to local ecological conditions. Ecotypes are made up of populations which are not uniform genetically, but are characterised by the frequency with which different alleles of genes occur within that population (Holden *et al.* 1993, 28; Simmonds 1979, ch. 1).

Changes in environmental conditions can be catastrophic for a given ecotype lacking the diversity necessary for adaptation. In most wild plants, a degree of diversity is maintained within the ecotype through genetic recombination *via* cross-fertilisation. As long as the available genetic diversity allows new adaptive combinations to be generated, so the possibility of new ecotypes adapted to new environmental conditions arises. Ultimately, a species comes to be composed of a series of ecotypes, the variation usually arising from differences in the frequencies with which particular genes are found in the populations rather than in the genes themselves. Similar considerations apply to cultivated crops under conditions of domestication. It is not so simple to talk of these activities as imposing 'additional' selection pressures since some selection mechanisms which would operate in the absence of human intervention are effectively selected against by human beings.[5] The factors determining the development of a particular agro-ecotype, or landrace, became intimately connected with human decision-making.

As human beings moved around the globe, they took crops with them, exposing them to changed selection pressures, leading to the emergence of new landraces.[6] This led to the development of diversity in cultivated plants and associated 'weeds'.[7] On the other hand, the wild relatives of domesticates dispersed unassisted by human intervention. Where varieties were in contact with related wild species, hybridisation and introgression continually led to gene flow between species. New species were probably domesticated along the way, and in some cases, geographical movement caused the attention of cultivators to switch their attention away from the crop once it had reached the limits of its adaptive range, and towards a 'weed' associated with the crop.

Agricultural technologies as experiments

Farmers have continuously, and quite consciously, experimented in their fields with crops.[8] Most observers have overlooked this process, emphasising instead innovations that have originated in laboratories and field stations of research organisations and corporations established specifically for that purpose. Yet, '[P]robably, the total genetic change achieved by farmers over the millenia was far greater than that achieved by the last hundred or two years of more systematic science-based effort' (Simmonds 1979, 11). In suggesting that developing country farmers were keen to utilise new technology if only it were available to them, a view deemed radical at a time when developing country farmers were visualised as both backward and conservative, Theodore Schultz (1964, ch. 2) implied that the technology they were employing was not changing. Similarly, Jackson and Ford-Lloyd (1990, 3) in suggesting that artificial selection by humans was 'probably of largely unconscious nature', seem to devalue the contribution of early farmers in developing crop species.

Probably, Cox and Atkins (1979, 67–8) are right to suggest some traits were selected for consciously, others by default. Nevertheless, the fact that seed selection has been practised all over the globe for millennia supports the view that one does not need a rigorous understanding of genetics to recognise value in such practices, and as Biggs and Clay (1981, 323–4) point out, farmers work in continually changing environments.[9] Without experimenting with crop varieties, most of the plants which today we recognise as crops would never have evolved to their current form.

Diversity in decline

Since the mid-nineteenth century, an increasing amount of research into plant breeding has taken place in organisations set up partly for that purpose. As plant breeding became more successful, producing cultivars matching the needs of an increasingly market oriented agriculture (expansion of which was assisted by breeding), so there began a decline in the genetic diversity of crops in the field. Holden *et al.* (1993, 29) suggest that crop genetic diversity probably peaked around the middle of the nineteenth century. The double bind which faces those who have sought to replace landraces, or diverse crop mixtures, with uniform 'improved varieties' is that the genetic diversity on which their activity depends may be lost as a consequence of their very success. This is one way in which 'genetic erosion' occurs.

The value of landraces for the future of agriculture was recognised by two German scientists, von Proskowetz and Schindler, at the end of the nineteenth century (Vellvé 1992, 25). Nikolai Vavilov (1951) then laid the basis for work on centres of origin and diversity of crop plants, suggesting that the genetic diversity of a given crop coincided with its centre of origin. An exact correspondence has been refuted, and today Vavilov's localised centres are discussed in terms of centres and non-centres (Harlan 1971) which cover significant geographical areas.[10] However, the idea that some areas of the world exhibit great diversity for particular crops continues to be widely accepted, and is of crucial importance to breeders today as genetic erosion gathers pace, even in these areas.[11] Loss of diversity has taken place both between and within species, and as a result, it can be argued, in the range of techniques employed in agriculture: 'the mosaic appears to be disappearing' (Van der Ploeg 1992, 20, also 21–6). As Wilkes notes, a shift from growing crops for consumption to growing crops for sale has resulted in:

> the dropping of many legume crops (20–35 per cent protein) which have been classic protein sources in the human diet for more fertiliser/water-yield responsive cereals (7–14 per cent protein). The current situation is a rapid reversal of eight millennia of *in situ* evolution where the cultivators ate the crops they cultivated. (Wilkes 1977a, 313)

Today, of more than 200 000 plant species currently known to scientists,[12] only 30 are used to supply humans with the bulk of their food (Holden *et al.* 1993, 13; Sattaur 1989, 41). If one was to compile a list of the top five crops in every country, only 130 species would be named (Fowler and Mooney 1990, xi). It should come as no surprise that there are many 'unknown' crops ideally suited for production and consumption purposes in the regions where they have been selected.[13] Of those crops that are widely grown, the extent of diversity exhibited within the species is also narrowing. This entails risks in terms of the food supply's vulnerability to catastrophic losses, though as with many other ecological problems, quantitative estimates of such risks are subject to great uncertainty. Even if the probability of such epidemics occurring were low (which we do not know), they may have severe consequences: 'Loss of genetic diversity – silent, rapid, inexorable – is leading us to a rendezvous with extinction – to the doorstep of hunger on a scale we refuse to imagine' (Fowler and Mooney 1990, ix). Given the prevailing uncertainty, it is equally valid to depict optimists as complacent as it is to brand the authors of the above words as scaremongers. Such is the state of knowledge about the system which feeds many of us.[14]

Estimates as to the extent of genetic erosion tend not to take on a reliable quantifiable form, partly because by the time the problem began to be addressed, much diversity had already disappeared. Although the loss of varieties is nothing new, it is the pace of recent changes that is without precedent:

> Since World War II, for example, virtually all of the local wheat varieties in Greece, Italy and Cyprus have been abandoned and most of the indigenous sorghum (*Sorghum bicolor*) races of South Africa disappeared after the release of high-yielding Texas hybrids. (Plucknett *et al.* 1987, 11–12)

But if the fact that genetic erosion is occurring is beyond doubt, the manner in which it occurs is disputed. Even those who emphasise the contributions made by traditional agriculture through crop selection have tended to fall into the trap of believing that the use of modern varieties always implies loss of traditional varieties. That they do so is surprising given that many seek to highlight the wisdom of indigenous practices. Why would such wise cultivators simply discard varieties they have so carefully selected and take a leap into the unknown? The work of Brush and others (Bellon 1991; Bellon and Brush 1994; Brush 1991, 1992, 1994; Brush *et al.* 1992; Dennis 1987) suggests that in some cases they do not:

> The dismissal of *in situ* conservation is based on an extrapolation from temperate or from atypical regions of the tropics. This reasoning rests on conventional ideas about modernisation, which envision a linear replacement of earlier stages of agricultural development by more advanced ones ... we should re-examine our assumptions about the loss of diversity, thus opening the plausibility of on-site conservation. (Brush 1991, 155–6)

How new varieties are introduced and used in agroecosystems varies from place to place and from farmer to farmer. It is not impossible for modern varieties to augment diversity, or lead to the production of novel populations (Wellhausen 1978; Dennis 1987; Brush 1992). Where this occurs, it reflects *diversity's value in use*. Even where replacement does occur, the value of discarded varieties may subsequently become apparent.

The risks of genetic uniformity

Why is genetic erosion a problem? In traditional agricultural systems, epidemics are rare, unnoticed, undocumented, or all three. This is partly due to the genetic diversity, and hence, flexibility retained in the agricultural system. Genetically uniform systems lack this flexibility (Sydor 1976; Perrin 1977; Simmonds 1979, 262; Thurston 1992, 9). The more uniform the crop over space and time, and the less frequent are environmental conditions deleterious to particular pests and pathogens (they are most infrequent in the tropics), the more likely it becomes that an epidemic will occur, and occur over a large area. Modern agricultural systems are characterised by their uniformity. Ecologically, they are artificial in the extreme, and the level of inter- and intra-specific diversity is, in terms of the system's ecological vulnerability, worryingly low

(Cox and Atkins 1979, ch. 6). The possibility that pest epidemics will emerge increases due to the strong selection pressure exerted in favour of pests which can overcome any resistance that the crop may have had.[15]

In 1970, nearly a quarter of the US maize crop was wiped out by southern corn blight owing to the uniformity of the cytoplasmic DNA of hybrids making each equally susceptible to the fungus *Helminthosporium maydis* (see Chapter 5; also Allaby 1973,181; NRC 1972, ch. 1; Clunies-Ross 1995, 5–6; Raeburn 1995, 127–45). A subsequent National Academy of Sciences investigation concluded that, not just in maize, United States agriculture was, 'impressively uniform genetically and impressively vulnerable' (NRC 1972, 1).[16] Among those alarmed by developments was Jack Harlan, whose father Harry had warned as long ago as 1936 that:

> When new barleys replace those grown by the farmers of Ethiopia or Tibet, the world will have lost something irreplaceable. When that day comes then our collections, constituting as they do but a small fraction of the world's barleys, will assume an importance now hard to visualise. (Harlan and Martini 1936, 317)

The significance of Harry Harlan's comments lay in his awareness that genetic variation is what breeders require to develop new varieties, and lost varieties may contain alleles of genes whose importance cannot be estimated in advance.

Well before Harlan made these observations, the same uniformity had led to:

- the potato famine in Ireland (Clunies-Ross 1995, 5, 11–12; Raeburn 1995, 121–4);
- the French wine industry being struck by powdery mildew in 1848 (Raeburn 1995, 137);
- the destruction of the coffee industry in Ceylon (then the world's largest producer) in the late nineteenth century (Allaby 1973, 180; also, on Ethiopia's coffee, Robinson 1996, ch. 21); and
- two wheatless days a week in the US in 1917 due to red rust (Allaby 1973, 180).

Since the corn blight episode, uniformity has led, amongst others, to problems with rust in the UK wheat crop in 1972 owing to widespread planting of two wheat varieties (supposedly resistant to rust); orange wheat blossom midge in the UK in 1993 (Clunies-Ross 1995, 36); rice blast in South Korea in 1980 when 75 per cent of rice lands were sown with modern varieties of similar pedigree (forcing the country to import rice) (Plucknett *et al.* 1987, 16); Russian wheat aphid in a number of countries including the US (Raeburn 1995, 47–58); and growing problems in 1990s US and the UK with the fungus responsible for the Irish potato famine, *Phytophthora infestans*.[17] Of all these episodes, only the case of potatoes in Ireland can be said to have created a famine. Yet the potential exists for widespread crop losses, and in a world where global food stocks are shrinking, the likelihood of such events occurring in the future may be increasing.

The standard approach to dealing with vulnerability associated with genetic uniformity has been to breed for pathotype specific (vertical) resistance to prevent losses due to a particular ecotype of a pathogen. For the last fifty years or so, this 'disadaptive breeding' (Simmonds 1979, 266) has been the approach of choice for formal agricultural research. Yet vertical resistance exerts the strongest selection pressure in favour of 'matching' pathotypes,[18] limiting its useful life. Because the search for vertical resistance makes the degree of horizontal (or pathotype non-specific – see below) resistance difficult to detect, the level of horizontal resistance tends to be lower than is desirable, resulting in a boom-bust cycle. This is the so-called vertifolia effect. Breeders partake in what has become known as a varietal relay-race in their efforts to pre-empt the devastation of crops by pests and pathogens.

In situations where vertical resistance is matched, or where vertical resistance cannot be found (because it may not exist),[19] the use of pesticides is unavoidable. The use of pesticides, in turn, has led to development of new ecotypes of old pests and pathogens which are resistant to chemical attack, as well as the emergence of new pests and pathogens that thrive in the absence of their natural enemies (Clunies-Ross and Hildyard 1992, 60–3; Dudley 1987, ch. 8; Bull 1982).[20] When spraying to kill some pests, particularly those whose vectors move easily between fields, farmers create positive benefits for themselves, and a positive externality for other farmers in that they help to contain populations. Yet they may also incur costs to themselves in terms of health impacts, and lost productivity as a result, as well as two types of negative externality. The first relates to environmental pollution (including human health – see Antle and Pingali (1994)). The second relates to the fact that pest resistance builds up at a rate determined in part by the volume of the active ingredient used (Kishor 1992, 6). All farmers are adversely affected in the medium term through yield loss owing to existing resistance being overcome. Uniformity is, therefore, a problem with at least two related dimensions; vulnerability, leading at best to market instability and loss of earnings, and at worst, to widespread hunger; and associated environmental pollution.

Alternative strategies to vertical resistance include:

- Breeding for pathotype non-specific (horizontal) resistance – horizontal resistance, though more difficult for plant breeders to achieve, is rather more durable.[21] Those favouring vertical and horizontal resistance can be said to be adopting, respectively, Mendelian or biometrical approaches to the problem (Robinson 1996, part 1). The problems associated with vertical resistance are increasingly well recognised and a shift is occurring towards horizontal resistance strategies (Simmonds 1979, 274; Burdon and Jarosz 1989, 293). However, horizontal resistance strategies are not without problems of their own, one of which is that it is not known how effective horizontal resistance strategies will be in the absence of agrochemical use.
- Diversifying the varietal make up of a crop in any given region – by planting varieties incorporating different vertical resistance genes in neighbouring fields, the possibilities of disease epidemics occurring are reduced. The method is most effective where an early variety is the source of reproduction for a pathogen of a late variety. Similarly, where a parasite is known to reside in a particular region in a given season and then spread in subsequent seasons, growing varieties with different resistance genes along its path may slow down, or prevent, development of related disease (NRC 1972, 64–5). In a slightly similar vein, Wilkes (1992, 51) talks of the use of 'genetic firebreaks', bands of diverse germplasm employed to limit the progress of any epidemic.
- Varietal rotation. Varieties incorporating different resistance mechanisms are grown once every few years and rotated with others.
- Incorporating, in a given field, diversity within varieties, between varieties, or between species.[22] At one end of this spectrum are multilines and mixtures, in which different vertical resistance genes are either backcrossed into a single genetic background or used in distinct varieties (Browning and Frey 1969; Wolfe 1981; Wolfe 1985; ABN Bulletin 1992; Clunies-Ross 1995, 23–4). At the other end are the diverse systems in home gardens, and of permaculture and agroecology (see below), where both inter- and intra-specific diversity are encouraged.

One can appreciate that these different recommendations require different responses from breeders, farmers and policy-makers. The methods vary in terms of their genetic flexibility. This may be a consideration that increases in importance if climates change quickly (in historical terms), causing 'environment erosion' of any resistance (vertical or horizontal) present in monocultures. Furthermore, use of diversity within the agroecosystem may offer the best prospects for eliminating the use of synthetic pesticides, though some advocates of horizontal resistance do so on the basis of its performance in this respect.

Risks associated with a narrow genetic base are reduced by widespread agrochemical use (with attendant consequences for the environment),[23] and by substantial investments by public and, increasingly, private organisations whose aim it is to keep hold of the batton in the varietal relay-race, either through developing new chemicals, or new varieties.[24] This places a considerable burden on these organisations to continually come up with the goods in the face of unexpected and unannounced attacks from pests and pathogens, as well as (possibly) changing climate. As pathogens overcome vertical resistance conferred by individual genes or become immune to chemical attack, and as genetic erosion continues in crop plants, weedy relatives of crops are assuming greater importance. Harlan (1976, 330) has been moved to comment that 'wild relatives stand between man and starvation.'[25]

The suggestion that genetic erosion is a problem does not inevitably lead to the recommendation of a re-direction of research in favour of practices that maintain, enhance, or encourage diversity *in situ*. Many advocate a more comprehensive, better described, and better co-ordinated system of *ex situ* conservation in genebanks (Arnold *et al.* 1986). Those who support this approach have confidence in *ex situ* conservation to provide adequate back-up for a breeding system which they implicitly trust to overcome any future problems.[26]

This position is increasingly being undermined by well-documented limitations of genebanks, leading to growing recognition that *ex situ* storage of germplasm is inadequate of itself to ensure a secure genetic foundation of agriculture into the future.[27] This has led some to advocate acknowledgement, and some form of reward system, for the role played by various farmers and communities around the world in maintaining crop genetic diversity *in situ*.[28] Within this school of thought, there is a difference in emphasis between those who advocate the conservation of genetic diversity *in situ* for use in future breeding programmes, and those who recognise diversity as a desirable property of agricultural systems in itself. The former view lends itself open to criticism on grounds that it can be taken to imply that *in situ* diversity would inevitably disappear unless prevented from doing so.[29] But the latter view, exemplified by the work of Brush and others, immediately suggests a need for reappraisal of the objectives of formal agricultural research. If diversity is of value in use, why do researchers not base their investigations more firmly on the incorporation of diversity at the inter- and intra-specific level? If farmers are capable of experimenting with novel germplasm, why do *ex situ* genebanks fail to promote an active exchange of their resources with farmers (Crucible Group 1994, 25–7)?

Is high-external-input agriculture the only way to feed the world?

Introduction

It is often stated, indeed, it is frequently taken as axiomatic that modern high-external-input agriculture (HEIA), unashamedly uniform genetically, constitutes the only possible way to feed the world's large and growing human population. There are many who believe that to retain genetic diversity, in whatever way, within the agricultural system necessarily implies an economic loss, or at least, a loss in production. Supporters of this view point to the achievements of the Green Revolution in Asia and parts of Latin America. Lipton and Longhurst (1989, 1) write: '[h]istory records no increase in food production that was remotely comparable in scale, speed, spread and duration'. This most probably is the case, yet the same could probably be said for the accompanying narrowing of the genetic base, and indeed, the 'scale', 'speed' and 'spread' factors each have negative ecological manifestations.

Many who discuss the Green Revolution tend to limit their number of counterfactuals to one, a continuation of the *status quo*. Is it to be understood that agricultural research could only work effectively if it concentrated on monocultures? Could production increases not have been achieved in any other way? Is it impossible to construct plausible counterfactuals paths for change which left agriculture less vulnerable in terms of its genetic inflexibility, less dependent on crop protection chemicals, and yet with the potential (in the sense that it meets the sufficiency criterion) to feed the world? It seems almost heretical to write these words, but even on grounds of productivity alone, it may be true to say that (and this may require qualification with respect to ecological regions) agroecosystems incorporating greater species diversity are the best performers in terms of output per unit area and/or economic performance (see below). At the same time, they can offer a number of environmental benefits relative to HEIA (depending on the system) which may include energy conservation, reduced agrochemical use, soil conservation and resource use efficiency.

Hence, there are three important, if for the most part obvious points to be made with respect to the position that HEIA is the only way to feed the world:

- modern agriculture does not have, as one of its aims, feeding the world. It is principally geared to making profits;
- farmers have a number of reasons for adopting one or other technology. Concern for global food supplies and their distribution does not weigh significantly in such considerations;[30] and
- nobody has, and nobody could, prove the truth of what is being stated. The choice with regard to approaches in agricultural research is far wider than is often understood, and suggesting that one or other is 'the best' (let alone 'the only') is likely to prove problematic, especially when so few of these approaches have been researched thoroughly.

There is reason to believe that more will be known about any given way of doing agriculture as more research is conducted upon it, raising the possibility for what we might term increasing returns to advocacy. In other words, if through gaining an early lead in the command of research resources, a particular approach improves its position relative to alternative approaches (I leave open for the moment discussion of what 'improvement' might mean), it is more likely to command research funds in the future. Whilst Plucknett (1994, 354–5) criticises those who argue the need for research on marginal lands (as opposed to the most productive lands), Lipton (1994, 611), mindful of the over-simplifications implied by researchers in dividing land into advanced or backward areas, asks the question:

> are the 'lead areas', often no longer the most promising (especially after much research with a long period of diminishing returns), but rather the areas that have been favoured by AR [agricultural research] in the past, and continue to be favoured because that is what scientists know how to do best?

Two questions need to be answered; why has one particular way of doing agriculture come to be so widely embraced the world over; and what are, what has happened to, and what might be the alternatives?

Comparing farming systems

In order to understand why it is at least somewhat odd that a particular way of doing agriculture has come to be accepted so broadly, it is worth considering some of the factors that might impinge upon the decision to farm in whichever way. The problems encountered when trying to compare farming systems are many and varied. They relate to:

- *Location specificity* – the only satisfactory way to compare two farming systems is not available to us. We cannot compare different systems at the same time on the same piece of land. Because the quality of land will be affected by past activities, a comparison conducted sequentially on the same piece of land will introduce errors. If the comparison is not to take on the form of an experimental test (and it is generally appreciated that research station performance differs widely from that in the field) one faces the problem of finding working farms which are not only similar in agroecological characteristics, but also in other respects considered below.
- *Motivations for farming* – farming systems are part of wider social systems and are embedded within a social context. Not all farmers have the same motivations, particularly if farming is not the primary source of income. Even where income maximisation processes are of importance, they may still take on different forms (see Low 1992). Many farmers will seek to optimise performance with respect to multiple criteria. Reintjes *et al.* (1992, 30–2) consider productivity, security, continuity and identity to be important objectives, any combination of which could constitute the guiding rationale for farming. Problems that might arise from a desire to control (if this were possible) for variation in motivations include the possibility that choice of farming technique may be, indeed almost certainly is, endogenous to these motivations.
- *Managerial capabilities* – that one farm 'does better' than another in a given comparison could simply be regarded as proof of the superiority of one farmer's management skills. But it is difficult to dissociate the issue of management from that of the farmer's objectives, and indeed, those of the researcher. One might also ask whether the adoption of one or other farming system should itself be considered an issue in measuring management capabilities. Furthermore, the adoption of one or other system might *require* of farmers that their management capability improves (Lockeretz 1989). Perversely, this might raise questions as to the wisdom of farmers deciding to practice more managerially demanding styles of agriculture. Lastly, the

different types of farming systems probably require completely different types of managerial skill, if that is the correct phrase. These considerations serve to caution us against treating technologies and techniques as isolated parts of what is actually a complex whole, in which the ability and resources available to manage the farm are often conditioned by the characteristics of the farm household. How the household functions as a unit, and how it interacts with the wider community, will affect the types of decision taken regarding farm management.

- *Duration of comparison* – farming has always been a risky enterprise, and although in different farming systems the principal source of those risks (and possibly where the burden of risks falls) is altered, it continues to be so. Issues of seasonality, year-to-year variation, and secular trends or even breakdowns in one or other variable in the system are of crucial importance. Most of these can only be revealed through a comparison which takes the longer-term into account. On the other hand, such studies of a longer-term nature as have been done have taken the form of experiments which, due to the degree of control that is desired, tend to ossify the production systems under investigation. This prevents any evolution in the system that might be desirable or necessary over time. Given that, for many systems, design and its improvement are of supreme importance, such comparisons lose relevance over the longer term.
- *Differences in systems* – farming systems may be so different that they simply are not comparable in any meaningful sense of the word. The sort of alternative systems considered above incorporate greater diversity on the farm, leading almost inevitably to a more diverse output than is characteristic of modern, highly-specialised farms. How can one really compare the output of such different systems, particularly when one also considers the many other factors mentioned above which mitigate against any easy comparison? Research methodologies which use the manipulation of a restricted number of variables as the basis for comparing organic and high-input agriculture have probably missed the point of, for example, organic agriculture altogether (Lampkin 1994b, 31). They suggest an inability to appreciate that in such agriculture, what is sought is system integration through the interaction of components, and this is a goal rather than an instrumental means to a particular end.
- *The comparison to be made* – to the extent that a focus on one or more indicators inevitably fails to consider all the dimensions of the farming system, any comparison will be less than complete, and

> therefore prone to biased conclusions. If such comparisons are attempted, it seems impossible to dissociate the comparison from the motivations of researchers. Additionally, if a comparison is made on the basis of chosen indicators, for the comparison to be a fair one, the farmers being compared should indicate an equal weighting of these indicators in their subjective assessment of farm performance.

Clearly, comparison of alternative farming systems is extremely difficult. Yet for all these difficulties, the homogenising tendencies of the expanding agro-food system are difficult to deny, still more so to resist (Van de Ploeg 1992, 21–5). Why do so many farmers apparently chose independently to farm in a broadly similar way the world over? Have the options open to them been inadequately (re)presented? To state that alternative options have received inadequate attention from formal agricultural research is partly to answer the question, 'why are these options called alternatives?' Consideration of this issue is important in the light of the genetic vulnerability inherent in the prevailing practice of agriculture, and the environmental problems this creates.

Schools of alternative agriculture

As applied to agriculture, my use of the term 'alternative' is culturally biased. In some regions, modern high-input agriculture is simply unfeasible and 'alternative' agriculture is all that is available. Furthermore, several alternative movements have referred to the validity of indigenous technologies which have, it goes without saying, pre-dated such movements by some considerable time.[31] Harwood (1990, 6) comments that the roots of alternative agriculture are to be found in the discourse evolving at the turn of the twentieth century juxtaposing holism and reductionism. Alternative agriculture's development parallelled that of modern agriculture, leaning on the insights of the growing number of soil scientists, and influenced by Darwin, who had a great interest in the biological attributes of soil. In the early twentieth century, several books were written which alluded to the complex interrelationships between components of the farming system that had been ignored or downplayed by agricultural scientists. The work of King (1927) and Howard (1940) on highly productive systems of agriculture in Asia helped reinforce this message.

However, alternative movements seem to have been severely affected by the Second World War. For practitioners of biodynamic agriculture, based on Steiner's anthroposophy, the impact came through ideological channels, as the National Socialists banned their activities.[32]

Harwood (1990, 9) cites a number of works which appeared in the 1940s and 1950s, a common theme of which was 'the increasing environmental harm and resource degradation brought about by 'modern' farming methods. Yet, for Harwood, the 1960s were a transition period of 'narrow focus', which saw the rise of modern agriculture and the parallel demise of alternatives in agricultural scientific discourse. Harwood's view is corroborated by the experience of different nations in the post World War Two era, during which the state has played a crucial role in the promotion of modern agricultural techniques.

Most interesting to note in this context is the way in which the prosaic critique by Rachel Carson (1965) was greeted by Harwood and his colleagues:

> there was little or no debate during those years, in the biological sciences at least, on development direction. The success of current technologies was so overwhelming that it stifled serious debate of alternatives. The alternative farming 'schools' were practically non-existent and certainly in disrepute. I remember clearly graduate school discussions in 1964 about the 'crackpot Rachel Carson and her whistling in the wind' against the great benefits of DDT. In looking back on those heady years, I wonder about our arrogance and narrowness of vision. I also wonder, parenthetically, if many of us still remain intellectually in the comfortable era of the early 1960s when we trained. (1990, 10)[33]

Since the early 1970s, a resurgence in awareness of the environmental impact of modern agricultural practices has taken root. In this context, many alternatives began to experience a revival, and others emerged. A re-appraisal of the relative validity and value of farmers' knowledge and that of research scientists was underway.[34] This stemmed partly from the appreciation of the environmental impact of modern agriculture, particularly in the tropics, but also from its failure to provide useful technologies and techniques for farmers cultivating marginal lands in the developing world. It seemed that agricultural researchers could cope with conditions where the degree of control over resources came close to that achieved in the laboratory, or in the countries from which they often came, but when faced with more complex environments, they had much to learn from cultivators though they were seldom willing to do so.[35]

It is difficult to delineate the different schools of alternative agriculture. Embracing most of the alternatives, and not necessarily excluding the use of external inputs, is what is known as low-external-input and sustainable agriculture (LEISA), an overview of which is given by Reintjes *et al.* (1992). They compare high-external-input agriculture (HEIA) with 'erosive forms of low-external-input agriculture (LEIA)', thus hoping to show that external inputs may in some cases be sufficiently convenient and important options that they should not be ignored. LEISA, then, is:

> ... agriculture that makes optimal use of locally available natural and human resources (such as soil, water, vegetation, local plants and animals, and human labour, knowledge and skills) and which is economically feasible, ecologically sound, culturally adapted and socially just. The use of external inputs is not excluded but is seen as complementary to the use of local resources and has to meet the above-mentioned criteria. Neither the conventional Western agricultural technology nor any alternative technology is completely embraced or condemned. The attempt is made, rather, to draw lessons from past experiences in agriculture in developing and industrialised countries and to merge them into a process of technology development which leads to LEISA. (Reintjes *et al.* 1992, xviii–xix)

Organic agriculture is often used, like LEISA, as a sort of umbrella term embracing a number of alternatives:

> A hypothetical High External Input farmer who wants (has) to become more 'environmentally friendly' will reduce inputs. This is what many would call 'integrated agriculture.' There are different grades, usually starting with Integrated Pest Management (IPM), adding Integrated Nutrient Management (INM), and making a real change when the farm becomes an Integrated Farming System (IFS). Continuing the quest to become more environmentally friendly and more sustainable, the farmers will move forward to LEISA when optimizing the IFS. The farmer will end up with an Organic Agriculture System, which to the conventional farmer's surprise, needs very few external inputs. (UNDP 1992, 3–4)[36]

Organic agriculture is the alternative that is best known and most practised in Europe and the United States. Many countries have certification bodies that award symbols for the produce of farms that qualify by dint of their following certain codes of practice. The certification bodies together form the International Federation of Organic Agricultural Movements (IFOAM), which has drawn up minimum production standards to which all members, in principle, subscribe (IFOAM 1992, 25–64). Typically, organic produce retails at a premium over other produce.

Organic agriculture has been gaining ground in recent years. A report prepared for the United States Department of Agriculture in 1980 noted that:

> Contrary to popular belief, most organic farmers have not regressed to agriculture as it was practised in the 1930s ... Most of the farmers with established organic systems reported that crop yields on a per-acre basis were comparable to those obtained on nearby chemical-intensive farms. (USDA 1980, xii)

And similar conclusions were reached in a report prepared nine years later by the National Academy of Sciences (NRC 1989). As Clunies-Ross and Hildyard (1992) point out, organic agriculture's roots early in the twentieth century mean that it addresses itself less directly to wider problems of modern society. The same cannot be said for permaculture, developed by Bill Mollison in Australia in the 1970s, which is more a philosophy than a way of doing agriculture. Its originator describes it as:

> the conscious design of agriculturally productive ecosystems which have the diversity, stability and resilience of natural ecosystems ... The philosophy of permaculture is one of working with, rather than against, nature; of protracted and thoughtful observation rather than protracted and thoughtless action; of looking at systems in all their functions, rather than asking only one yield of them; and of allowing systems to demonstrate their own evolutions. (Mollison 1988, ix–x)

This philosophy is distilled into design principles (Mollison 1988, 15–16).

Permaculture has no official global networking centre, though it is making headway in countries on all continents. There is a loose network of practitioners, many of whom are 'graduates' who have attended courses at Institutes of Permaculture. The aim of permaculture practitioners is to continuously improve on the design of their agro-ecosystems so that they become self-sustaining and self-maintaining.

The emphasis is not on any one species, but on multiple interactions between different species and the landscape. It is notable how permaculture practitioners tend to describe species in terms of their multiple outputs and inputs rather than in one-dimensional terms. With the emphasis on knowledge-based design, there is no stigma attached to either 'scientific' or 'traditional' knowledge. Ultimately, experimentation provides the tests of the usefulness of a species. For this reason, and those mentioned above, permaculture has proven readily transferable across nations and climates.

There is much in permaculture that is common to agroecology. Though based on the work of Klages and Papadaksis before the Second World War, only since the 1970s has there been a major expansion in the agroecological literature. This reflects the resurgence of interest in indigenous knowledge systems and the growing awareness of environmental impacts of human activities. Indeed, Hecht (1987, 1) is of the view that 'decentralised, locally developed agronomic knowledge is central to the continuing performance of these [agricultural] production systems', a view shared by Van der Ploeg (1992, 37) and Richards (1985, 141) among others. She observes the processes by which indigenous knowledge systems were destroyed or devalued, and the biases which led to their being overlooked by conventional agricultural science, a point developed further by Norgaard (1987). Observing that the odds were stacked against the 'rediscovery' of agroecology, she attributes this to:

> an unusual example of the impact of pre-existing technologies on the sciences, where critically important advances in the understanding of nature resulted from the decision of scientists to study what farmers had already learned how to do. (Hecht 1987, 4)[37]

Hecht also writes:

> Loosely defined, agroecology incorporates ideas about a more environmentally and socially sensitive approach to agriculture, one that focuses not only on production, but also on the ecological sustainability of the production system ... At its most narrow, agroecology refers to the study of purely ecological phenomena within the crop field, such as predator/prey relations, or crop/weed competition. (Hecht 1987, 4)

Agroecology is perhaps best associated with the works of Miguel Altieri and Stephen Gleissman. It is well-advanced in Latin America, where a coalition of non-governmental organisations (NGOs) has been formed under the banner of the Consorcio Latinoamericano sobre Agroecologia y Desarrollo (CLADES) in Chile (Altieri 1992). Other alternatives exist which have emerged/experienced a revival in the 1970s/1980s. For example, Fukuoka's work (1978; 1985), extending the concept of natural farming, has been an inspiration to many in the alternative movement. Fukuoka himself was inspired by the sight of healthy rice plants growing in uncultivated road verges, leading him to ask why so much labour was employed doing what could be done naturally.

As one moves progressively away from HEIA and monocultures, one finds diversity being strategically employed in the design and evolution of agroecosystems. Diversity can enable adaptation to the particular growing season in question. Richards (1989, 40) makes the point that farmers adapt crop mixes in accordance with 'sequential adjustment to unpredictable conditions', and do not plan their planting of crops in a once-off pre-season activity. Agriculture becomes a performance. The contrast with HEIA, with its scheduled plantings of genetically uniform monocrops, and by-the-calendar sprayings could hardly be more stark.

Performance of alternatives – the benefits of diversity

It is generally accepted that most of the alternatives considered above perform rather better than HEIA in terms of their environmental impact.[38] They are also more efficient in terms of calorific output.[39] Abstracting for the moment from the social and cultural relations of those on the farm, the reason why alternatives are viewed less than favourably by formal agricultural research is that they are supposedly less productive per unit of land than HEIA. Indeed, the supposedly superior productivity of HEIA is often used to suggest that, though damaging to the environment in the direct sense, it actually prevents degradation of land less suited for agriculture.[40]

In what follows, I will seek to set out the case for diversity in terms of the agroecosystem's yield.[41] The comparative literature in this respect is sketchy, partly for the reason which preoccupies me in this work, namely that there are limited funds devoted to alternatives.[42] In addition, researchers studying alternatives are more concerned with improving their own practices than with comparisons which they feel to be of limited value, and where comparisons are made, they often concern dimensions other than yield which are equally worthy of concern. I will admit of some selective citation, but feel justified in this given the fact that alternatives have not benefited from anything like the concentration of research devoted to HEIA. Had they done so, there might be more 'positive' examples with respect to yield. Broadening our criteria for judging the value of one or other system might make the body of evidence even more convincing.

Intercropping

I use Vandermeer's (1989) definition of intercropping, that is, cultivation of two or more species in such a way as they interact agronomically. Agroecosystems have, historically, incorporated diverse elements in the field and there appear to be advantages in doing so. There are several reasons why farmers choose to undertake intercropping, including yield; resource (soil nutrients, water and sunlight) use efficiency; plant complementarity;[43] pest and disease reduction;[44] weed suppression; spreading labour costs; insurance against individual crop failure; and dietary diversification (Altieri 1987a, 73–5; 1987b; Dover and Talbot 1987, 32–42; Thurston 1992, 159–60; Reintjes *et al.* 1992, 81–101). In Africa, 98 per cent of all cowpeas, the continent's most important legume, are grown in combination with other crops. Beans grown with other crops account for 90 per cent of bean production in Colombia, 73 per cent in Guatemala, and 80 per cent in Brazil (Dover and Talbot 1987, 32). These examples show how polyculture systems make use of the nitrogen-fixing property of leguminous plants, a major source of protein, to fertilise companion crops (Wilkes 1977a).

The yield performance of polycultures is usually expressed in terms of land equivalent ratios (LERs), or relative yield totals (RYTs). The LER/RYT of a polyculture can be understood as the amount of land planted to monocultures that would be required to grow the same amount of material as is grown on one hectare of polyculture. An LER greater than 1 indicates 'overyielding'. Trenbath (1974) found that around 20 per cent of 572 experiments that he carried out on crop mixtures were more productive than monocultures, a statistic biased downward by the fact that the work was experimental, and did not concentrate on mixtures in widespread use (Gleissman, cited in Dover and Talbot 1987, 33).

Amador (1980, cited in Gleissman 1986) compared yields of polycultures composed of corns, beans and squash with monocultures planted at four different densities in Cardenas, Tabasco, in Mexico (Table 1.1). This polyculture system, with LER much greater than one, has been widely practised in Latin America for centuries (often with additional crops). Yet most formal agricultural research on corn has focused on monoculture.[45]

Table 1.1 Yields and total biomass of maize, beans and squash (kg/ha); planted in polyculture as compared to several densities (plants/ha) of each crop in monoculture

Crop	*Monoculture*				*Polyculture*
Maize					
Density	33 300	40 000	66 600	100 000	50 000
Yield	990	1 150	1 230	1 170	1 720
Biomass	2 823	3 119	4 478	4 871	5 927
Beans					
Density	56 800	64 000	100 000	133 200	40 000
Yield	425	740	610	695	110
Biomass	853	895	843	1 390	253
Squash					
Density	1 200	1 875	7 500	30 000	3 330
Yield	15	250	430	225	80
Biomass	241	941	1 254	802	478
Total polyculture yield					1 910
Total polyculture biomass					6 659
*Land Equivalent Ratios (LER)**					
Based on yield					1.73
Based on biomass					1.78

* LER = sum [(yield or biomass of each crop in polyculture)/(maximum yield or biomass of each crop in monoculture)]

Source: Amador (1980), in Gleissman (1986).

Liebman (1987, 117) cites many examples of polyculture systems that overyield. He goes on to indicate that profitability in one study of various systems was between 42 and 149 per cent greater for polyculture systems than for monoculture. Also, Mutsaers (1978, cited in Dover and Talbot 1987, 34) did work on maize and groundnut mixtures in Cameroon, which overyielded whether fertilised or unfertilised. The above evidence suggests that Hobbelink (1991, 140) may well be correct in commenting that:

> Perhaps the most important misconception about these complex farming systems is the claim that they tend to produce less than monocultures. They might produce quantitatively less of one and the same crop, but generally the combinations yield far more.

Agroforestry

Agroforestry is not new, though to read much of the development literature, one might falsely deduce that it was 'discovered' in the 1980s. The integration of trees into agroecosystems is a practice as old as agriculture itself (Rocheleau *et al.* 1989, 14–15). The problems associated with monocropping, and the limitations of transferring such systems to the tropics, have led to a reassessment of the importance of trees in agroecosystems. Not unusually, farmers have sought to incorporate within their agro-ecosystems trees that perform multiple functions, including improvement of soil fertility and prevention of soil erosion, as well as other benefits (see Farrell 1990).

Alley cropping is a form of agroforestry in which rows of crops are grown between rows of trees and shrubs, usually fast growing leguminous species.[46] There appear to be several factors affecting the yield of plants grown under tree canopies (Farrell 1990, 178–82). Although soil properties appear to be improved in many, if not most cases, light is often a limiting factor in promoting crop growth. One reason why *Acacia albida* is believed to increase crop yields is that as well as its water holding and soil improving characteristics, it sheds its canopy for the understory growing season (see Felker and Bandurski 1979; Poschen 1986; Table 1.2).

Table 1.2 Effects of *Acacia albida* trees on soil characteristics and crop yields in Africa

Nutrients returned annually to the topsoil	*Amount (kg/ha)*
Element	
Nitrogen	186
Phosphorous	4
Potassium	76
Calcium	222
Magnesium	39
Increases in soil quality and yield under Acacia crowns	*Per cent increase*
Item	
Total soil nitrogen	33–110
Organic matter	40–269
Cation exchange capacity	50–120
Millet yields	37–104
Sorghum yields	105

Source: McGuahey (1986).

Dover and Talbot (1987, 46) cite an example where *Leucaena* leaf litter contributed some 200 kg of nitrogen per hectare, enabling a yield of 2.8 tonnnes per hectare of alley-cropped maize to be achieved without synthetic fertiliser. Behmel and Neumann (1981) discuss a project at Nyabasindu in Rwanda where an agroforestry system was built on the foundations of the knowledge of farmers in the area. Mixtures in this system provided 54 per cent more calories, 31 per cent more protein, and 62 per cent more carbohydrates than monocultures, as well as producing more fuelwood than was needed.

Multistory cropping and home gardens

Multistory cropping as practiced in home, or forest gardens is a form of polyculture/agroforestry deserving special attention. This is due not simply to their importance in tropical agriculture, but also to the incredible diversity they incorporate, and their yields. Javanese home gardens account for some 17 per cent of the country's agricultural land, and one survey of 351 such gardens revealed 607 species in use. The layered canopies can intercept as much as 99.75 per cent of sunlight (Dover and Talbot 1987, 32–3).[47] Similar systems are found in the Philippines. In one study by Clawson, a 416 m^2 plot produced more than 2 tons of a diverse range of products (see Table 1.3). The equivalent yield of maize alone is more than twice the Philippines national average for the 1989–91 period, and close to half that of the United States,[48] yet this constitutes little over 5 per cent of the edible biomass produced in this plot.

Gleissman (1990, 166) reports on the diversity of productive home gardens and their multi-functional roles. He makes the point that:

> The ability of an agroecosystem to respond to different factors or conditions in an environment, to meet the needs of the inhabitants for a great diversity of products and materials, and to respond to external socioeconomic demands are all very important components of a sustainable system. At the same time, we have become more aware of the need to find systems to lessen dependence on expensive, imported agricultural inputs, as well as to limit the environmental impact of the ways in which we farm. What we are learning from agroforestry systems, such as home gardens, is that there are practical, time-tested ways to combine all of these characteristics into manageable, productive, and sustainable farm units. Trees are the key elements in the functioning of the system.

Table 1.3 The productivity of a polyculture plot in Quezon City, Philippines, 1984

Crop	*Annual yields (kg, 416 m² plot)*	*Equivalent yields (kg per ha)*
Upper story		
Banana (*Musa paradisiaca*)	186	4471.15
Papaya (*Carica papaya L.*)	195	4687.50
Third story		
Cassava (*Manihot esculenta C.*)	184	4423.08
Maize (*Zea mays*)	125	3004.81
Sugar cane (*Saccarum off.*)	210	5048.08
Okra (*Hibiscus esculentus L.*)	24	576.92
Second story		
Taro (*Colocasia esculenta S.*)	200	4807.69
Arrowroot (*Marantha arudinacea*)	50	1201.92
Chile Pepper (*Capsicum annuum*)	8	192.31
Ground layer		
Swamp Cabbage (kangkong) (*Ipomoea aquatica*)	200	4807.69
Sweet Potato (*Ipomoea batatas*)	600	14 423.08
Squash (*Cucurbita maxima D.*)	75	1802.88
Total	2057	49 447.12

Source: Clawson 1985b, cited in Altieri (1987b).

Indigenous knowledge and varietal management

Altieri (1991, 96) cites an experiment in Peru where farmers were assisted in reconstructing raised field farms along the lines used by the Incas 3000 years ago:

> These 'waru-warus', which consisted of platforms of soil surrounded by ditches filled with water, were able to produce bumper crops in the face of floods, droughts and the killing frosts common at altitudes of almost 4000 metres.

Initial tests showed that yields outstripped those achieved on chemically-fertilised farms. Yields of 10 tonnes per hectare compared with the regional average of 1–4 tonnes. Silt, sediment, algae and plant and animal remains decay in the canals and can be applied as manure to the raised beds.

Hecht (1989) compares the Kayapo system of agriculture with colonist and livestock agriculture in Brazil and finds them superior in virtually every respect, including yield. A well-known project known as the Yurimaguas project is also compared with the Kayapo system, this being based on, or so Hecht suggests, a US Land Grant College style of agronomic investigation. Again, the Kayapo system seems to be superior. Clay *et al.* (1978) cite an example of indigenous varietal management in Bangladesh. Deepwater varieties of rice have been selected over time which yield around 4 tonnes of rice per hectare without the use of chemical fertilisers. These varieties exploit the nitrogen fixing property of algae in the water. The yield compares extremely favourably with any available alternative for this environment, and indeed with yields obtained in other environments in Bangladesh.

Salazar (1992) reports the experiments conducted at a rice station set up with support from the US and Belgian branches of Oxfam in which a traditional cultivar was compared in a variety of situations with a variety, IR42, developed by the International Rice Research Institute. The traditional variety gave comparable yields of rice, as well as superior production of straw. He gives other examples where, in the absence of chemical fertilisers and pesticides, traditional varieties outperform modern varieties.

Nature farming

Aggarwal (1990, 462) gives an account of changes that occurred in his village, Rasulia in Madhya Pradesh, once the community began to address problems of soil fertility decline and rural poverty. The community stopped using, 'chemical fertilisers and poisons. Some plants protested mildly, but then accepted the change. Mexican hybrid wheat tried, but was not able to do without chemicals. This, we found, was true of all highly engineered seeds.' The community devoted a small plot of their least productive land to the method at first, slowly increasing the area given over to the system. Despite teething problems,

> Some of our salient achievements are: yields of up to 20 quintals of paddy per acre; highly respectable yields of all other food crops except wheat; higher total production than under the previous, chemical-assisted system; a six to eightfold increase in net profits, and, most important of all, vast improvements in the health and fertility of our soil. (Aggarwal 1990, 463)

A similar story of a Thai farmer, Kumdueang Phasi, converting to Fukuoka's methods can be found in SEASAN (1992, 200–204). Kumdueang noticed no change in yields, but that the soil quality improved immeasurably, leading him to believe that future yields on the nature farming plot would improve relative to the chemically treated land which, he noted, continued to deteriorate. An example of nature farming was widely reported in Japan's daily newspapers in 1993. In a cold summer that wiped out many conventional rice crops, nature farming rice, in the words of Dr Hirokazu Nakai, 'managed to survive, comparable to a strong man bending in the face of adversity' (WSAA 1993/4, 1). The rice varieties grown were varieties long disregarded by modern Japanese farming.

Biodynamic agriculture

Schilthuis (1994, 60) compares two farms, the conventional and biodynamic, and finds the biodynamic outyields the conventional farm by more than 25 per cent (see Table 1.4).

Another interesting example comes from Perlas (1993):

> Lorenzo Jose, a small rice farmer in the Pampanga Province [Philippines], became one of the government's early Green Revolution heroes by producing a yield of over 8 tons of paddy rice per hectare on his 1.6 hectare plot. Yet less than ten years later, Mr. Jose found his soil so depleted that he had to apply four times more chemical fertilizer to

Table 1.4 Farming costs and yields: conventional and biodynamic (based on the annual report of the ministry's accounting service)

Costs and yields	*Tallhof farm*	*Comparable conventional farms in the district*
Expenses for fertilisers or materials for preparations and straw (DM/ha/year)	7.70	147.00
Yields: grains (kg/ha/year)	3600.00	2900.00
milk (kg/cow/year)	4399.00	3376.00
Bought-in concentrate (DM/cow/year)	35.00	225.00
Hectares per worker	10.80	9.70
Income per hectare (DM)	1800.00	1111.00
Income per worker per year	18750.00	10760.00

Source: Adapted from Schilthuis (1994).

> maintain his early yields. His soil had become sticky and difficult to plow. To control infestations of increasingly chemical resistant insects he had to continually increase insecticide applications. Wild fishes and snails, important protein sources, began to disappear. Returns no longer covered costs and his debts mounted. He was more prone to illness. His skin was itchy and wounds healed slowly.

Having asked for, and been denied assistance from government services to assist in converting to organic agriculture, Mr Jose sought the assistance of a non-government organisation, the Centre for Alternative Development Initiatives (CADI), and they began to implement biodynamic practices:

> In the first year of large-scale experimentation, one farmer who used the full spectrum of Ikapti [CADI's] technology harvested 6.5 tons per hectare, three times the regional average. One third of participants had yields that exceeded the Masangana 99 target and well over twice the provincial average. Nearly all had yields in excess of the average for chemical farmers. The enhanced flavor and aroma of the biodynamically grown rice brought premium prices, while input costs were substantially lower, resulting in net profits in some instances more than two and a half times those of typical chemical farmers. (Perlas 1993)

These examples show the relevance of biodynamic agriculture to developed and developing world, temperate and tropical climates alike.

Organic agriculture

An increasing number of studies have been carried out comparing organic and conventional farming in the developed country setting.[49] Most of these, to the extent that they consider yields, do so in the context of a comparison of financial performance of the two systems. Over time, the debates over comparisons have become considerably more sophisticated as interest has grown in organic farming. Many of the problems mentioned above in comparing systems are recognised (yet persist regardless of the increase in data available).

On the whole, yield performance of organic agriculture is not as great as in modern HEIA. Whilst this is not universally the case, most have been led to conclude from these studies that the gap between yields does appear to increase with the intensity of use of external inputs both by crop, over time, and by country/region (Padel and Lampkin 1994). Others believe that size of grain also is an important factor, with organic farms' yield performance relatively better for small grains than for large. As a result, relative yield performance tends to be better for small grains and in regions/countries where synthetic chemical input use is less intense. The latter is an important consideration when one considers the changing policy context of developed country farming. Various forms of farm subsidy have promoted a high level of input use, but these are now being scaled back, partly (or so it is stated) owing to environmental concerns. This is one reason why interest in organic farming, even amongst those practising HEIA, is growing.

Table 1.5 gives a flavour of the performance indicators in terms of yield. As mentioned above, this gives an unfavourable picture of organic farming's financial performance since many of the outputs are awarded premium prices (in Pretty and Howes (1993, 18–19), the 'sustainable' farms in the UK yield, on average, some 15 per cent less, but gross margins are greater with few exceptions). Whether or not these premia would persist under more widespread conversion to organic farming is a key factor influencing whether farmers decide to adopt organic practices. Many (especially developed country) organic farmers (remembering that this term embraces many types of farming) do not employ great intraspecific diversity in the field, and consequently, their place as an alternative in the context of this thesis is somewhat questionable.

In the developing country context, 21 case studies of 'organic farming' are considered in UNDP (1992). The case studies are varied, and include cases of agroforestry. With the exception of two cases, diversity increased on the farm. The evidence regarding yields led the study to conclude that: 'yields obtained under organic farming are clearly higher than in traditional farming ... When compared with HEIA systems, yields are lower [with one exception]' (UNDP 1992, 170). But the benefits to farmers are invariably greater than the systems with which they are compared (one of only two systems where this was not the case was, interestingly, the instance in which organic farming outyielded HEIA). The overall conclusion of the study is as follows:

> The question, then, is whether organic agriculture presents an attractive alternative to current non-sustainable practices. The answer is a qualified 'yes' ... both in high potential areas and marginal lands, organic agriculture offers agronomically feasible solutions for problems of environmental sustainability. The feasibility includes a positive microeconomic effect for most of the producers involved, and possibly leads to positive effects on regional and national economies as well. (UNDP 1992, 183–4)

Table 1.5 Yields from organic agriculture as compared with conventional counterparts

UNITED KINGDOM

(i) Soil Association in conjunction with C.U. Dept. of Land Economy (Figures are for Winter Wheat)

	case study 1		*case study 2*		*case study 3*		*Average (Soil Assoc.)*		*Average (Camb. Survey)*		*Soil Assoc. as % Camb Survey*	
	1972	*1973*	*1972*	*1973*	*1972*	*1973*	*1972*	*1973*	*1972*	*1973*	*1972*	*1973*
Yield (cwt/acre)	33.4	33.7	36.2	38.3	32.2	31.9	33.9	34.6	38	34.8	89%	99%
Gross output (£/acre)	71.48	136.3	74.9	155.9	63.7	118.2	70.02	136.8	62.9	69.6	111%	197%
Total variable costs	3.61	8	10.8	13.4	5.2	6.31	6.53	9.23	12.8	15.5	51%	60%
Gross margin/acre	67.87	128.3	64.1	142.5	58.5	111.89	63.49	127.57	50.1	54.1	127%	236%

Source: *Quarterly Review of the Soil Association*, March 1977.

(ii) Study at Haughley, Suffolk, carried out at Pye Research Centre – carried out by C.U. Dept. of Land Economy

	Organic section		*Chemical section*		*Organic yield*
	Yield (t/ha)	*Gross margin (£/ha)*	*Yield (t/ha)*	*Gross margin (£/ha)*	*as % of chemical*
Winter wheat	5.29	593	5.7	449.4	92.8
Spring barley	5.04	500	4.24	322.7	118.9
Oats	5.17	499	4.64	353.8	111.4

Source: Widdowson (1987).

Table 1.5 continued

(iii) Yields on wholly organic farms as reported by the C.U. Dept. of Land Econonomy compared with yields from conventional farms as reported in Agriculture in the UK, MAFF, 1991

	Organic	*1987*	*1988*	*1989*	*1990*	*Organic as % conventional average 87–90*
Wheat		5.99	6.23	6.74	6.97	
Winter wheat	3.73					57.5
Spring wheat	3.24					50.0
Barley		5.04	4.67	4.88	5.21	
Winter barley	3.09					62.4
Oats	3.59	4.57	4.55	4.46	4.96	80.5
Potatoes – early	17.03	23.6	24	20.9	27.4	81.5
Potatoes – main	18.98	39	39.6	37.4	37.3	50.7

Source: Author's calculations using Murphy (1992) and MAFF (1992).

(iv) Yields from organic farming in England and Wales

Year	*Yield of winter wheat*		*Organic as % of conventional*	*Yield of spring barley*		*Organic as % of conventional*
	organic	*conventional*		*organic*	*conventional*	
1977	4.9	4.8	102.1	3.7	4.6	80.4
1978	3.0	5.6	53.6	2.0	4.0	50.0
1978	3.7	5.6	66.1	3.7	4.0	92.5
1978	4.3	5.7	75.4	4.3	4.4	97.7
1978	4.3	5.3	81.1	4.9	4.0	122.5
1978	4.7	4.6	102.2			
1978	4.8	4.9	98.0			
1979	3.5	n/a	n/a	3.0	4.0	75.0
1979	4.9	5.3	92.5	3.2	n/a	
1979	4.8	5.4	88.9	4.2	4.3	97.7
1979	5.3	5.1	103.9	3.7	3.8	97.4
1979	5.8	5.1	113.7			

Source: Vine, A. and Bateman, D. (1981) (in Lampkin 1985).

Table 1.5 continued

SWITZERLAND

(v) Organic agriculture compared with conventional farms in Switzerland

	Organic	*Conventional partner*	*Regional average*	*Organic as % conventional*	*Organic as % reg. average*
Wheat (t/ha)	3.9	4.5	4.7	86.7	83.0
Rye (t/ha)	4.4	*	4.5	*	97.8
Maize (t/ha)	4.4	4.7	4.9	93.6	89.8
Oats (t/ha)	4.2	5	4.9	84.0	85.7
Barley (t/ha)	3.9	4.5	4.5	86.7	86.7
Potatoes (t/ha)	31.1	31.4	36.3	99.0	85.7
Milk (l/cow)	4517	5111	4912	88.4	92.0
Milk (l/for ha)	8609	10669	11254	80.7	76.5

Source: Steinmann, R. (1983) (in Lampkin 1985).

(vi) Performance of organic and biodynamic farms in Baden-Württemberg, Germany

	biodynamic farms only 1971–74 average biodynamic	*conventional*	*biodynamic as % of conventional*	*organic and biodynamic farms 1983 survey of 200 holdings average*	*range and no. in sample*	*conventional 1983*	*altenative as % of conventional*
Winter wheat	4.54	4.09	111.0	3.3	(1.0–5.3, 145)	4.7	70.2
Spring wheat	4.08	4.07	100.2	2.8	(1.0–4.6, 52)	3.9	71.8
Winter barley	4.05	4.22	96.0	3.5	(1.0–5.0, 28)	4.8	72.9
Spring barley	3.33	3.59	92.8	2.6	(0.7–4.2, 21)	3.7	70.3
Oats	3.9	3.66	106.6	3.2	(1.2–5.0, 36)	3.9	82.1
Rye	*	*	*	3.2	(0.8–5.2, 52)	3.8	84.2
Carrots	*	*	*	40.5		42.3	95.7
Early potatoes	*	*	*	13.8	(7.0–20.0, 20)	18.5	74.6
Main crop potatoes	22.8	22.7	100.4	16.5	(5.8–40.0, 120)	22.6	73.0
Beetroot	*	*	*	29.9	(5.0–62.5, N/A)	32.6	91.7

Sources: MELU (1977) (in Lampkin 1985) and Bockenhoff, E., *et al.* (1986).

Table 1.5 continued

UNITED STATES

(vii) Comparison of organic and conventional farms in 1974 and 1975 in the US mid-West

	Organic		*Conventional*		*Organic yields as % conventional*			*Percentage organic less than conventional*
	1974	*1975*	*1974*	*1975*	*1974*	*1975*	*Average*	
Soyabeans (bu)	35	32	38	29	92.1	110.3	101.2	0
Wheat (bu)	28	28	38	29	73.7	96.6	85.1	16
Corn (bu)	74	74	94	76	78.7	97.4	88.0	13
Oats (bu)	58	56	60	60	96.7	93.3	95.0	5
Hay (tons)	4.5	5	3.9	3.4	115.4	147.1	131.2	–30

Source: Lockeretz, W. *et al.* (1975).

(viii) Comparison of organic and low-input farm yields (t/ha) with those on conventional farms in Nebraska and N. Dakota

Crop	*State*	*Period*	*Organic*	*Low input (a)*	*Conventional*	*Org./Low input as % conv.*
Maize	Nebraska	1979–1991	5.0	n/a	5.3	92.0
Maize	S. Dakota	1985–1988	n/a	4.6	5.3–5.6	82–87
Soyabeans	S. Dakota	1985–1988	n/a	1.5	1.5–1.8	85–105
Spring wheat	S. Dakota	1985–1988	n/a	2.8	2.7–2.8	100–105

(a) No synthetic fertilizers or pesticides used, but not labelled 'organic.'
Source: Sahs *et al.* (1992) and Mends *et al.* (1989) (in Anderson (1994)).

Table 1.5 continued

(ix) Other case studies from the United States

Crop yields from case study farms (tonnes/ha)

Farm, period and crop	*Organic*	*Conventional (a)*	*Organic as % conventional*
From NRC (1989)			
Spray Brothers, Ohio, 1981–1985			
Maize	9.1–9.4	7.0	130–135
Soyabeans	3.2	2.3	140
Wheat	3.0	2.9	105
Oats	2.9	2.3	122
Kuztown farm, Pennsylvania, 1978–1982 (b)			
Maize	6.8	5.4	127
Alfalfa hay	7.4	6.7	110
Corn silage	32.0	32.0	100
Rye	1.8	2.0 (c)	92
Soyabeans	2.6	2.0 (c)	126
Wheat	2.5	2.5	97
Pavich and Sons, California/Arizona, 1986			
Grapes	16.8	13.4	125
From Matheson *et al.* (1991) (all from Dryland Northern Plains and Intermountain Northwest USA).			
Gould and Partners			
Winter wheat	1.9	2.3	82
Quinn Farm			
Winter wheat	2.4	2.4	103
Thomas farm			
Wheat	2.0	1.6	125
Sunflowers	1.4	1.2	114
From Cavigelli and Kois (1988), Kansas farms.			
Tangeman farm, 1984–1986			
Wheat	1.7	2.4	74
Soyabeans	2.2	1.9	114
Maize	5.1	5.3	96
Oats	2.3–2.5	1.6	141–152

Table 1.5 continued

(ix) Other case studies from the United States (*continued*)

Crop yields from case study farms (tonnes/ha)

Farm, period and crop	*Organic*	*Conventional (a)*	*Organic as % conventional*
Bennett farm			
Wheat	1.7	2.2	82
Soyabeans	3.0	2.4	122
Vogelsberg farm, 1983–1986			
Maize	6.0, 5.7(d)	4.3, 2.9(d)	141, 198(d)
Soyabeans	2.0	1.7	115
Oats	2.6	2.1	126
Alfalfa/red clover	9.0	6.9 (e)	129
Wheat	1.8	2.4	75

(a) County average unless otherwise stated
(b) Not certified organic
(c) State average
(d) Drought conditions
(e) County average for alfalfa only

Source: Anderson (1994)

CANADA

(x) Comparison of low-input and conventional farms in Ontario, Canada

	Conventional	*Low input*	*Low Input Relative to conventional (%)*	*Organic*	*Organic Relative to conventional (%)*
Maize yield (t/ha)	5.5	6.5	118	6.4	116
Winter wheat (t/ha)	2.9	3.3	114	3.3	114

Only two sample points were used for each category.
Ten-year provincial average yields were 6.4 t/ha for maize and 3.8 t/ha for wheat.
Source: Henning (1994).

Conclusion

The above account has served to do two things; it has shown that modern agriculture, based as it is on genetic uniformity, substitutes diversity for uniformity, makes crops vulnerable to pest attack, and is at the root of environmental problems associated with pesticide use; and it has also shown that alternative ways of doing agriculture are at least worthy of greater attention from formal agricultural research than they currently receive, and that, notwithstanding the difficulties encountered in comparing farming systems, it may not be universally true to say that HEIA gives yields superior to any other system. Why, given the ecological problems associated with it, has formal agricultural research promoted high-input agriculture so successfully for so long? It seems reasonable to suggest that formal agricultural research should be more willing to understand the diversity of approaches to agriculture rather than single-mindedly adopting, as the technique of choice, one based on the purchase of chemical inputs and the use of genetically uniform seeds. Who knows what might be achieved if more resources were directed to research on alternatives?

The question underlying my attempt to understand the determinants of the path of technological change in agriculture is 'what are the factors that appear to mitigate *against* greater emphasis on agricultural techniques that embrace, rather than destroy, genetic diversity?' This question is at the heart of any attempt to understand why environmentally damaging chemicals have been employed in agriculture for so long since the function of these chemicals is, in large part, to compensate for the vulnerability to crop losses that genetic uniformity brings.

I suggest in this book that the reason why the incorporation of diversity in agriculture has been given so little attention by formal agricultural research has historical roots. Decisions made in the past influence the menu of choices on the basis of which future decisions are taken. It is critical to understand that there are always choices to be made, though in some cases choices are made less consciously than in others. Viewed in this light, the development of agricultural research needs to be understood as the tracing out of a path through time. As time changes, so do the nature of the choices to be made. What, then, determines the path traced out by technological change in crop agriculture? Were there choices open to those who developed the technologies, or those responsible for their development, and, if so, what of the roads not taken?

Notes

1. For a discussion of this process, see Rindos (1984), Cox and Atkins (1979, ch. 3), Reed (ed.) (1977), Harlan (1975), Sauer (1952) and Anderson (1967).
2. Whilst it is tempting to contrast human selection with natural selection, to do so would suggest that human beings are not part of nature, and that there is no place for humans in 'the wild'. Climate change provides a good example of the difficulties in separating 'natural' and 'human' selection pressures.
3. See Norgaard (1987, 1988a,b, 1994, ch. 3), as well as Rindos (1984, 1989) and Holden *et al.* (1993, 12).
4. Adaptation is not explained solely through genetics; 'some variability is genetic, some is environmental. The genotype (or genetic constitution) determines a certain potential for development, environment determines the developmental track adopted and the phenotype is the outcome' (Simmonds 1979, 67).
5. The classic example is selection for non-shattering plants, though this is not universal (Fowler and Mooney 1990, 15). For other examples, see Holden *et al.* (1993, 15, 28), Fowler and Mooney (1990, 15–16) and Simmonds (1979, 14–19).
6. The way in which diversity is expressed in these landraces depends very much on the mating system of the crop in question. The evolution of crop plants may have increased the tendency towards inbreeding, which promotes local adaptation at the expense of long-term flexibility. Outbreeders maintain flexibility, but the possibility of deleterious recessive alleles concurring implies that less fit individuals will exist in any population.
7. The perception of a plant as a crop or as a weed is purely subjective (Fowler and Mooney 1990, 12). Datta and Banerjee (1978, 309) list 124 'weed' species used in West Bengal rice fields. Chacon and Gleissman (1982) describe farmers in Tabasco (Mexico) who differentiate only in terms of 'good' and 'bad' plants, it being possible for the same plant to be either depending on where and when it appears.
8. On farmer experimentation, see Johnson (1972), Biggs and Clay (1981), Chambers *et al.* (eds) (1989) and Scoones and Thompson (eds) (1994). On experimentation with varieties, see Clay *et al.* (1978), Clawson and Hoy (1979), Hernandez X (1985), Richards (1985, 1986, 1994), Dennis (1987), Maurya (1989), Bellon (1991), Brush (1992), Brush *et al.* (1992) and Bellon and Brush (1994).
9. See Wilkes (1977c). There is a fascinating discussion of the relative contributions of 'conscious' and 'unconscious' selection in Rindos (1984, ch. 1), terms first used by Darwin to differentiate the work of breeders and others. That Darwin should have seen fit to make such a distinction even prior to the rediscovery of Mendel's work seems premature given that, more than one hundred years later Simmonds (1979, 27) was referring to the scientific knowledge employed by breeders as 'more or less secure', and their methods as 'tolerably well understood.' In a section 'Biology and Intent,' Rindos (1984, 85–99) makes disparaging noises concerning the contributions of early cultivators. The strict interpretation he demands of intent makes a nonsense of the notion of experimentation, the nature of which must surely be that *the outcome cannot be intentional, but the act invariably is.*

10. On centres of origin and diversity, see Vavilov (1951), Zohary (1970), Harlan (1971), Hawkes (1983).
11. On genetic erosion in cradles of diversity, see Frankel (ed.) (1973).
12. There are many plants 'new' or 'unknown' to science which have been known, and used for years by people now confined to the remote parts of the globe.
13. On the possibility of raising the profile of some of these crops, see Plotkin (1988) and Sattaur (1988). Many such crops are medicinal. For the 1960–85 period, 25% of all prescriptions dispensed in the US contained active principles still extracted from higher plants, and around 80% of people in the developing world depend on a form of medicine which relies almost exclusively on plants (Farnsworth 1988, 83, 91).
14. There is actually no means to *prove* the truth of a position which states that the food supply will *not* succumb to such a catastrophe. The only proof possible is that the opposite is the case. Unfortunately, this proof arises through the hypothesised catastrophe acquiring the status of fact.
15. For the purposes of this thesis, I will use the term pest to imply all biological organisms which are deemed unwanted or undesirable in a given farming system (i.e. including insects, rodents, microbial organisms and 'weeds'). Insects resistant to pesticides have long been a problem, but herbicide resistant weeds are a relatively recent, though no less tractable phenomenon. There were 107 such weeds world-wide as of the late 1980s, 81 of which were resistant to more than one chemical formulation (LeBaron 1991).
16. Garrison Wilkes, in trying to update the 1972 National Academy of Sciences study on the situation in the United States (the USDA stopped monitoring the situation shortly after the study was published), concluded that in virtually all cases matters had worsened (in Raeburn 1995, 148).
17. Doyle (1985, 206), noting the spread of the Russet Burbank as the potato of choice for MacDonald's french fries, anticipated this situation in 1985 and suggested that the company should be culpable, if not liable, should a virulent organism begin to affect the potato crop.
18. It is generally accepted that a gene-for-gene relationship exists between pathogens and their hosts, functioning like biochemical locks and keys (Robinson 1996, ch. 5).
19. Allelic forms responsible for vertical resistance may have been lost due to genetic erosion. Several cases are known where resistance to a particular pest has been found in populations that are almost extinct (see refs. in note 25 below).
20. The former problem is most often discussed with reference to pests and pathogens, yet the same considerations apply with increasing force to weeds (see LeBaron 1991).
21. Horizontal resistance is sometimes called 'durable' resistance because some believe horizontal and vertical resistance to be at opposite ends of a continuous spectrum (Dempsey 1992, 15).
22. Since fields vary enormously in their size, this could be taken to be the same as option (ii) in the limiting case. What is implied here is inter- and intra-specific intercropping.
23. These practices, entailing as they do an element of risk all of their own, alter the source of risks and change the distribution of their burden. Risks

are pushed underground into groundwater supplies, and are widely distributed in the food supply.

24. It is widely acknowledged that the costs of so-called maintenance research are rising (Plucknett and Smith 1986; Hayami and Otsuka 1994, 34). Yet the character of maintenance research, and thus, probably its cost, relates to the strategy by which past yield increases have been achieved. Rising costs of maintenance research are related to past reliance on vertical resistance strategies.
25. On the usefulness of sources of diversity in this respect, see Holden *et al.* (1993, ch. 3), Fowler and Mooney (1990, ch. 4), Burdon and Jarosz (1989), Cox and Atkins (1979, 108), Myers (1979, 61–2) and Plucknett *et al.* (1987, ch. 8). The importance of wild relatives as a source of insect resistance is also used by Chang (1989) to argue in favour of large germplasm collections for each crop.
26. This confidence may well be misplaced. See, for example, Fowler and Mooney (1990, 161–72); Vellvé (1992, ch. 3); and Raeburn (1995, ch. 2).
27. For criticisms of genebanks, see GRAIN (1992, 5–7); Vellvé (1992, ch. 3). On the need to supplement *ex situ* with *in situ* conservation, see Vaughan and Chang (1992, 371–2), Brush (1991, 153–4); Holden *et al.* (1993, ch. 5), Cohen *et al.* (1991a, b), and Brush (1994, 4–14). There are crops (mainly tree crops and tubers) which are simply not amenable to easy *ex-situ* storage. Biotechniques (*in vitro* storage) may offer new possibilities in this regard.
28. For example Altieri and Merrick (1988), Nabhan (1985), Prescott-Allen and Prescott-Allen (1982), and Brush (1994).
29. Hence, Williams (1988, 246); 'Landraces cannot be conserved by growing them in primitive agricultural conditions; it is neither practical nor can it be justified morally.'
30. The first two points here raise thorny questions concerning whether the problem of feeding the world can actually be compartmentalised in a manner that allows one to dissociate the question of food sufficiency from the way food is produced. Even if it were true that modern (or for that matter any other) agriculture was the best way to produce more food, what if the meeting of the 'sufficiency' criterion precluded the meeting of the necessary distribution criterion?
31. For the history of alternatives, see Harwood (1990), Rodale (1990), Schilthuis (1994) on biodynamic agriculture; Tate (1994), Lampkin (1994a), Hecht (1987) on agroecology, and Higa (1989).
32. In biodynamic agriculture, sowing and planting is dictated by cosmic cycles (Schilthuis 1994, 79–83), a practice carried out in other agricultural societies around the world (Salas 1994, 57–69). Theodore Schultz (1964, vii) saw such practices as backward. Referring disparagingly to the the approach of policy makers to assisting agricultural growth, he wrote, 'policy makers are about as sophisticated in this matter as farmers who once upon a time planted crops according to the face of the moon.'
33. Carson's *Silent Spring* (1965) was not just derided by graduates in agricultural colleges. The book also attracted the vitriol of the pesticide industry (see Graham (1980), and in a similar vein, Van Den Bosch (1980)). Carson's work actually failed to achieve any concrete changes in policy

towards pesticides until it was established that DDT and dioxin affected not just animals, but humans too (MacIntyre 1987, 577).

34. The key works of King (1927) and Howard (1940) both marvelled at the ingenuity of 'traditional' farming practices. Others had done so even earlier (see Dogra (1983)).
35. See Thurston (1992, 4–8).
36. The term 'integrated agriculture' has different meanings in different countries. In Asia, particularly Thailand, the term is used to define a type of farming which is closer to agroecology/permaculture (see below) (see, e.g. SEASAN 1992).
37. This is close to the concept of reflexive modernisation discussed by social theorists such as Beck (1992) and Giddens (1991).
38. Some perform less well in certain respects. Organic monocultures are notoriously 'leaky' in terms of nitrate leaching.
39. There are few comparisons of the environmental impact of farming systems. Several attempts to gauge the costs of HEIA have been undertaken (see Pearce and Tinch 1998, Pimentel *et al.* 1993 and Dinham 1998). A substantial literature on energy output/input analysis in agriculture now exists. For a historical review, see Martinez-Alier (1990). Otherwise, see Berardi (1978), Cox and Atkins (1979, ch. 24), Freedman (1980), Stout (1980), Pimentel *et al.* (1983), Pimentel *et al.* (1990) and Dazhong and Pimentel (1990).
40. See Graham-Tomasi (1991).
41. This should not be construed as acceptance of a ranking of agricultural systems in terms of their yield alone. However, the rarely questioned assumption that HEIA is the only way to feed the world deserves to be closely examined. Evidence regarding the productivity of alternatives incorporating diversity might be a useful strategy in persuading research administrators to allocate greater resources to these alternatives. Furthermore, since variable input costs of alternatives are invariably lower than for modern counterparts, and outputs, whether in developing country local markets, or through developed world labelling and local marketing schemes, can attract premium prices, even yields which are lower for alternatives can bring higher gross margins to farmers.
42. Reintjes *et al.* (1992, 21) suggest that there is a wealth of 'grey' literature on LEISA, but that 'most of the experiences of innovative farmers and fieldworkers throughout the world have not been documented, although much has certainly been spread by word of mouth.' Their Appendix C3 (228–32) contains a list of about 100 publications on sustainable agriculture in the tropics.
43. The nitrogen-fixing property of legumes is a well known example. Also, allelopathy studies chemical interactions between plants. See the volume by Rizvi and Rizvi (1992), especially the chapter by Rizvi *et al.* (1992).
44. It is not true to say that all polycultures reduce the incidence of all pests. However, those that are in use appear to perform better than monocultures more often than not. See Thurston (1992, ch. 19).
45. See Chapter 6.
46. A good deal of alley cropping work has suffered from the same shortcomings as other work in formal agricultural systems, many of which are highlighted in later chapters. Principal among these being has been its lack of

relevance to farmers owing to a failure to work within a participatory approach, despite the fact that the high cost and long duration of trials, and the need for adaptive experimentation in terms of crop configurations and management, seem to make farmer participation in research particularly important in alley cropping. Furthermore, there has been a tendency to concentrate on the use of one or two supposedly 'miracle trees', one of which, *Leucaena leucocephala* has proven particularly susceptible to pests (see Carter 1995).

47. See also Freeman and Fricke (1984) and Thurston (1992, ch. 20).
48. Based on CIMMYT (1992).
49. See, for example, Lockeretz *et al.* (1984); Oelhaf (1978, 195–225), Parr *et al.* (1983), Wookey (1987, ch. 9), Widdowson (1987), NRC (1989), Lampkin (1985, 1990, esp. ch. 13); Lampkin (ed.) (1990, papers 1,3,6,8,9), Stanhill (1990), Murphy (1992, esp. ch. 8), Pretty and Howes (1993, ch. 3); Lampkin and Padel (eds) (1994, esp. chs. 5–12).

2
Technological Change in Agriculture: Orthodox Views

Introduction

Chapter 1 sought to show how the technological transformation of agriculture has persistently narrowed the genetic base on which agriculture rests. The question arises as to whether a theory can be formulated which attempts to account not just for the occurrence of change, but also the direction which change follows. Such a theory must appreciate that a variety of different solutions may be more or less appropriate for solving any given solution. It must therefore concern itself *not just with the changes that actually occur, but with why alternatives do not occur.*

This chapter begins with a brief historical discussion of the views of Smith, Marx, Malthus and Ricardo. It then reviews two of the more widely discussed theories of technical and technological change in agriculture, those of Ester Boserup, and of Yujiro Hayami, Vernon Ruttan and Hans Binswanger. In the context of critiques that are not intended to be comprehensive, they are shown to suffer from significant shortcomings, both on their own terms and by the criterion mentioned above.

Classical economists and Marxism

The classical political economists tended to accord technological change a more central role in their theories than their neo-classical successors, especially those of the first half of the twentieth century. For Adam Smith, agriculture seemed to have properties which prevented it from being subject to the type of division of labour that took place in industry:

> the ploughman, the harrower, the sower of seed, and the reaper of corn are often the same. The occasions for those different sorts of

> labour returning with the different seasons of the year, it is impossible that one man should be constantly employed in any one of them. This impossibility of making so complete and entire a separation of all the different branches of labour in agriculture is perhaps the reason why the improvement of the productive powers of labour in this art, does not always keep pace with their improvement in manufactures. (Smith 1976, 16)

In other words, the concentration of so many tasks in the hands of one person meant that the these tasks were not so susceptible to the sorts of division of labour that Smith famously analysed in the case of a pin factory.

Marx's view was rather different. In a particularly prescient passage in the *Grundrisse* (Marx 1973, 527–8), he writes:

> If agriculture rests on scientific activities – if it requires machinery, chemical fertiliser acquired through exchange, seeds from distant countries, etc., and if rural, patriarchal manufacture has already vanished – which is implied in the presupposition – then the machine-making factory, external trade, crafts, etc. appear as *needs* for agriculture ... Agriculture no longer finds the natural conditions of its own reproduction within itself, naturally, arisen, spontaneous, and ready to hand, but these exist as an industry separate from it ... This pulling away of the natural ground from the foundations of every industry, and this transfer of the conditions of production outside itself, into a general context – hence the transformation of what was previously superfluous into what is necessary, as a historically created necessity – is the tendency of capital.

By the second half of the nineteenth century, it had become clear to Marx that science was beginning to serve the process of commodity production. He postulated that, 'Invention then becomes a business, and the application of science to direct production itself becomes a prospect which determines and solicits it' (Marx 1973, 704). Ironically, therefore, it was Marx, not Smith, who envisioned the beginnings of the process by which tasks within agriculture would become specialised, and concentrated in the hands of different specialists. Crucially, Marx also observed that such specialists would often reside off the farm.

It was Malthus' *Essay on the Principle of Population* (1976a) which sparked off a debate, which has waxed and waned in intensity ever since, concerning the relationship between demography and food supply.[1] Malthus' key hypothesis was that whilst increases in agricultural production followed an arithmetic progression, population increased at a geometric rate. Consequently, population would, at some point in time, overtake the capacity of food supplies to provide for it (Malthus 1976a, ch. 2). Checks on population growth would then come into play.[2] It is often pointed out that Malthus ignored the capacity of technological change to stave off the ultimately tragic scene which he supposed would befall humanity. This is not to say that he did not consider technological change at all. Nobody writing on agriculture at the time of Malthus could fail to recognise the contributions made to productivity by new methods of farming, and Malthus was no exception.[3] But he held, as an article of faith, that food production would increase by constant amounts per unit of time (arithmetically) rather than, as is typically estimated today, in constant percentages (geometrically). In addition, Malthus' argument seemed to consider the United Kingdom as a closed economy in which trade in food products was not given adequate consideration.

This last criticism could not be levelled at David Ricardo (1973), whose *Principles of Political Economy* owes much to Malthus' theory of rent (ch. XXXII) and makes several references to his *Essay*. It is striking in the work of Smith, Malthus and Ricardo that the level of wages are deemed to be so closely tied to the price of basic commodities, especially corn. Ricardo uses this concept to suggest that even when food prices rise, profits do not necessarily increase since wages rise also. This represented a constraint on profits which Ricardo was anxious to see removed.

An area of concern, therefore, for classical economists, most clearly expressed by Ricardo, was the idea that the progress of invention would prove incapable of offsetting the effects of diminishing returns to incremental increases in capital and labour as applied to the land.[4] The theory of diminishing returns actually originated with Turgot in 1767, and was first applied to problems in agriculture by Malthus and West in the debate over the Corn Laws in 1815.[5] Ricardo and his colleagues elaborated two cases within the agricultural sector in which decreasing returns set in – those of the Extensive and Intensive margins. In the former, the theory suggested that newer lands brought into cultivation would tend to be those with poorer soils and that therefore, for a given amount of capital and labour, production would be less than on lands already cultivated. In the latter case, it was believed that intensification on a given plot of land would, beyond a certain point, be characterised by smaller increments in increased production for a given increase in capital or labour (Ricardo 1973, 56).

In cases where the supply of land was inelastic, a condition which would have characterised Britain at the time of Ricardo's writing, diminishing returns in the case of the Intensive Margin were assumed to constitute the major constraint on economic growth. Ricardo used his theory to advocate repeal of the Corn Laws, and a freer trade regime in general.[6] Only in this way, or so he argued, could Britain's intersectoral terms of trade be prevented from turning against industry.

It is now widely accepted that the classical model downplayed the possibilities for offsetting diminishing returns in agriculture through changes in technology. For a given technology, one can understand the relevance of diminishing returns at the Intensive Margin, but with changes in technology, the efficiency of transformation of inputs may be increased, new inputs may be developed, and the constraints on growth represented by the onset of diminishing returns at the Intensive Margin can be pushed into the future. Related to this, Hayami and Ruttan (1971, 31) have criticised the classicals' view of the agricultural sector as separate from other sectors of the economy, though clearly, this could not be said of Marx.

Boserup's 'conditions of agricultural growth'

Outline of Boserup's theory

Ester Boserup responded to Malthus' theory in her well-known work, *The Conditions of Agricultural Growth* (1965). Whilst Malthus hypothesised that inelastic growth in the supply of food determines the rate of population growth that a given society can sustain, Boserup's approach is:

> the opposite one [to that of Malthus] ... that the main line of causation is in the opposite direction: population growth is here regarded as the independent variable which in its turn is a major factor determining agricultural developments. (1965, 11)[7]

She reviews the development of agriculture, taking as the key variable the frequency of cropping. Agricultural development (and somewhat awkwardly, cultural development also) are deemed to proceed in a linear fashion in which cropping frequency passes through phases of forest fallow, bush fallow, grass fallow, and eventually, annual cropping and multiple cropping (15–17)

Perhaps the key part of her argument consists of her attempt to show that in passing through these stages, output per person-hour declines whilst output per person remains roughly constant (41, 43–55). This is taken to imply that rational cultivators will not change their practices unless they are forced to do so by increased population density. Thus, population growth is viewed as the spur to increased frequency of cropping (for which, read agricultural development).

Criticisms of Boserup's theory

Throughout the discussion, Boserup implicitly assumes that cultivators seek to maximise food output per person-hour consistent with the prevailing demographic situation in a world where higher output per unit of land always entails lower output per unit of labour. It is not clear that this is the only motivation of cultivators. Neither is it clear that this specifies a technique in the way that Boserup seems to suggest (see Chapter 1). Furthermore, the fact that output per person-hour on today's most modern farms is higher than at any time in the past would appear to fly in the face of her assumption that as output per unit of land increases, the input of human labour immutably rises. The substitution of fossil fuel energy for human labour through mechanisation, and the use of chemical inputs for soil fertilisation and crop protection, has enabled greater output for lower levels of labour input. Yet it will not suffice to assume that if one simply imputes calorific equivalents for the tools and machines used, the theory will hold good since Boserup shows herself to be aware of the labour-saving implications of new tools (ch. 2, 112–14), both those fashioned on the farm, and off it. Indeed, Boserup writes that output per man-hour 'increases by leaps and bounds when industrial methods are introduced in rural communities in already industrialised countries,' yet is moved to conclude only that such increases are unlikely 'in underdeveloped countries.' The implications for her thesis as a whole are not explored.

Population pressure is given a privileged status as scarcity's cause.[8] One can understand why, when so much of her work deals with centuries past, Boserup did not look beyond population in this respect. Had she done so, she might have appreciated that other factors would provide the same sort of spur to innovation that she suggests population pressure provides. War seems an excellent example of a situation in which intensification is driven by circumstances in which, if anything, the demographic trend is likely to turn downwards.[9]

Although critical of Malthus' position (59–61, 116–18), Boserup's position is much closer to Malthus' than she would like to admit (though she effectively does so on p. 41). Her work implies that some agricultural communities have the requisite techniques to enable them to adapt and some do not. Those that succeed have these techniques 'hidden up their sleeves' and they employ these techniques and the complementary technologies only when the prevailing demographic situation requires it. Yet in order for them to both know of, and know not to use earlier, the assumed techniques, these techniques would have to have been experimented with. The alternative (probably wrong) interpretation of Boserup's work suggests bursts of innovative activity in perfect adaptation to the demographic situation, in which case, Boserup's view has much in common with Social Darwinism (which, Hodgson (1993) and Rindos (1984, esp. ch. 2) are both at pains to point out, has little to do with the natural selection of Darwin himself).

Sauer (1969, 21–2) was probably right in proposing that:

> People living in the shadow of famine do not have the means or time to undertake the slow and leisurely experimental steps out of which a better and different food supply is to develop in a somewhat distant future ... the inventors of agriculture had previously acquired special skills ... that predisposed them to agricultural experiments.

Thus, when Boserup herself admits that communities have, in the past, fallen foul of a Malthusian trap (1965, 41), it is surely incumbent on her to explain why their population growth did not lead to the development of appropriate technologies. If she cannot, then as she herself states, circumstances will arise when, '[t]he population would then have to face the choice between starvation and migration' (41).[10] Malthus was not always wrong.

Boserup's theory tells us little about how technologies are actually developed and who develops them. In the modern era, formal agricultural research, private and public, has been a key agent in transforming agriculture in many parts of the world, in conjunction with changes in technology elsewhere in the economy. Any theory of technological change which wishes to be of contemporary relevance must explain the development of, and roles played by, formal agricultural research organisations. Furthermore, such a theory must account fully for the impact of changing relations of production and exchange in increasingly integrated, if highly distorted, agricultural markets. These are important in determining the attitude of producers to technological change. The basis for deciding whether or not to adopt different technologies is likely to change as opportunities for specialisation in agriculture increase. Population growth as a driving rationale for technological change is likely to become far weaker as the markets into which producers sell become more integrated over space and time, weakening the influence of local demography on producers' decisions.[11] The roles of communication, transport, and other non-agricultural technologies are critical (and Boserup addresses these in part in a later work (1981, ch. 10)).

It might be supposed, therefore, that Boserup's theory begins to lose relevance from the mid-nineteenth century onwards, the era of research organisations designed specifically to carry out scientific research in agriculture.[12] These have facilitated the institutionalisation of the application of the methods of science to the understanding of agriculture. Yet her theory can retain validity if one adopts a more global view of the population inducement to technological change. The emergence of agricultural research organisations could be viewed as a response to a need to increase production on a world-wide basis so as to meet the needs of this expanding population. This is something closer to what is proposed by Hayami and Ruttan, to whom we turn below. Even so, if one accepts this to be the case, the question of motivations for technology adoption looms large. Illuminating in this respect is Boserup's chapter 7 in which she considers why certain technologies (promoted by colonial administrators) are rejected by cultivators. Had this part of her work been taken to its logical conclusion, she might have appreciated that a whole host of factors affect the decision whether or not to adopt a given technology, the demographic situation being only (sometimes) one.

An obvious criticism of Boserup's thesis is that technological capabilities do not appear to be correlated with population density. Yet even if agricultural productivity appears not to be correlated with population density at, say, a national level, one should be careful about drawing conclusions from such statements regarding the capacity of a country to innovate in agriculture. Technologies in use are often ones chosen, for one or other reason, from a range of options. It may be that options which have not been pursued so vigorously, or which are not so visible, do exist and are, where they are used, highly productive. The example of home gardens in densely populated regions of Southeast Asia (see Chapter 1) are an excellent example of highly productive agricultural systems in a densely populated region that have hardly gone noticed by researchers for whom the absence of monocropping has often been taken to imply backwardness. It is as if home gardens are not actually to be considered as a form of agriculture.

On the surface, Boserup's work appears to be sensitive to ecological constraints on agricultural production. The motivation for change is, effectively, an acknowledgement of the fact that the carrying capacity of the ecosystem as understood in the context of the prevailing technologies and techniques, is being, or is about to be, exceeded. However, Boserup's theory adopts far too linear an approach to ways in which production can be increased to accommodate growing populations. Her thinking is constrained by the chronological sequence into which her theory is squeezed. Also, Boserup ignores the wider ecological consequences of intensification. This can be the cause of declining productivity in the longer-term, since the persistent use of a given technology on any piece of land alters its properties. This could, in turn, quicken the onset of the Malthusian trap, more especially since soil erosion is less visible than population growth.

Boserup's work reflects the prevailing modernisation paradigm of the time in which she wrote. Two decades later, far from suggesting the superiority of field crop agriculture, agricultural scientists and researchers were alerted to the numerous possible ways of doing agriculture in the wake of environmental damage and the limitations to the transfer of North American and European techniques to resource poor farming areas of the developing world. Boserup might have sympathised with the fact that some of these methods seek to maximise the use of the sun's energy (and other 'secondary' energy cycles) so as to economise on human labour inputs, which, as long as food needs are met, is the key aim of cultivators in her theory.

Induced innovation theory

Outline of the theory

The roots of the theory of induced innovation are to be found in what David (1975, 33) calls a 'fatefully vague passage' in Hicks' *Theory of Wages* (1932, 124–5), in which he suggested that high labour costs might bias the direction of invention or innovation in a labour-saving direction within privately owned firms.[13] In the first edition of *Agricultural Development* (AD1) (1971), Hayami and Ruttan (H&R) adapted the theory to the case of the agricultural sector. Recognising that much agricultural research is undertaken by the public sector, H&R sought to extend the theory into the public domain. Their model attempted:

> to include the process by which the public sector investment in agricultural research, in the adaptation and diffusion of agricultural

> technology and in the institutional infrastructure that is supportive of agricultural development, is directed toward releasing the constraints on agricultural production imposed by the factors characterised by a relatively inelastic supply. (AD1, 54)

Binswanger (1978a) made an effort in *Induced Innovation: Technology, Institutions, and Development* (I.I.) to tighten the theoretical basis of induced innovation and also in I.I., Ruttan (1978) sought to provide a theory which endogenised institutional change within the economic framework. Recognising the fact that much new technology resulted from the work of public sector organisations, this theory tried to account for the establishment of these organisations. The second edition of *Agricultural Development* (AD2) (1985) saw H&R attempting to bring these various new insights together in a tighter theoretical framework.

The theory of induced innovation is part of a wider theory of agricultural development.[14] H&R's point of departure is the work of Theodore Schultz. Consistent with his view of peasant farmers as 'poor but efficient' (Schultz 1964, 36–52), Schultz suggested that peasant farmers lacked the incentive to increase output since the marginal productivity of labour was low, and the lack of profitable technologies on the horizon constituted a disincentive to save for future investment. He advocated the need for new technologies for the agricultural sector so as to remove the constraints under which farmers were functioning. Investment was required to make these available to farmers, leading Hayami and Ruttan to label his model the 'high-payoff input model' (AD2, 60). H&R, though in broad agreement with Schultz's recommendations, were concerned that Schultz's work was silent on the matter of how new technologies could be generated for the agricultural sector.[15] For H&R, the key questions are what would determine the amount of resources allocated to agricultural research, given that these resources are not traded in the marketplace, and the relationship between technological and institutional change (AD1, 2–3, 39–43; AD2, 2–3, 61–2)?[16] Their theory, essentially two inter-dependent theories of induced technical change, and institutional change:

> attempts to make more explicit the process by which technical and institutional changes are induced through the responses of farmers, agribusiness entrepreneurs, scientists and public administrators to resource endowments and to changes in the supply and demand of factors and products. (AD1&2, 4)[17]

Induced technical change hypothesises that technical change occurs in such a way as to economise on factors which are becoming relatively more scarce and hence, more expensive. In their view, research administrators and research scientists can be viewed as responsive to economic indicators in the same way as can a profit maximising firm (AD1, 57–8; AD2, 88). The authors suggest that not all technical changes are induced through a demand-driven process, some being supply-driven as a result of exogenous advances in the state of science and technology (AD1, 59; AD2, 89). Induced institutional change is postulated to occur in response to similar signals as induced technical change, the two being seen as connected.

Problems of definition

Two problems of definition deserve to be addressed before we proceed any further, both of which are important for the criticism that ensues. In AD2 (94–5), the definition of institution employed by H&R is such as to encompass the concepts of institution and organisation. This seems to be inconsistent with their own view that:

> Institutions are the rules of a society or of organisations that facilitate co-ordination among people by helping them form expectations which each person can reasonably hold in dealing with others. (AD2, 94)

Bromley (1989, 22–3) has pointed out that there is a problem in maintaining the earlier definition, bringing together, as it does, both institutions, understood as the rules and conventions that facilitate co-ordination among members of society, and organisations, such as government research bodies, other government agencies, banks and the like.[18] It makes sense to maintain the distinction between an organisation and an institution (as in this book), if only because it is through a given institutional framework that organisations are given their meaning.

Koppel (1995b, 58) criticises the breadth given to the definition of an institutional change by H&R on other grounds, embracing as it does the Chinese Revolution as well as changes in tenancy in a village in the Philippines. Under Bromley's interpretation, it would not be inconsistent to follow H&R's line in this respect. The criticism of Koppel probably stems less from the organisation/institution distinction than from the perception of such changes *only* as institutional changes, their other dimensions being ignored. The significance of this will be made more clear below.

The second problem relates to the definition of the terms 'technological change' and 'technical change' (and innovation). Surprisingly, in much of the literature on technological and technical change, the terms are rarely defined, and quite often, they are used interchangeably. Precisely because so little caution is taken in defining the terms, it is not possible to tell whether the terms are being used carelessly, or whether implied definitions allow for such fluidity.[19] Binswanger (1978a, 18–19) uses the following definitions:[20]

> [technical change] ... *changes in technique* or production at the firm or industry level that result both from research and development and from learning by doing.
> [technological change] ... the result of the application of new knowledge of scientific, engineering, or agronomic principles to techniques across a broad spectrum of economic activity. (My emphasis)

The definition of technical change is curiously circular given the lack of a definition of 'technique'.[21] However, if it were made clear what was understood by 'technique', then surely the italicised phrase would suffice.

Both definitions pay attention to the source of change and its level of applicability (firm, industry, economy etc.), but they lose sight of what it is that is changing. Technique and technology are left undefined, which would not be so gross an emission were it not for the awkwardness of the definitions of technical and technological change, which imply something more than, and other than, respectively, 'change in technique' and 'change in technology.' As a result, we are left with intuitively contradictory notions. Suppose a producer selects one input from a choice of two (or more) to perform a particular task one year and subsequently makes use of the other since the first appeared to change output per worker. It might appear that we have an instance of technical change.[22] But in fact, 'those *shifts in individual factor productivities that result in choices among known techniques* or from changes in commodity mix brought about by changes in relative prices of factors or products' are excluded from the definition of technical change (Binswanger 1978a, 19, my emphasis), so this is not an instance of technical change. Something has changed, but these definitions admit of the possibility that it is neither technology nor technique.[23]

Whether a change may be considered 'technological' apparently depends on whether or not a technical change arising from 'the application of new knowledge of scientific, engineering, or agronomic principles' affects a 'broad spectrum of economic activity.' Therefore such change is always, chronologically speaking, 'technical' before it is 'technological.' When it becomes 'technological' depends on the breadth of the spectrum of economic activity that one accepts as characterising such changes (raising, incidentally, the problem of how this would be measured). These definitions are so divorced from what is intuitively understood by the words themselves that, unsurprisingly, the author is unable to sustain coherent use of them. Yet H&R defer to Binswanger's theoretical treatment, raising questions as to exactly what they understand by these terms.

In this book (in line with Ashford 1994, 276; Becker and Ashford 1995, 220, though they too emphasise 'the new') the word technology denotes, broadly speaking, inputs, whether they be tools, chemicals, seeds or knowledge. Technique is understood as a way of doing something (the way in which technologies are used either in isolation or in combination). Technological and technical change are defined as, respectively, changes in technology and technique. Innovation is the first practical use of a particular technology, whilst invention is the act of discovery of what may subsequently become an innovation.

The internal logic of induced innovation

What, given these definitions, does induced technical change explain? It becomes clear that the attempt by induced innovation to endogenise the process of technical change in the economic system is somewhat less inclusive than it first appears. To retain consistency with their definitions, it might have been more honest to call the theory one of 'the impact of inducement mechanisms on the factor bias of research and development portfolios of research organisations seeking to develop *new* techniques.' The theory is not concerned with choice from among existing techniques, or innovation by farmers, or the diffusion of techniques already known. The emphasis is on the new, on formal research, and on the funding thereof.

Binswanger (1978a, 26) introduces the concept of an innovation possibility curve (IPC)[24] as follows:

> At a given time there exists a set of potential production processes to be developed. This set of processes may be thought of as determined by the state of the basic sciences. Each process in the set is characterised by an isoquant with a relatively small elasticity of substitution, and each of the processes in the set requires that a given quantity of resources be developed to the point where the process can actually be used. The IPC is the envelope of all unit isoquants of

> the subset of those potential processes which the entrepreneur might develop with an exogenously given amount of research and development expenditure.

The scientific frontier is defined as the IPC with no budgetary constraint imposed (Ruttan *et al.* 1978, 46). Hence, we are given a picture of a set of IPCs corresponding to an exogenously determined budget being generated over time as scientific knowledge advances, each IPC a hyperbola moving progressively closer to the origin over time (see Figure 2.1).

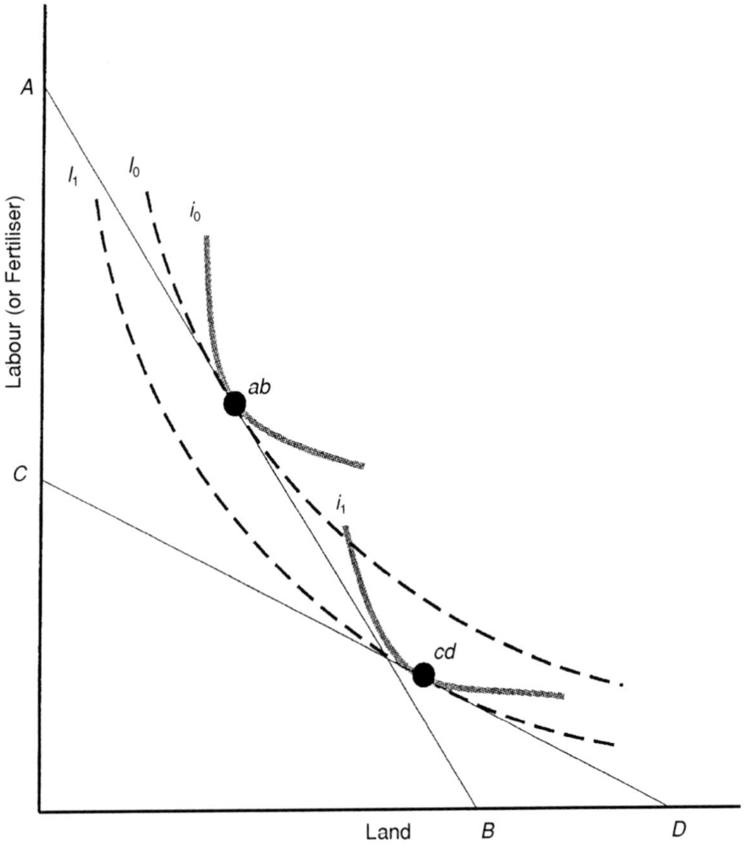

Figure 2.1 Representation of induced technical change in agriculture

Source: Adapted from Hayami and Ruttan (1985, 91)

Induced innovation does not completely endogenise the inward movement of the IPC over time:

> We do not argue, however, that technical change is wholly of an induced character. There is a supply (an exogenous) dimension to the process as well as a demand (endogenous) dimension. In addition to the effects of resource endowments and growth in demand, technical change reflects the progress of general science and technology. (AD2, 89)

Resource endowments influence the choice of techniques to be developed from a given IPC. The role of demand is presumably twofold, on the one hand bringing about changes in relative factor prices, and on the other affecting the resources devoted to exploration, both in terms of finance and new organisations. This would determine the position of the IPC relative to the scientific frontier and contribute to, though it would not account for all of, its forward movement. Therefore, induced technical change concerns itself primarily with determining, at a given moment in time, where on the IPC a research organisation should be functioning (the hypothesised ideal being determined principally by factor price ratios) and the IPC's position relative to the scientific frontier.

There are three major problems with this conception of technological change. The first relates to the fact that the scientific frontier does not always 'lead' the process of technological development. Indeed, as Rosenberg has often pointed out, technological developments often indicate fruitful possibilities for scientific research. The second problem concerns the lack of an adequate discussion of the influence of any inducement mechanisms on the scientific frontier itself. Why, if markets can influence the direction of technical and institutional change, should they not influence the direction of science itself, particularly if, as the previous point indicates, H&R are of the view that science leads technological development? Part of this book concerns the influence of a variety of factors, sociological, institutional, political, and financial, on the direction of science itself (see Chapters 3 & 4), yet H&R seem happy to accept the view that scientific progress either remains uninfluenced by such factors, or that the direction of advance in 'basic science' is irrelevant to the subject under discussion. The Mertonian ethos is deemed to prevail in the field of scientific endeavour. Figure 2.2 seeks to show the significance of focusing on one area of science rather than the whole universe of knowledge. H&R's IPC is effectively a subset of all possible ideas, represented in Figure 2.2 by a

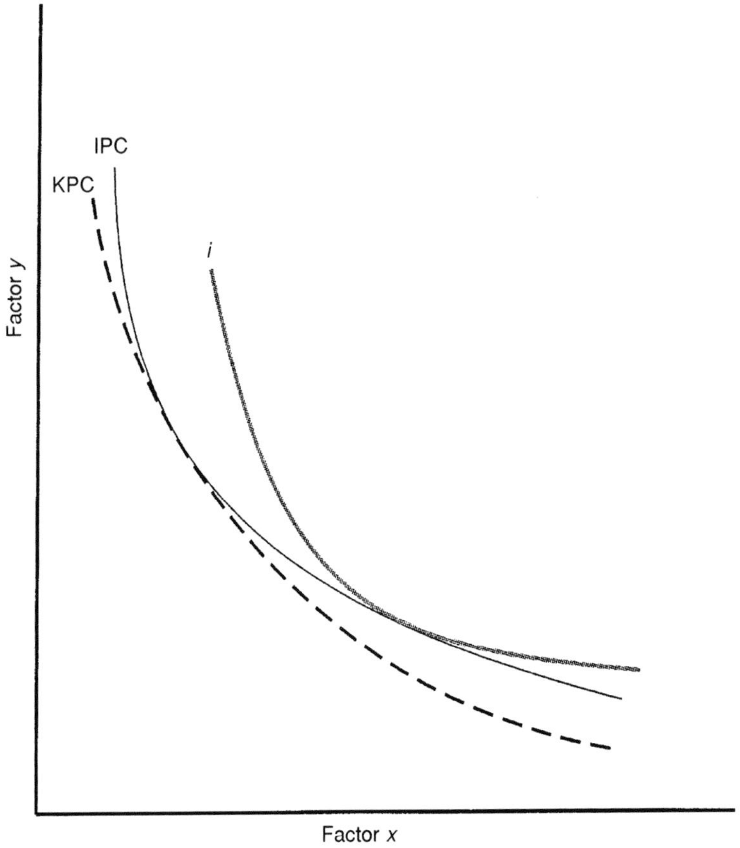

Figure 2.2 Innovation possibility curve within the knowledge possibility curve

knowledge possibility curve (KPC). The IPC represents the effect of selecting particular scientific approaches (for reasons other than finance) in preference to others. Lastly, there is actually very little discussion by H&R on how new techniques come into use. It is one thing to develop a technology, quite another to have it used.

Untestable hypotheses

Recognising what we have implied above, i.e. that induced innovation deals with research and development, not the use of technologies, it becomes clear that induced innovation theory has not always been true to itself in the sense that it has often purported to be explaining much more than, logically, it does. Indeed, tests on induced innovation are performed by determining the factor bias of technical change *in production*. Consistency with what is explained ought to imply testing *not what is used in agriculture, but what is being researched in research stations*. If it is assumed that the two are the same, there are good reasons to doubt the validity of this assumption. H&R hypothesise that:

> technical change is guided along an *efficient path* by price signals in the market, *provided that the prices efficiently reflect changes in the demand and supply of products and factors and that there exists effective interaction among farmers, public research institutions, and private agricultural supply firms*.
> Farmers are induced, by shifts in relative prices, to search for technical alternatives that save the increasingly scarce factors of production. They press the public research institutions to develop the new technology and also demand that agricultural supply firms supply modern technical inputs that substitute for the more scarce factors. Perceptive scientists and science administrators respond by making available new technical possibilities and new inputs that enable farmers profitably to substitute the increasingly scarce factors, thereby guiding the demand of farmers for unit cost reduction in a socially optimal direction.
> ... Given effective farmer organisations and a mission- or client-oriented experiment station, the competitive model of firm behaviour can be usefully extended to explain the response of experiment station administrators and research scientists to economic opportunities. (AD2, 88, my emphasis)

This, it should be stressed, is their *hypothesis*: it is not a theory, and it holds explanatory power only to the extent that the hypothesis can be tested and validated. It does suggest a mechanism by which the efficient path *might* come to be followed, and it seems rather more than a suggestion.

In any case, quite what is meant by the term 'efficient' is not addressed by H&R. Binswanger (1978b, 91–2), on the other hand, provides some clues:

> a society can gain in efficiency (that is, increase its economic growth) ... technologies are said to be efficient if they lead to an optimal increase in production measured by factor and goods prices that are assumed to reflect true factor and good scarcities.

From these definitions, the efficient path *will* be followed if technical change responds optimally to price movements which in turn reflect changes in supply and demand of products and factors. The spotlight thus falls upon whether this is likely to occur, and if efficient interaction is likely to be sufficient to make it occur. The test of the inducement hypothesis must therefore take place at the level of the process being investigated, yet H&R fail to do this.

It seems difficult to believe that the movements in market prices are reflected in the work of research and development organisations in this way. There are a number of reasons for this:

1. Market prices for land, labour, machinery and fertiliser fluctuate on differing time-scales. Furthermore, the fact that machinery and land are essentially capital goods (purchased infrequently) whilst the cost of labour and fertiliser are variable cost items (purchased annually) means that the nature of farmers' demand for different items will be far from uniform.
2. Few if any research organisations are in the business of shifting personnel from seed research into agricultural engineering, or indeed firing some of those working on the former, and hiring more of those working on the latter so as to reflect changing demands of farmers. More plausibly, research and development carries on in each sphere independently of movements in relative factor prices. In connection with this, induced innovation theory has been criticised for assuming that farmers are, at any one time, seeking to save on the use of one, relatively more scarce factor, as opposed to any of them (Salter 1960, 43–4; David 1975, 33).
3. Much research and development will have a long lead time. Even if relative prices do change, the ability of research and development to respond 'efficiently' must be questioned.

Changes in relative prices certainly alter the balance in farmers' investment decisions concerning whether or not to invest in certain technologies that they know to exist. But this is not the issue which is being addressed in the above hypothesis. The hypothesised interaction runs from the farmers' response to prices, to the scientists and researchers, then back to the farmer through the medium of 'price relevant' technologies and techniques. As the hypothesis stands, and given the definitions of efficiency above, the hypothesis is a little circular: if all the conditions are fulfilled, and prices guide technical change (which they do if the conditions are fulfilled), then technical change would always be efficient according to Binswanger's definition. The hypothesis is tautological, 'true in all possible worlds'.

Regarding institutional change, H&R embellish the (supposedly) Marxian perspective that technical change drives (revolutionary) institutional change with the view that factor endowments and product demand also play a role. In addition, they hypothesise a supply side to institutional change, which has a distinctly neo-classical hue:

> It is useful to think of a supply schedule of institutional innovation that is determined by the marginal cost schedule facing political entrepreneurs as they attempt to design new institutions and resolve the conflicts among various vested interest groups (or suppression of opposition when necessary). We hypothesise that *institutional innovations will be supplied if the expected return from the innovation that accrues to the political entrepreneurs exceeds the marginal cost of mobilising the resources necessary to introduce that innovation.* (AD2, 107, my emphasis)

The hypothesis employs marginalist concepts, but these would only be relevant if what was being discussed was how much, or how many institutional innovations are being supplied. Institutional innovations are not divisible, nor are they homogeneous. It is therefore meaningless to talk in any sense of a supply schedule for institutional change, a point made clear by considering what might be meant by the resource use elasticity of supply of institutional change.

What is at issue is whether or not one or other institutional change occurs. It seems highly unlikely that there is a knowable (even within fairly large margins of error) 'cost' associated with a given institutional change. If there was, the relevant analytical framework would be a simple cost-benefit analysis, not a marginalist interpretation. Even then, the idea that the costs and benefits of institutional change can be as easily determined as for, say, a power project seems fanciful given the multiplicity of interests that potentially have a stake in supporting or resisting the change. These can and do radically alter the strategies designed (as with power stations). The hypothesis, as well as being incorrectly stated, presumably as a result of the wishes of the authors to make their concept of institutional innovation more plausible to followers of the marginalist revolution, is not testable.

Nevertheless, the authors have carried out tests of what they perceive to be the induced innovation hypothesis. These have mostly been in the form of an econometric test, and they have not been without problems. Ruttan *et al.*'s (1978) test came out with some very strange coefficients, especially when testing the relationship between land per worker and the relative price of land with respect to labour. For many countries, there was a positive correlation. The authors suggest that opportunities for technology transfer, and a possible 'exogenous labor-saving bias in the process of technical innovation' (AD2, 63) may be to blame.[25] Existence of the latter would, of course, leave induced innovation theory still-born, but this point is not treated with the seriousness it deserves.[26]

Power, vested interests and efficiency

It seems that H&R's occasional appreciation of the role of power and vested interests in institutional change is confined to the development of their theoretical perspective on institutional change. In their empirical analysis, they rarely waiver from a belief that it is factor prices and economic opportunities that are the only important issues in explaining institutional change. However:

> To the extent that the private return to the political entrepreneurs is different from the social return, the institutional innovation will not be supplied at a socially optimal level. Thus the supply of institutional innovation depends critically on the power structure or balance among vested-interest groups in a society. If the power balance is such that the political entrepreneurs' efforts to introduce an institutional innovation with a higher rate of social return are adequately rewarded by greater prestige and stronger political support, a socially desirable institutional innovation may occur. But if the institutional innovation is expected to result in a loss to a dominant political block, the innovation may not be forthcoming even if it is expected to produce a large net gain to society as a whole.
>
> It is also possible that socially undesirable institutional innovations may occur if the returns to the entrepreneur or the interest group exceed the gains to society. (AD2, 107–8)

Thus, power and vested interests are part of H&R's theory. Note also that the marginalist language has disappeared to be replaced by the language of cost-benefit analysis (lending support to our criticism above).

Bromley (1989, 23–25) examines H&R's presentation of their case study of a change in tenancy and employment relationships in a Filipino village (AD2, 98–103; Ruttan and Hayami 1984, 207–13). He opines that H&R seem blind to the power relations in the community which underpin institutional change, these relations resulting from the prevailing institutional set-up.[27] Though they suggest 'the process [of institutional innovation] is much more complex than the clear-cut, two-class conflict between the property owners and the propertyless as assumed by Marx' (AD2, 96), it does not appear to be so in the case of the village which they discuss. Their principle concern is with 'efficiency,' a term which they themselves do not define, but which for Binswanger, equates with economic growth, and following price signals. This raises questions not only about H&R's view as to the role of agriculture in the economy, but also about whether such a line is not an attempt to dress up a normative position in positivist language. Important questions, eloquently raised by Bromley (1989, 25), seem to have been swept under the carpet somewhere in the leap between the elaborating of a hypothesis, and its 'testing':

> [induced innovation] is a model of institutional change that seems to ignore as mere transfers the distributional implications of new institutional arrangements created by sheer imbalance of economic power. In attempting to condone the new institutional structure, this model fails to deal with the fundamental analytical issue: efficient with respect to what? By endogenizing institutional innovation in this manner one is left precisely where conventional welfare economics leaves us – able to comment on changes that seem to be efficient, but unable to comment on the important distributional issues that are at the core of institutional innovation.

Bromley (1989, 23–5) also discusses what he regards as a potential flaw in H&R's induced innovation hypothesis, which is to say that disequilibria in market processes, which are the motivation for institutional change, are the outcome of prevailing institutional arrangements. In a similar vein, resource endowments, in which property rights are central, are defined by institutions: 'To say that institutions change in response to new resource (or factor) endowments is to say that new institutions appear in response to new institutions – not a very interesting prospect' (Bromley 1989, 26–7). But this *is* an interesting prospect, and it suggests why H&R fail to understand the centrality of power and vested interests in institutional change. Institutional change is a path-dependent process. The prevailing institutional set-up does, as Bromley says, determine income streams, knowledge streams, resource and factor endowments, and *ergo*, economic power. To the extent that economic power can and does influence institutional change, future possibilities are connected to the present via the prevailing institutional framework (affecting the distributional outcomes).

There is no mention of the influence of power in technological development itself, although it could be claimed that this comes via institutional channels (AD2, 361–2). Ruttan (1983, 7) makes an effort to convince critics that induced innovation theory is consistent with interest group and structuralist models. In this defence, he implies that the structure of income distribution and its effect on prices is significant for considerations of efficiency. This disappears from AD2, receiving limited discussion only in the context of extremely polarised rural societies (AD2, 361–2). Yet if changes are guided by factor prices which diverge significantly from those reflecting true factor scarcities, Bromley's question, 'efficiency with respect to what?', resurfaces with considerable force, all the more so if the path of institutional change simply entrenches the prevailing power distribution.

Although Binswanger (1978b, 92) makes some reference to the issue of distribution in affecting relative prices, he does not see that this should alter the concept of (technical) efficiency in any way. Indeed, he goes some way to justifying the prevailing price system, and by implication, the extant distribution of income, through suggesting that since these are what determine trading opportunities for producers, if research responds to price signals, efficient outcomes will result. The ecological implications of considering only the existing factor prices in setting research agendas are considered below.

Several authors have written on the biases that power brings to the process of technological change, notably de Janvry and associates.[28] An extremely critical view of the induced innovation hypothesis comes from Burmeister (1987, 1988), who uses the concept of 'directed' rather than 'induced' innovation. Examining the South Korean experience, he notes that the state basically imposed new technologies on farmers, 'inducing' them to use new varieties through administering prices of complementary inputs. In more recent work (1995, 43), he points out that a correlation between behaviour in terms of technology adoption and factor price movements can be due to administered prices adjusted by state officials to elicit the observed response. He thus highlights a major problem with the testing of induced innovation (and many other) theories through econometric tests. Correlation with relative price movements says little about the direction of causality between technological change and prices, yet the direction is important if the aim is to verify the existence of the hypothesised inducement mechanism.[29]

Agriculture, environment, efficiency and resource use

H&R refer to the primary resources in agriculture as land and labour (AD1&2, 4). Though enabling a straightforward adaptation of induced innovation theory to agriculture, this choice reflects a narrow view of agriculture. They use only two partial productivity measures, those with respect to land and labour.[30] Astonishingly, they measure productivity *per male worker only*, the reason given being, 'so as to preserve the international comparability of the data' (AD2, 450). There are other measures of productivity that might be equally important, for example, productivity with respect to cash input, or energy input (Altieri 1987a, 41)?[31] With greater imagination, measures of productivity per unit of non-rainfall water, or measures relating productivity and ecological integrity in other ways, such as genetic diversity, might be developed.

Consider the emission of pollutants from agriculture as use of 'the environment', and as a factor of production.[32] Currently, this use is free, or priced low, with few exceptions. Economic, and indeed, induced innovation theory would expect the development of agricultural technologies to have followed a path that uses more of the environment than is desirable. If a technology 'leads to an optimal increase in production measured by factor and goods prices' in circumstances where prices do not reflect scarcity values, though it may not be economically efficient, we might regard it as financially the most profitable. If, as environmental economists do, we distinguish between 'economic values' and prices, it begins to appear as though induced technical change endogenises technical innovation within the price, not the economic system. Similarly, and H&R seem to recognise this, institutional change is likely to be induced not by considerations of economic efficiency so much as possibilities for private profit, or the desires of the state.

Figure 2.3 illustrates what this implies for induced technical change as proposed by H&R (AD2, 91) and Ruttan *et al.* (1978, 47). At any one time, H&R consider two dimensions (land and labour, or land and fertiliser) of what is a multi-dimensional problem. As their IPC moves inward, we are left in the dark as to its ecological implications. For them, at any given moment in time, a producer may be using a technique defined by its position in the x–y plane. If we consider 'use of the environment,' the producer's activity is defined by a point in three-dimensional space. It should be stressed that there may be an extremely large number of these. The closer to the x–y plane the producer's operating point, the lower the use of environmental services. Industrial agriculture has tended to move its point of operation along the z-axis (increasing use of the environment), represented by the move from *ab* to *cd* in Figure 2.3. But the problem of reducing the environmental impact of agricultural techniques may not be so simple as moving back towards the x–y plane parallel to the z-axis. This is because:

(i) knowledge of how to do so (or genetic resources) may have been lost over time, or may never have been available; and/or
(ii) the technology exists but the institutions which support the prevailing way of doing agriculture actually prevent, or constitute disincentives to, such a change.

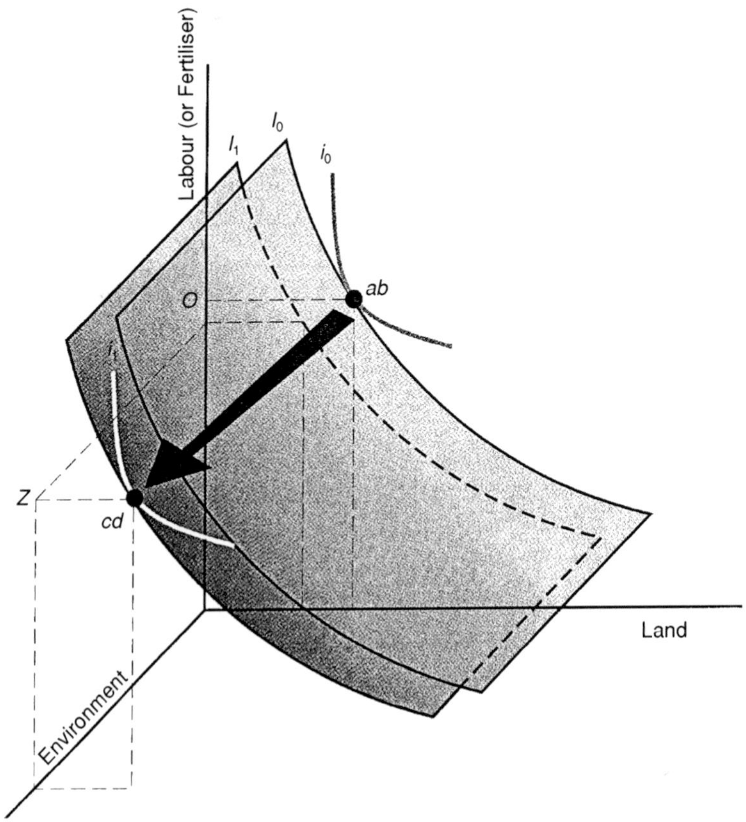

Figure 2.3 Integrating the environment into the theory of induced technical change

Both cases imply that even where there is considerable agreement as to the ecological impact of prevailing practices, pointing out alternatives, however desirable they may appear, may not be sufficient to alter prevailing practices. The truth content of this statement is likely to increase as more organisations, providing more varied services, become engaged in an established 'way of doing agriculture.' At more mature stages, technologies in use are those that both support and are supported by other technologies.

H&R's work pays scant attention to the environmental impact of new technology. It is not even considered in AD1. This is quite surprising given their opening gambit that:

> a common basis for success in achieving rapid growth in agricultural productivity is the capacity to generate an *ecologically adapted* and economically viable agricultural technology in each country or development region. (AD1&2, 4, my emphasis)

They are more concerned with adaptation of institutions in order to realise 'the growth potential opened up by new technical alternatives' (AD1, p. 4) than they are with the effects of new technologies on agro-ecosystems. There is no appreciation of the possibility that the goals of realising growth potential, and of achieving an ecologically adapted technology, may come into conflict. Such an analysis might have forced them to ask whether there might be forces, or even elements of the institutional framework, which skew the development of agricultural technology in favour of growth at the expense of ecological adaptation. Indeed, if factor prices do not account for the environment, this would be expected in an 'enhanced' induced innovation theory.

Consistent with the hypothesised inducement mechanism, the short section by way of a response to the ecological critique of green revolution agriculture in AD2 states:

> We do not regard the ecologically oriented criticisms as a fundamental challenge either to the long-term validity of the seed-fertiliser revolution or the induced innovation hypothesis. Rather, we see the environmental spillover impacts of the new technology inducing a focus on the part of both ecologically and agriculturally oriented agricultural scientists toward the invention of cropping, pest management, and farming systems that are both ecologically and economically viable. (AD2, 297)

Hidden in this statement is a clear admission that the seed-fertiliser revolution is not ecologically adapted, at least not yet. In their view, the ecological damage incurred so far will induce agricultural researchers to produce ecologically sound agricultural techniques in the future. There is no thought given as to whether this is a realistic proposition within the prevailing institutions, nor whether the institutions necessary to bring this about are likely to emerge, and if so, how? What interests would induce the required changes in institutions or the emergence of the new ecologically sound technology?

It was suggested above that such changes might not be so easily forthcoming. AD1 was written in 1971 and ecological critiques of HEIA were already appearing. AD2 was written in 1985, yet ecologically adapted techniques, which H&R suggest might have been induced in the meantime, had had such a marginal impact that H&R themselves are forced to write of their coming into existence in the future. If fourteen years of criticism had produced such marginal results, why should the hypothesised inducement mechanism be expected to work in the future? Most recently, Ruttan and Hayami (1995b, 182–3) have conceded that, '[t]he design of institutions capable of achieving compatibility among individual organizational and social objectives remains, at this point an imperfect art.' At best, this implies that what they hoped would be induced has not been.[33] Alternatively, one can make the point that, interest groups prevent prices from reflecting the true costs of their use, making it impossible for even a well-functioning inducement mechanism to carry technical change down a path that is efficient (in the sense of overall economic efficiency), and suggesting that interest groups can bias induced innovation through affecting the evolution of prices.

For H&R, the non-agricultural sector has a crucial role to play. In their view, 'A continuous stream of new technical knowledge and *a flow of industrial inputs in which the new knowledge is embodied are necessary* for modern agricultural development.' (AD2, 5, my emphasis) Knowledge flows *to* the farm from outside it. A major concern with H&R's work is that they consider HEIA as the only viable path for agriculture to follow. In their international comparisons of agricultural productivity, one test of the induced innovation hypothesis seeks to demonstrate that land productivity and fertiliser input per hectare are positively correlated, but the Appendix on statistical data reveals that fertiliser input per hectare is taken from statistics that only reveal commercial fertiliser use, thus ignoring the nutrients applied by many farmers which do not fall into that category. Nor is there any attention paid to the role played by animal traction in augmenting labour productivity. In addition, Binswanger's work (1978b,106–14) on research resource allocation among commodities assumes that the world of agriculture should be/is one of monocropped fields.

Elsewhere, Ruttan *et al.* (1978, 58) adopt the role of evangelists for HEIA:

> In order to make the productivity ratios described by the 1970 metaproduction function accessible to farmers in countries whose productivity ratios fall to the right of that function, technical and scientific knowledge about improved crop varieties, animals, chemicals, and equipment must be made available through investment in experiment station and industrial capacity.

It is not seriously considered that agriculture could be performed without industrial inputs, and there is no mention in their books of techniques other than HEIA. In their AD1 and AD2, H&R's model of rice varietal adoption is based solely on the relative price of rice and fertiliser. The obvious limitations of such a simplification are rarely noted.[34] In a recent paper, Hayami and Otsuka (1994, 20) themselves point out (in a footnote) that other factors, such as water control and cultural practices are relevant, the point being that the fertiliser response curve would look very different under systems where there was variation in these factors. They go on to state: 'It is a reasonable assumption, however, that the rise in marginal productivity of fertiliser due to the development of the [modern varieties] raises the marginal productivities of these complementary inputs' (Hayami and Otsuka 1994, 20). Whether or not these complementary inputs (and the credit needed to purchase them) are absolutely necessary, not to mention available, is never seriously questioned. The ecological implications of widespread planting of genetically uniform monocultures in tropical countries where the climate favours year-round existence of a pathogen are that vulnerability to epidemics is about as high as it can be. The use of crop protection chemicals becomes all but a necessity, notwithstanding efforts to incorporate pest-resistant genes (c.f. Chapter 1).

Underlying their support for this model of agriculture is a belief that the way to relieve constraints to agricultural growth in all countries is through the application of science generated in research organisations. These have by and large underestimated the value of cultivators' knowledge and practices. It is assumed that countries vary only in the distance they have travelled down this road, and in the factor-saving bias that is revealed in their pursuit of HEIA. Apparently all cultivators need the same thing: '[a] critical assumption in [the induced innovation] approach is that the technical possibilities available to agricultural producers in the different countries can be described by the same production function' (AD1, 89).

Information: its availability and flow

Information is neither obtained, nor always supplied, without cost. Typically, new developments may be made known to farmers either through what we might call '*in situ*' transmission (that is, through word of mouth and observation of neighbouring fields), through journals of the trade or of farmers' organisations, or through extension agents, who may represent public or private interests (or both). None of these receive any significant treatment in AD1&2. H&R appear to have stuck with the assumption employed by economic theorists of technological change which holds that knowledge of all available techniques is at the fingertips of all potential users. Clearly, this is not true. How new technologies and techniques are publicised among potential users is critical, and the type of information which reaches farmers may be decisive in increasing the popularity of one or other technology over others which might be considered competitors.

The issue of information flow is of significance in considering the incentives to which H&R believe scientists and research administrators respond. Their view will be worth quoting at length, this being:

> the critical link in the inducement mechanism. The model does not imply that it is necessary for individual scientists or research administrators in public institutions to consciously respond to market prices, or directly to farmers' demands for research results in the selection of research objectives. They may, in fact, be motivated primarily by a drive for professional achievement and recognition. It is only necessary if there exists an effective incentive mechanism to reward the scientists or administrators, materially or by prestige, for their contributions to the solution of significant problems in society. Under these conditions, it seems reasonable to hypothesise that the scientists and administrators or public sector research programs do respond to the needs of society in an attempt to direct the results of their activity to public purpose. Furthermore, we hypothesise that secular changes in relative factor and product prices convey much of the information regarding the relative priorities which society places on the goals of research. (AD1, 58)

Note that in this model, no direct contact with farmers is required. Their demands, or those of society (it is not clear which, even though they may not be the same) are supposedly reflected in market prices. Does this legitimise a one-way transfer-of-technology research system? The claim made that agricultural research institutions reflect the economic rationality of farmers serves to maintain the theory's overall consistency, but it begs the questions, 'How?', and, 'Which farmers?' If it is assumed that these organisations arose solely in response to the demands of farmers, this is incorrect, and if in addition it is assumed that their goal is to serve *all* farmers' needs, this is a serious error. In any case, scientists most often seek prestige within their particular discipline, and the conditions for attaining that prestige are rarely coupled to societal needs, a term which defies simple definition. Administrators in public sector organisations are as, if not more likely to respond to the needs of those who finance their work as they are to the needs of society (see Chapter 4).

Conclusion

Both the theories of Boserup, and H&R and Binswanger, are, principally, theories of agricultural development in which technical change plays a key role. Both theories fail to consider technologies other than those that have come into widespread use. Both also are imbued with techno-optimism, and faith in human ingenuity to continue to stave off the Malthusian spectre. Yet neither comes to terms with the environmental implications of agricultural intensification, and the possibility of Malthus' ghost re-appearing through the ecological back door is denied, at least in H&R's work, through appeal to the power of market forces to induce the appropriate technology for the situation in (or even out of) hand.

H&R and Binswanger's theory tries to explain the whole process of technical and institutional change through appealing to inducement mechanisms driven by changes in relative factor prices and economic opportunities. Setting aside the very important questions of definition and testability, in contradistinction with Boserup, it ignores the farmer since the proponents of induced innovation focus on agricultural research organisations as the locus of innovations. It supports a particular way of doing agriculture and oversimplifies the issues relating to institutional innovation, including questions of efficiency, distribution, and the environment.

H&R's theory is frustrating since the interpretation of the theory allows everything that happens in the development of agricultural technologies to be 'induced' by markets, however skewed the interpretation has to be. Indeed, there is little thought given as to how changes occur beyond being induced by market-type forces, any deviance from what they suppose to be the best way forward for agriculture being explained away through reference to 'cultural endowments'.[35] This diminishes the theory, and in telling us how change is induced, what is left out of the explanation is often more revealing than what is contained within it. Yet perhaps the most powerful support for the induced innovation theory comes from the failure, or non-emergence, of institutions to encourage the development of a more sustainable agriculture. It is a sobering thought that the entrenchment of powerful interests in the field of agri-business is likely to make such a transition likely only when enlightened self-interest on the part of these organisations compels such a change.

This of course has implications for any recommendations which H&R might consider making as to how agricultural development can be made to pursue an 'optimum path'. Presumably, an optimum path, where it is not being followed, requires a change in institutions. But who or what will induce the changes required to transform the sub-optimal institutions into optimal ones? And doesn't the belief that one particular organisational/institutional framework is the optimal one for all countries imply a normative judgement which makes a nonsense of the complexity of the problems involved?[36] With apologies to Adam Smith, too many of the hands at work developing new technologies for agriculture are, in the theory of induced innovation, obscured by the hidden hand. If an attitude of ambivalence towards environmental issues in agriculture could be sustained in the early 1970s, H&R's complacency in this respect was misplaced in the 1980s, and unacceptable in the 1990s. Sustainable development, a malleable catch-all designed to suit the needs of those who speak it, has as one central concern environmental quality. The second one is equity, within and across generations. These concerns raise issues of a normative nature which cannot be addressed by some blasé appeal to the strength of inducement mechanisms inspired by existing institutions and prices which reflect the prevailing distribution of property rights.

Issues of sustainability cannot be divorced from questions of technology and technique since these largely shape the practice of agriculture. It is surprisingly rare to find commentators writing about theoretical aspects of technical and technological change in the context of sustainable agriculture. It is critical that this task is undertaken so as to understand why; (i) it has taken so long for sustainability to become the buzz-word that it is today; and (ii) whether sustainable development will remain, as it has largely done thus far, a 'sermon without deeds' (Bressers and Huitema 1996, 2).

Notes

1. This debate has helped demonstrate the complexities involved in the morally and intellectually challenging debate inspired by Malthus' work. Any respectable neo-Malthusian debate concerns itself not only with food and demographic trends, but also with natural resources and technology, all of the plausible counterfactual futures which these could generate, and the ethical questions raised by each.
2. Malthus' work was written in response to the more optimistic views of the day (i.e. the aftermath of the French Revolution) and suggested that improving the lot of the poor was a fruitless exercise as long as they maintained their propensity to rapidly reproduce. It is untrue to say that Malthus was unconcerned for their plight.
3. Malthus' views on agricultural improvement are best discerned from the summary view of his work, published in 1830 (Malthus 1976b).
4. Anticipating Hayami and Ruttan's (1971) distinction between mechanical, and biological and chemical types of technology, Ricardo (1973, 42) noted that; 'improvements in agriculture are of two kinds: those which increase the productive powers of the land and those which enable us, by improving our machinery, to obtain its produce with less labour....If they did not occasion a fall in the price of raw produce they would not be improvements; for it is the essential quality of an improvement to diminish the quantity of labour before required to produce a commodity.' Like many after him, Ricardo seemed uninterested in product quality and the environment.
5. See Ricardo's original Preface (Ricardo 1973, 3).
6. See especially Ricardo (1973, chs. II, V, VI and VII).
7. This is an argument that stretches back in time to de Candolle (1886) who argued that agriculture was not 'chosen' but a necessary adaptation in response to population increase.
8. One can argue that Boserup's theory is a precursor of induced innovation theory (see below), though one restricted in its concerns to population pressure.
9. The importance of the Second World War in the history of United Kingdom agriculture would be difficult to overstate. It was the compact forged during the War years between the Government and the National Farmers' Union that led to intensified agricultural production through subsidised use of modern inputs by farmers (see Lowe *et al.* 1986). Part of the significance of wars relates both to reduced opportunities for trade, and (for example in the US during the First World War) a concern to provide for allied countries.
10. Depending on what one takes to be the labour implications of migration (Boserup 1965, 66), one could argue that this possibility is already exhausted by the time matters reach such a critical stage, in which case, starvation is the only plausible future.

11. This will be all the more true when the effective demand for agricultural output is weakest in countries with high population growth rates. Several studies have attempted to show how population growth increases income inequality (e.g. Lindert 1978).
12. Public sector organisations were established in Germany, the US and Japan in the 1850s, 1860s and 1880s respectively (Hayami and Ruttan 1985, ch. 8).
13. At the time, it was felt that technical change was inherently biased in a labour-saving direction, raising concern for the implications for employment and income distribution. Interest in Hicks' work was renewed in the late 1950s, and more especially in the 1960s (see Fellner (1961) and Ahmad (1966)).
14. This forms part of a wider theory of technologically generated economic development along lines suggested by Simon Kuznets (1966).
15. In fact, Schultz (1964) was less silent than one might infer from reading AD1 and AD2. Chapters 9, 10, 11 and 12 of his work are entitled, respectively, 'A Supply and Demand Approach'; 'Suppliers of New Profitable Factors'; 'Farmers as Demanders of New Factors', and 'Investing in Farm People' (Schultz's use of the term 'new factors' is broadly synonymous with 'technology').
16. H&R ignore 'informal' innovation in their theory (informal is defined in Chapter 4).
17. A notable omission from this list are philanthropic organisations, such as the Rockefeller Foundation, which have played such a major role in the promotion of agricultural research around the world (see Chapter 6).
18. Perhaps H&R's inconsistency explains why Ruttan (1988a, 85) has downplayed the importance of what he calls a 'distinction without a difference' by suggesting that the rules by which organisations function are themselves a product of another organisation's decisions.
19. An interesting discussion as to why this is the case is found in Winner (1977, 8–12).
20. I have used the definitions employed by Binswanger because it was he who fleshed out the theoretical basis of the H&R model of induced technical change in agriculture. H&R hardly bother to define either. There is an extremely awkward definition of technical change hidden away on page 86 of AD2, given in reaction to the criticisms of induced innovation levelled by Salter (1960); 'We regard technical change as any change in production coefficients resulting from the purposeful resource-using activity directed to the development of new knowledge embodied in designs, materials, or organizations.' This is different to Binswanger's definitions and appears to exclude any serendipitous discovery by virtue of its not being a 'purposeful resource-using activity'. The language is deserving of attention, the definition being phrased in terms of numbers in a formula. A section on 'changes in technology and resource endowments' (AD2, 98) uses the terms 'technical change' and 'technical innovation', but not technology or technological change.
21. Eliasson (1988, 151) is one of many scholars who have found themselves employing circular definitions of this nature; 'I shall try to be consistent in my use of the term "technology" as the knowledge of techniques, and "technical change" as the result of the application of new technology.'

22. Imagine, for example, using one of two (or more) pesticides which appear to affect the health of labourers, subsequently inducing a different choice (e.g. Antle and Pingali 1994).
23. The reason for Binswanger's, and his followers' confusion is that they are trying to distinguish between 'factor substitution', and 'technical change' (see Salter 1960, 43–4; Binswanger 1978a, 25–6; AD2, 86). The former is defined as a choice from existing techniques, and the latter, the development of new ones. This exercise in semiotics enables a change in technique to escape definition as technical change. Indeed, technical change from this definition has nothing to do with the choice of technique, only concerning itself with the research and development leading towards a *new* technique (I would call this innovation).
24. He leans on Ahmad (1966, 347).
25. The most likely explanation for the relationships being counter-intuitive is given in Kislev and Peterson (1981, 563–4), who note that induced innovation theorists have failed to adjust the price of machinery for changes in quality.
26. Kislev and Peterson (1981, 562) point out that factor bias need not result from price changes if the innovation possibility curve itself shifts in a biased manner.
27. H&R do this elsewhere in their analysis. The Enclosure Acts in England in the fifteenth/sixteenth and eighteenth centuries 'can be viewed as an institutional innovation designed to exploit the new technical opportunities opened up by the innovations in crop rotation, utilizing the new fodder crops (turnip and clover), in response to the rising food prices' (AD1, 59–60; AD2, 97).
28. See de Janvry (1978); de Janvry and and Le Veen (1983); de Janvry *et al.* (1995); and last, but by no means least, Sanders and Ruttan (1978).
29. I would suggest that one could make similar arguments to Burmeister's in almost every other country where prices have been manipulated by the state. Burmeister (1995, 42) makes the point:

    > the induced innovation framework implies more than just the congruence of new technology with aggregate factor endowments and relative factor prices. Evidence must also be adduced at the processual level – that is, one must analyse how agricultural research decisions are made, implemented, and adopted. This analysis should complement studies of technical bias and research allocation patterns.

30. One approach that has been suggested for measuring sustainability in agriculture is through trends in total factor productivity (Harrington 1992a, b). An acknowledged problem in this respect is what should be included as partial productivity measures. In general, however, any attempt to derive an overall index founders on the rock of finding a suitable metric to aggregate different partial productivity (or any other) measures.
31. H&R express the view that agriculture, formerly a resource based activity, is now a science based activity. This is very much at odds with the findings of those engaged in research on agricultural energetics which suggests that the use of science-derived inputs in agriculture makes production far more resource intensive than it has ever been in the past (c.f. note 39, Chapter 1).

32. For the purposes of this exposition, I use 'the environment' though as will become clear, I could disaggregate this by medium such as soil, air, water, etc.
33. Ruttan saw his earlier (1971) article on technology and the environment as an early move towards induced innovation theory (see Ruttan and Hayami 1995a, 34). This bears all the hallmarks of the natural resource economists' belief in limitless possibilities for substituting for increasingly scarce resources. Whilst, like H&R, Binswanger (1978b, 126–7) supposes that inducement mechanisms may work to improve matters in both respects, as regards exhaustible resources, he adds that;

    > Innovation possibilities may be constrained, and for certain exhaustible resources it may not be possible to invent substitutes or to reduce their use substantially in most applications (127).

    Apparently disregarding the seriousness of what this might imply (both for the matter under consideration and the nature of the production functions used by Binswanger elsewhere), he glibly concludes that, 'the induced innovation perspective leads to a more optimistic assessment of future growth possibilities than the antigrowth proponents would predict'.
34. Not only do these include the fact that other factors, such as taste, cooking quality, aesthetics, drought-tolerance, earliness, etc. affect adoption decisions, but the rice:fertiliser price ratio is not moving immutably in one direction.
35. Ruttan states;

    > the capacity of a nation to respond to the demand for technical change is strongly conditioned by its cultural endowments – by its ability to organise and sustain institutional changes conducive to research in agricultural science and the development of agricultural technology, for example (Ruttan 1988a, 98; also AD2, 96).
36. Bromley makes a similar point rather differently. He points out that economically efficient solutions, and Pareto optimal solutions are specific to a given institutional context. This is not what is being implied by H&R whose notion of 'optimal' seems to transcend the various (infinite) institutional permutations.

3
The Determinants of the Path of Technological Change in Agriculture: An Unorthodox View

Introduction

This chapter seeks a closer understanding of how various influences are brought to bear in the development, by formal agricultural research, of new techniques and technologies in agriculture. In particular, it sheds light on why some technologies are chosen rather than others. An explanation is offered for the possibility of becoming locked-in to a particular way of doing agriculture. In so doing, an evolutionary metaphor is employed, from which perspective, different influences act as sources of selection pressure. These lead to the adoption of a path followed by research which determines which technologies may emerge from what are presumably myriad possibilities.

First, the conceptual basis for this view is elaborated. Following this, using concepts of lock-in (to a particular technological trajectory) and lock-out (of alternatives), it is suggested that research tends to follow trajectories of innovation based around a fairly stable, though nonetheless evolving, techno-economic mode. Chapter 4 builds on this synthesis.

Evolutionary economics and the study of technical change

Schumpeter and the revival in evolutionary economics

Evolutionary perspectives in economics are very much in fashion. Evolutionary thinking in economics is nothing new, and has a notable pedigree.[1] Renewed interest reflects the belief that the keystones of economic orthodoxy, its assumptions concerning the rationality of actors and preoccupation with equilibrium, are losing their capacity to support the edifice founded upon them. Dismissing disequilibria as temporary aberrations in a world of economic equilibrium, has become increasingly untenable in a world defined by, as Giddens puts it (though he specifies the capitalist world), 'chronic economic mutation and technological innovation' (1979, 223).

Yet concern with change need not necessarily imply an evolutionary perspective. How one chooses to define an evolutionary process is no trivial matter since even within biology, from which a growing number of social sciences have borrowed the term, considerable debate surrounds the concept (leading Hodgson (1993, ch. 3) to set up a taxonomy of meanings of the term 'economic evolution'). Evolution can be characterised by three properties:

- the potential for continual change;
- processes of changes are irreversible; and
- successive states are linked to earlier ones through a more or less well understood mechanism.

The latter states, however, are not *determined* by earlier ones.[2] Darwinian natural selection is one type of evolution which occurs through the two-stage process of natural selection from continually emerging, undirected variation.[3] The properties mentioned above imply nothing concerning the desirability or otherwise of evolutionary processes. Evolution is not necessarily the hand-maiden of progress, more especially since the concept of progress is a metaphysical one.

Much of what is termed evolutionary economics addresses itself to change, but not necessarily to evolution. The recent interest in this area has been provoked by the fact that economic analysis in the first half of this century, if not the second also, has been characterised by equilibrium analysis in which technologies are considered as given. As Freeman (1988, 2) notes, mainstream economic theory demoted technical and institutional change, 'to the status of "residual factors" or "exogenous shocks" even though they were at one time subsumed within the general framework of classical "political economy."'[4] Even where attempts were made to endogenise technology within the economic system, the process by which technological change occurs has been implicitly kept outside the considerations of such theories. Technologies, for all we were told, fell out of the sky instantaneously to assist firms in adjusting to new factor prices.[5]

Nelson (1987, 2–7) suggests several influences responsible for reviving interest in the economics of technological change in the 1950s, though these were not recognised as parts of a coherent investigation.

Among these was the legacy left by Joseph Schumpeter. It is often stated that he alone among major twentieth century economists accorded technological change a central place in the economy. Schumpeter is also accorded a central role in evolutionary economics, but both matters are disputed. For whatever reasons, and perhaps they are to be found in his personal background,[6] the main focus of Schumpeter's analysis of change is the entrepreneur, but the proximate cause of change is new combinations, or innovation. In his *Theory of Economic Development* (Schumpeter 1983, first published 1911), it is the entrepreneurs who are the source of endogenous change in the economy, and they introduce 'new combinations' by forming new enterprises, or 'partially new combinations', implemented within existing organisations.[7] The term 'new combinations' is given a broad interpretation, and in *Business Cycles*, it is made almost synonymous with the term innovation, which is defined as a variation in the production function (Schumpeter 1939, 87).[8]

In his most famous work, *Capitalism, Socialism and Democracy*, Schumpeter (1976, 132), having already noted that new combinations and new enterprises are not the sole cause of change, takes another step: 'innovation itself is being reduced to routine. Technological progress is increasingly becoming the business of teams of trained specialists who turn out what is required and make it work in predictable ways.' This is but part of an argument in which he ponders the expansion of (economic) rationalistic approaches to various spheres of human existence (Schumpeter 1976, chs. 11 and 12). Such a process, he suggests, will lead ultimately to the disappearance of the entrepreneur and the death of capitalism since innovation is based on intuition, and seizing opportunities. Neither is possible in a fully rationalised world.[9] But Schumpeter's focus on the entrepreneur distracts him in large degree from considering technological change in any depth. Though he is closely associated with theories of technical change, '*technical change, in the strict sense of the development of new technical knowledge and possibilities, and the diffusion of knowledge are almost wholly absent from his exposition*' (Heertje 1988, 82, my emphasis).

Although Schumpeter uses the term evolution to describe change in capitalist economies, he rejected a biological analogy. His concept of evolution is closer to that of Marx in that the evolution he describes is that of a system, the capitalist economy:

> The essential point to grasp is that in dealing with capitalism we are dealing with an evolutionary process. It may seem strange that

> anyone can fail to see so obvious a fact which moreover was long ago emphasised by Karl Marx ... Capitalism ... is by nature a form or method of economic change and not only never is but never can be stationary. (Schumpeter 1976, 82)

Hodgson (1993, 150) suggests that, like Marx, Schumpeter's concept of evolution bordered on revolution. His famous passage in *Capitalism, Socialism and Democracy* suggesting that capitalists survive by coping in the midst of a 'perennial gale of creative destruction', continues with the statement that, 'the problem that is usually being visualised is how capitalism administers existing structures, whereas the relevant problem is how it creates and destroys them' (Schumpeter 1976, 84).

It is noteworthy, therefore, given the influence of Marx on Schumpeter (Part I – four chapters – of *Capitalism, Socialism and Democracy*, are devoted to, and full of praise for, Marx), that so many economists who purport to be working in an evolutionary framework cite Schumpeter and not Marx as their inspiration. One reason for this may be that Schumpeter's work was based on that of Walras, and whilst there is some debate as to the extent to which Schumpeter admired or rejected Walras' theory,[10] the interpretation most often adopted by evolutionary economists (implicitly or otherwise) is that Schumpeter recoiled from Walrasian equilibrium analysis. By invoking the authority of Schumpeter, one tactically defers to someone who, though familiar with neo-classical formalism, rejected its central tenets.

For all the attempts of his followers to associate him with evolutionary theory and technological change, Schumpeter's system fails to do either adequately.[11] Indeed, the capitalist system, in Schumpeter's view, was heading towards stasis as a consequence of the rationalisation of the inventive process.[12] However, his appreciation of the need to consider and understand the motive forces for change in the economy does set him apart from other major twentieth century economists. It has been this legacy which has inspired much of a growing body of work which now seeks to incorporate technological change into an understanding of the economy, some of which is discussed below.

Variation, selection and inheritance

(Neo-) Darwinian evolutionary theory[13] is essentially a two-stage process involving both the production of random, or undirected, variation, and selection from that variation. This alters a population through greater reproductive success of advantageous variants (Gould 1980, 79). Advantageous variations can be passed on to progeny if there is a heritable basis for that variation. Whilst, as pointed out above, natural selection acts at the level of the phenotype, much interest and insight into natural selection has been gained from genetics. The commonly held misconception that natural selection acts only on the genotype, in conjunction with the renewed interest in evolution in the social sciences, has led many authors to explain evolution in the social sphere through analogies with genes (or genes themselves), selection mechanisms, and the basis for heritability of these 'genes', as well as with 'genetic' mutation.

Such analogies, however apt in specific circumstances, fail to convey the richness of evolutionary processes in social systems. For the purposes of our analysis, we need to be explicit about some of the issues which we feel have to be considered in any 'economic' understanding of technological change. The first is the nature of the individual (subject). The second is an understanding of the importance of institutions in structuring 'economic' behaviour and the individual's interaction with society. Hodgson's (1988, 10) definition of institutions is that they are the means through which structure is imparted to human societies, but since individuals also *constitute* institutions, the two actually co-evolve. The third is an appreciation of the process by which technologies are developed. This of necessity relates to the other two issues since it is people, working either alone, or within organisations, or, increasingly, within a number of organisations working more or less co-operatively, who create new technologies and innovations. Within the economy, therefore, it is likely that evolution occurs at different levels. There are probably different forms of 'selection', correspondingly varied 'units of selection' (and forms of variation within them), and related diversity in mechanisms of 'inheritance.' Whether 'mutation' is a useful term or not might depend on which of this plethora of mechanisms we are considering, but we might suppose that the socio-economic equivalents of mutation are serendipity, and its antonyms.

By way of example, loosely defined, culture is the outcome of the transmission of information through institutions and individuals to other individuals and institutions which affects, or potentially affects, the development of both. It would appear that some species have a capacity for culture and others do not, and that this capacity for culture *might be* genetically determined. However, to suggest that culture itself is genetically determined (as Wilson (1978, 167) implies) seems, as Rindos (1984, 55) points out, 'to fly in the face of the fundamental observation concerning cultural behaviour, namely, that in its various particular manifestations, culture is not genetically transmitted.' The transmission of culture depends on the way in which information is transmitted and received by individuals, not on genetics. Consequently, as has often been pointed out, cultural evolution can occur at a much faster pace than the evolution of populations in biology.

The transmission of cultural information does not, of course, guarantee that the behaviour of individuals will be modified in accordance with cultural norms. Indeed, if this were the case, cultures might lack the source of variation enabling them to evolve. Information is processed through cognitive structures of individuals who may subsequently either accept or internalise the values or norms of a culture, or choose to reject them. The situation rarely being so clear cut, matters are further complicated by the need to consider not only the effect of the subconscious, but also the tacit nature of much knowledge that individuals make use of over the course of their lives. As Sagoff (1988, ch. 3) has cogently argued, the framework for decision-making employed by any given individual changes as that individual assumes different roles (for example, citizen or consumer) within their life. Thus, in the social sphere, evolutionary analogies, if they are to be made, are likely to involve considerable complexity.

Co-evolution

Co-evolution has been described as:

> an evolutionary process in which the establishment of a symbiotic relationship between organisms, increasing the fitness of all involved, brings about changes in the traits of organisms [...] Co-evolutionary sequences frequently may be described as cooperation, but they do not depend on recognition by the organisms of the advantages involved. (Rindos 1984, 99–100)

Co-evolution is a concept that has been applied to issues in economics and agriculture by Norgaard (1987, 1988a,b, 1992, 1994). However, for reasons associated with the use of the term 'evolution' in the social sciences in the past, many scholars have been understandably shy of the term.[14]

In his theory of structuration, in which he seeks to build a bridge between a theory of action and the subject, and structuralist theory,

Giddens is understandably wary of the teleological functionalism which has characterised use of the term evolution in sociology. [15] However, the much discussed notion of reflexivity is strongly suggestive of society and individuals co-evolving. Where the analogy breaks down is in the fact that the individual is *part of society rather than being distinct from it*, but this coupling perhaps strengthens the case for a co-evolutionary analogy.

The importance of Giddens' theory of structuration is that it provides a means to understand the process by which institutions change. The supposedly Marxian view that technological change conditions institutional change shows an appreciation of the importance of both institutions and their evolution, but the direction of causality is claimed to be one way. Co-evolutionary views provide a promising approach to understanding the interaction between changes in either domain. It then becomes necessary to view technology, institutions and individuals as a coupled system, rather than independent areas of study.[16] The bewilderingly complex interplay of these elements drives economic change, or evolution.

Fitness Landscapes

The concept of a 'fitness landscape' was used as long ago as 1932 by Sewall Wright, building on his own earlier work on evolution in Mendelian populations. In his original work, 'contour lines are intended to represent the scale of adaptive value' (Wright 1933, 357). Kauffman (1989, 275–6) defines the fitness landscape as: 'the distribution of fitness values over the space of objects, proteins, morphologies, or otherwise.' Adaptive evolution within the landscape entails an uphill walk to a higher point in the landscape, representing an increase in the fitness of the object being studied. Hence, the notion of adaptive hill climbing. Wright noted that for any population, more than one peak could exist on a given landscape. Exactly which peak an organism begins to ascend is determined by early selection acting on the diversity generated through genetic recombination (and mutation), so that the adaptation process is path-dependent (see below).

Fitness landscapes need not remain stationary over time. Not only can external factors deform a landscape, but landscapes of co-evolving entities are coupled, with the result that a co-evolutionary adaptation in one entity may lower the fitness of another, the so-called Red Queen effect (Rosensweig *et al.*, 1987). This in turn increases the number of possible 'mutants' which can be considered to enhance fitness in the new landscape (there are more points on a hill that are higher up if one starts lower down).

Kauffman (1988) has sought to apply the concept of fitness landscapes to economic processes. The concept provides a useful way of visualising the evolutionary process of technological development. In the biological case, mutations and new genetic combinations, as well as changes in the landscape, alter the fitness of a species. Changes in the landscape are caused both by environmental changes and by changes in the fitness of other organisms. In the technological sphere, the fitness landscape selects out 'ways of doing things.' The fitness of a 'way of doing something' is determined by the landscape, which is defined by the socio-economic institutions in which the particular 'way of doing a particular thing' is to be introduced. The equivalent of mutation, or genetic recombination, is the process of learning about existing techniques and technologies, and applying the lessons learned. The landscape can be altered by changes in the way related activities are performed or, by changes in the institutions which constitute, and are constituted by, society. Owing to the complexity of the evolutionary process in social systems, the fitness landscape finds application in correspondingly diverse ways:

1. At the level of ideas and the shaping thereof. Ideas about how things can be done are selected through institutions of education, through socialisation (in society and academic disciplines), and through career prospects;[17]
2. At the level of specific technologies and their development. Although a great many ideas may be forthcoming as to how to approach a particular problem, some are more appropriate to the prevailing situation than others. The topography of the fitness landscape is shaped by organisational and disciplinary hierarchies (whose ideas are most likely to be carried forward?), economic (cost) considerations, and regulatory institutions, as well as the fit of the technology in question with those already in existence. Kauffman (1988, 127) speaks in terms of how well a product or service 'meshes into the existing economic web';
3. At the level of species. Both the species and the varieties within species that are used in agriculture are selected as agricultural systems and the societies of which they are a part co-evolve. Non-agricultural technologies promote specialisation through integration of markets over space and time, and as the seed industry itself becomes more market-oriented, so the institutions governing that market shape the choice of species and varieties within species (Norgaard 1988); and

4. At the level of ways of doing agriculture. The web of technologies increases in complexity over time, as more technologies are developed on the basis of their inter-relation with others. Subsequently, they co-evolve as the web itself evolves. Technologies tend, by the very nature of this process, to become inter-related. Thus, HEIA is characterised by a technological package, and choice as to the make-up of that package is restricted to that between brands rather than between technologies *per se* (where differentiation is rather narrow).

Fitness, both in biological, but even more so, in social systems must be understood as internally referential, since it is defined by the fitness landscape, which is constituted by other organisms, or innovations and institutions, and their interactions. This highlights the difficulty of talking in terms of '*evolutionary progress*', and indeed, with the exception of the most teleological of Lamarckian evolutions, in which evolution is directed towards a particular end (whence the phrase becomes tautological), the phrase is an oxymoron.[18]

Technological trajectories, regimes and paradigms

In formulating their evolutionary view of the economy, Nelson and Winter made several important contributions to the theory of innovation. Reflecting their dissatisfaction with demand-pull and supply-push theories, and acknowledging the significance of uncertainty in the innovation process, they suggested that employing (profit) maximisation procedures to the problem of selecting and screening research and development (R&D) projects was implausible (1977, 49–52). Those engaging in R&D employ a heuristic search process in which are employed 'proximate targets, special attention to certain cues and clues, and various rules of thumb ... While a costless maximizing algorithm would be preferred by decision makers ... good heuristics is the best one can hope for' (Nelson and Winter 1977, 53).[19] In their view, most organisations are confined to particular areas of R&D, and within each area, procedures followed may be quite similar from project to project. At the industry level, technological change occurs with an apparent inner logic, reflecting the fact that once a particular technology has matured, the heuristics associated with it enable payoffs to be realised by continuing development in the same direction (1977, 56). This direction is the natural trajectory. They hypothesise a link in some cases between the concept of a natural trajectory and a technological regime which relates to:

technicians' beliefs about what is feasible or at least worth attempting ... The sense of potential, of constraints, and of not yet exploited opportunities, implicit in a regime focuses the attention of engineers on certain directions in which progress is possible, and provides strong guidance as to the tactics likely to be fruitful for probing in that direction. In other words, a regime not only defines boundaries, but also trajectories to those boundaries. Indeed, these concepts are integral, the boundaries being defined as the limits of following various design trajectories. (Nelson and Winter 1977, 57)

Natural trajectories may exhibit complementarities. There may also be trajectories which affect several technologies such as mechanisation of manual operations and exploitation of latent scale economies.

Dosi (1982) has postulated the existence of technological paradigms analagous to the scientific paradigms of Thomas Kuhn (1970).[20] He too uses the concept of a trajectory:

We shall define a 'technological paradigm' broadly in accordance with the epistemological definition as an 'outlook', a set of procedures, a definition of the 'relevant' problems and of the specific knowledge related to their solution. We shall argue also that each 'technological paradigm' defines its own concept of 'progress' based on its specific technological and economic trade-offs. Then, we will call a 'technological trajectory' the direction of advance within a technological paradigm. (Dosi 1982, 148)

Later, Dosi (1982, 152) draws an analogy with what Kuhn termed 'normal science':

As 'normal science' is the 'actualization of a promise' contained in a scientific paradigm, so is 'technical progress' defined by a certain 'technological paradigm.' We will define a *technological trajectory* as the pattern of 'normal' problem solving activity (i.e. of 'progress') on the ground of a technological paradigm.

Dosi's main contribution comes not from these definitions, but in their elaboration. Borrowing from Lakatos (1970)[21], he suggests that technological paradigms entail positive and negative heuristics i.e. both which projects to pursue, and which to neglect. As a result:

Technological paradigms have a powerful *exclusion effect*: the efforts and the technological imagination of engineers and of the

> organizations they are in are focused in rather precise directions while they are, so to speak, 'blind' with respect to other technological possibilities. At the same time, technological paradigms define also some idea of 'progress'. Again in analogy with science, this can hardly be an absolute measure but has some precise meaning within a certain technology. (Dosi 1982, 153)

Thus Dosi recognises the importance of the question which arises as to how an established paradigm comes to be selected in preference to others. In this and later work, Dosi makes a somewhat vague distinction between *ex-ante* and *ex-post* selection. I believe this to be an important development since although Dosi does not tell us what determines when we shall call selection *ante* or *post*, it suggests that positive and negative heuristics can function even at the level of ideas, a possibility which Nelson and Winter (1982, 142) denied (c.f. note 17).

Dosi's technological paradigms have undergone a further mutation in the work of Perez. Dosi (1982, 160) suggested that the notion of technological paradigms was relevant to the revival in interest in long-waves in the economy first inspired by Kondratief.[22] Perez (1983) perceived that long waves are manifestations of successive 'technological styles' developing, and then reaching limits of, their potential. These she defines as:

> a kind of ideal type of productive organization or best technological common sense which develops as a response to what are perceived as the stable dynamics of the relative cost structure for a given period of capitalist development. (Perez 1983, 361)

She depicts the economic system as consisting of two main subsystems, the (fast-responding) techno-economic and the (slower-responding) social and institutional. Long waves arise due to successive phases in the evolution of the total system.

In more recent work with Freeman, Perez uses the term 'techno-economic paradigm' (Freeman and Perez 1988; also Roobeek 1995).[23] It is suggested that structural crises in the economy exemplified by downswings in successive Kondratiefs can be explained through recognition that a particular paradigm is reaching its limits in terms of the opportunities for profit that it presents. During this downswing, a new paradigm emerges which is incompatible with the institutional and social framework in which the previous paradigm developed. Structural changes are required if the new paradigm is to prosper and generate an upswing in the economy.

Although Freeman and Perez (1988, 47) suggest that the concept of the techno-economic paradigm 'corresponds to Nelson and Winter's concept of "general natural trajectories"', I would argue that the two are qualitatively different. The techno-economic paradigm, which has more in common with the notion of a regime developed by the French regulationist school (Boyer 1988), is used to describe the potential for economic growth. There is a greater appreciation of the importance of the social and institutional framework, and this becomes not just a 'selection environment', but equally, an environment that is itself selected by the possibilities contained in the new techno-economic paradigm.[24] It is the fit of the two that becomes important, not the fitness of a paradigm in a given environment. Technology and the social and institutional framework co-evolve.

The idea of the need to match social and institutional structures with new technologies within a 'techno-economic paradigm' can be usefully applied at the microeconomic level. Moreover, for the reason just given, there is no contradiction in talking about both techno-economic paradigms and technological trajectories, though there are problems with the use of the word 'paradigm' relating to its usage in the scientific context. There, new paradigms eventually displace the old, since co-existence is more or less impossible. In the case of technology, especially agriculture, there is no immutable law saying that two or more ways of doing agriculture should not co-exist, notwithstanding the pressures propelling things in one or other direction.[25] Henceforth, the term techno-economic mode will be used to denote a particular fit between technologies and socio-economic institutions, and a techno-economic trajectory will be understood as the way in which both of these develop, usually in a dialectical manner, over time.

Increasing returns, path-dependence and lock-in

Evolution was defined above as a process in which future states of the system being studied are dependent on past states. Path-dependency, or non-ergodicity, implies not simply that history matters, but that history plays a crucial role both in shaping the existing state of affairs, and determining the menu of futures which can become the present. In David's words, '*Historical* choices among techniques thus rule the future' (1975, 60, my emphasis).

The very notion of a trajectory implies a form of path-dependence, the path being determined by technicians' beliefs about what can be done within a particular paradigm. Of importance to our discussion is the fact that many innovations are incremental ones. They are improvements on existing designs, the need for which may become apparent either during testing, or through its performance in a market, or even owing to complementary innovations elsewhere in the economy. In sectors where patents are an important source of competitive advantage, much innovation may arise from competitors' desire to innovate around an existing patent, being incremental, but also sufficiently 'radical' to entail patentability in its own right.

The concept of path-dependence has been most closely associated with the work of Arthur (1988a,b, 1989, 1990, Arthur *et al.* 1987) and David (1975, ch. 1; 1985, 1987). Arthur's work leant on studies in the natural sciences and theoretical biology which dealt with 'auto-catalytic', or self-reinforcing mechanisms.[26] Systems that are out of equilibrium can exhibit non-linearities in their behaviour such that small fluctuations, or deviations from the norm, are amplified. In such circumstances, a system may begin to organise itself into one of a number of possible states, though exactly which form is beyond prediction and determined by the fluctuations themselves. The fluctuations drive the system's evolution.

In biological systems, one might suppose that the fitness of an organism was as, if not more, likely to decrease as a result of a random mutation as it was to increase. An optimal strategy might be one that minimised error-making through not mutating at all. However, under the operation of natural selection, fitter organisms survive more readily than the less fit. New combinations enable adaptive hill climbing on fitness landscapes, and despite persistent error-making, new combinations allow selection to drive evolution. Similarly, there are as many, if not more, technological flops as there are successes, yet the capacity to innovate generates fitter innovations. Innovation demands novel ideas which could not emerge in a system at equilibrium composed of like-thinking rational beings. Some behaviour of the 'non-average' kind is required.[27] To focus only on a core of average behaviour is to miss the motive force for change in the system under examination.

The motivation for Arthur's work was a fairly general enquiry into the phenomenon of increasing returns, or self-reinforcing mechanisms, in the economy. Arthur (1988b, 591) cites five sources:

1. learning by using;
2. network externalities;
3. scale economies in production;
4. informational increasing returns; and
5. technological interrelatedness.

Path-dependence is one of four properties of such mechanisms in economics noted by Arthur (1988a, 10), the others being the existence of multiple equilibria, possible inefficiency, and lock-in. One can illustrate the possibility of becoming locked in to a particular choice from a number of alternatives in a number of ways. Arthur (1988a, 13) uses the example of a company's research and development department:

> Suppose an economic agent – the research and development department of a firm perhaps – can choose each time period to undertake one of N possible activities or projects, A_1, A_2, A_3... A_N. Suppose activities improve or worsen the more they have been undertaken. A_j pays $\pi_i(n)$ where n is the number of times it has previously been chosen. Future payoffs are discounted at the rate β.
> ... if activities increase in payoff the more they are undertaken (perhaps because of learning effects), the activity that is chosen first, which depends of course on the discount rate, will continue to be chosen thereafter. The decision sequence 'grooves out' a self-reinforcing advantage to the activity chosen initially that keeps it locked in to this choice.
> Notice that at each stage, an optimal choice is made under conditions of certainty; and so there can be no conventional economic inefficiency here. But there may exist *regret*.

In other words, an initial choice on the basis of a discounted cash flow comparison can lock the research and development department into its initial choice, regardless of the nature of what the payoffs of alternatives might be as more is learned about them. Arthur goes on to suggest how 'regret' might surface, but since, in the examples he gives, no new information has been acquired, these seem more a comment on the practice of discounting than instances of true regret. Although, as Arthur (1988a, 14) argues, payoffs in future years may be higher with an unselected alternative, if these had been discounted properly under conditions of certainty, there would be no cause for regret (else why discount at all?)

The only way out of this corner is to acknowledge the importance of uncertainty related to the initial assessments of the projects (including their likely future payoffs, which implies some assessment of what will be learned) and to the choice of an appropriate discount rate. Uncertainty as a cause for regret must, after all, be prevalent when discussing alternative research projects whose outcome is, by definition, unknown. The possibilities for 'backing the wrong horse' (Cowan

1987, 1991) seem unlimited (how do we know in advance what is the 'best' technology?)

Arthur also shows that lock-in can occur through a stochastic process in which, under conditions of increasing returns, the random order of arrival of agents with differing 'natural' preferences for the technologies available in the market can determine the way in which the market is ultimately shared. In some cases, monopoly becomes inevitable, but exactly *which* technology achieves monopoly cannot be predicted in advance: '[t]hus the small events that determine the [times of entry and choice of the agents] decide the path of market shares; the process is non-ergodic or path dependent – it is determined by its small-event history' (Arthur 1989, 122). Due to one, or a sequence of, small events, a particular technology can become established in preference to another, ensuring its monopoly in the future.[28] For all the mathematical sophistication of Arthur's argument, his most illuminating comment is that defining '"chance" or "historical" events':

> Were we to have infinitely detailed prior knowledge of events and circumstances that might affect technology choices – *political interests, the prior experience of developers, timing of contracts, decisions at key meetings – the outcome or adoption market-share gained by each technology would presumably be determinable in advance* ... I therefore define 'historical small events' to be those events or conditions that are outside the *ex-ante* knowledge of the observer – beyond the resolving power of his 'model' or abstraction of the situation. (Arthur 1989, 118, my emphasis)

In Arthur's model, only the timing of arrival and choice of the adopters is outside the knowledge of the observer, and this determines market share outcomes. But other small historical events, such as those italicised above, also affect market share outcomes. They may also determine which technologies reach the market in the first place, imparting a form of selection operating well in advance of market-based decisions. They determine whether, and if so, in what form, a market shall exist, and may lock-out what become unexplored (or marginalised) alternatives. These are critical points for they suggest the necessity of a historically informed approach to the understanding of how things have become as they are. If our goal is to explore the viability of counterfactuals, only through the lens of history can we judge which counterfactuals to admit and which to leave aside.[29]

David's work, more than Arthur's, is informed by this view of the importance of history:

> it is sometimes not possible to uncover the logic (or illogic) of the world around us except by understanding how it got that way. A path-dependent sequence of economic changes is one of which important influences upon the eventual outcome can be exerted by temporally remote events, including happenings dominated by chance elements rather than systematic forces ... In such circumstances 'historical accidents' can neither be ignored nor neatly quarantined for the purpose of economic analysis; the dynamic process itself takes on an essentially historical character. (David 1985, 332; also David 1975 11–16).

Using the example of the standardisation of the QWERTY keyboard, he too refers to 'lock-in'. He suggests that the decision of typists to learn one or other keyboard is like Arthur's problem of technological choice. He adds that there may also be an element of self-fulfilling expectations at work. Citing important work by Katz and Shapiro (1985, 1986) as support, he suggests that:

> Although the initial lead acquired by QWERTY through its association with the Remington was quantitatively very slender, when magnified by expectations it may well have been quite sufficient to guarantee that the industry eventually would lock in to a de facto QWERTY standard. (David 1985, 335)

Over time, the cost of the typewriter came down but the costs of retraining typists remained high. As a result, the short run decision was made to adapt the machines to people rather than the other way round. With new hardware technology, the QWERTY arrangement has been proven slower than alternatives which led to mechanical problems on earlier typewriters, but the early standardisation means QWERTY survives.

In a later paper dealing with the setting of standards, David makes some important observations relating to the possibility of lock-in occurring. Firstly, there may be only brief 'windows in time' in which policy makers have scope to act before lock-in sets in. Secondly, this period is likely to coincide with the time at which policy-makers have least relevant information – they act as 'blind giants'. Thirdly, whatever decision is made, some investors will have sunk resources into paths to be no longer pursued – they become 'angry orphans' (David 1987, 210).

The ideas of lock-in and co-evolution have also found expression more recently in Bijker's work. Bijker (1995, 273) suggests 'the technical is socially constructed and the social is technically constructed'. He prefers to speak in terms of the study of sociotechnology. Technical artefacts rarely exist in isolation, but are usually part of a seamless web, a sociotechnical ensemble. Bijker distinguishes between three types of sociotechnical ensemble, characterised by distinct technological frames:

> a technological frame structures the interactions among the actors of a relevant social group. This is not an individual's characteristic, nor a characteristic of systems or institutions; technological frames are located between actors, not in actors or above actors. A technological frame is built up when interaction 'around' an artefact begins. (Bijker 1995, 123)

If these seem similar to Freeman and Perez's technoeconomic paradigms, the roots of both can be found in Kuhn's work. The difference lies in the inclusiveness of Bijker's concept which encompasses quite explicitly all social groups. Bijker borrows the notion of interpretative flexibility from the sociology of science, which holds that nature does not provide a determinate outcome to a scientific debate. In the technological sphere, such flexibility opens up political space in the process of technological development. Rejection of flexibility forces one to accept deterministic interpretations of the future. Acceptance, on the other hand, suggests the existence of plausible counterfactual technologies. Even so, many artefacts display an obduracy which makes it difficult, if not impossible, to entertain significant flexibility. Obduracy, Bijker's equivalent of lock-in, is greatest where a state of closure, the end of interpretative flexibility, has occurred, and artefacts have achieved the status of exemplars. Social groups have invested so much in the exemplar that its meaning becomes fixed.

It should be clear that all of these considerations regarding lock-in are extremely important if it subsequently becomes clear that technologies in use are creating problems which were not foreseen (raising questions as to the ability of institutions to foresee such problems). Any number of polluting technologies could be used as examples here; DDT; the 'drin' pesticides; nuclear power stations; petroleum-driven cars; and chloroflourocarbons (CFCs), to name but few.[30] Arthur (1988a, 13) borrows an analogy from physics in using the concept of a potential barrier to describe the accumulated economic advantage which lock-in confers on a chosen technology. An injection of energy, or an economic subsidy, may be needed to exit from lock-in. Whether or not this will occur in practice depends upon the source of self-reinforcement and 'the degree to which the advantages accrued by the inferior "equilibrium" are reversible or transferable to an alternative one' (Arthur 1988a, 16). Arthur cites learning and specialised fixed costs as presenting major barriers, and one could add, related to the latter, economic webs or interests which support the prevailing situation. If technology contributes to environmental damage, lock-in may commit us to many more years of degradation than would appear prudent, even where viable alternatives are available. Addressing the problem may not be quite so straightforward as showing people alternatives to current ways of doing things.

Formal agricultural research

Science, experimentation and technology

A significant proportion, though by no means all of agricultural research undertaken in formal organisations is carried out within a particular techno-economic mode. Chambers (1993, ch. 1; Pretty and Chambers 1993), in obvious reference to Kuhn's 'normal science', criticises 'normal professionalism' in agricultural research and extension. The questions arise as to what techno-economic mode this professionalism supports, and how it has emerged.

No techno-economic mode can be unequivocally adjudged 'the best'. To argue that it can, through appeal to the fact that the institutional framework supports it, is tautological if one accepts that institutions change with changes in technology. Furthermore, since markets are themselves institutions, to suggest that selection takes place through the workings of the market is to fall into the same trap. A problem with much evolutionary economics literature is that, although informed by biological, sometimes ecological metaphors, it employs these to conceptualise technological change and economic evolution through market-based selection, but there, the appeal to ecology ends. Some technologies that are 'successes' in the marketplace create significant environmental damage. The way in which institutions governing markets are structured is a potent force in the shaping of techno-economic trajectories. The fact that these institutions are not so structured as to engender respect for ecological constraints reflects a particular perspective on technology, or markets, or both. Indeed, the term used by economists to account for this phenomenon, 'externalities', is suggestive of more or less conscious efforts to exclude the 'side-effects' of technological and economic development from the field of view.

Various authors have sought to show how agricultural science has suffered from myopia as a result of the legacy of scientists such as Galileo, Bacon, Descartes and Newton.[31] Reductionism, a masculine approach, a mechanistic view of the universe, and positivism[32] are seen by these critics to be at the root of a failure to appreciate, variously, farmers' knowledge, women's knowledge, the importance of biological diversity, and the ecological context in which agriculture is carried out. It would, however, be making rather too much of the influence of the ideas of these men to suggest that the history of agricultural research has been founded uniquely on their ideas, and has suffered as a result. Important though they have been, the context in which these have been carried forward has also been of significance in shaping the development of science, as well as technology.[33] The success of Newtonian physics in not only describing, but also predicting the movement of macroscopic bodies, was an important event in the development of the modern world. From the seventeenth century, Europe was transformed from a society in which divine providence dominated life, to one in which empirical observation and the notion of progress founded on science and reason achieved ascendancy. Scientific experimentation was revealing aspects of the natural world which could be harnessed for the benefit of humanity.

Towards the end of the eighteenth century, England was already in the midst of the transition from an agrarian, craft-based society to one based on the factory and the machine. The triumphant march of scientific and technological rationality, especially in the wake of the French Revolution, prompted an extension into the social sphere of the rationalisation process.[34] All nations of Europe felt the effects of this break with tradition in the nineteenth century. Whereas in traditional societies, the present looked back to the past in creating the future, modern societies, aided by science and technology, began to write history anew. If science itself was reductionist, so it became the subject of its own desire to reduce a whole into constituent parts. From the mid-nineteenth century, the acquisition of knowledge was subjected to the same division of labour as Adam Smith's pin factory. With increasing disciplinary specialisation, it became less likely that any one person would achieve professional (for science was now becoming professionalised) competence in more than one discipline. Given the importance of science and technology to industrialisation, and thereby, to opportunities for profit, it was unsurprising that such specialisation in the process of knowledge production should have occurred.[35]

Such fragmentation is not necessarily undesirable, indeed, it may have been inevitable as the body of scholarship grew. But whether the *pattern* of such disciplinary fragmentation was desirable, and even if it was, whether it is appropriate today is open to question.[36] Furthermore, as this fragmentation developed, so there developed with it a hierarchy of disciplines, closely related to their scientific 'hardness.' Today, new problem areas give rise to new disciplines, even when such problems result from the application of disciplines already in existence. Progressive fragmentation enables established ways of doing things to remain beyond critical scrutiny, and for problems to be hived off to other disciplines as and when they arise. This is exactly what Beck means when he writes: 'Science is no longer concerned with 'liberation' from pre-existing dependencies, but with the definition and distribution of errors and risks which are produced by itself' (1992, 158).

There are three principal reasons, therefore, why, in a world in which the modern scientific view of the world is pre-eminent, negative externalities associated with the use of new technologies are likely to be significant:

1. The *reductionist view of the material world* prominent in much research. Reductionism is itself a self-reinforcing system. Technological developments increase the resolving power of instrumentation, enabling the examination of new elements, necessitating the development of instruments capable of their closer examination. Reductionism's assumption that the whole is *only* the sum of the parts simplifies analysis, but it risks losing much of the interesting behaviour in real world systems, such as the self-organising properties mentioned above. In the biological sciences, Darwin's theory of natural selection was originally proposed without knowledge of the mechanism of heredity. Indeed, Darwin appears to have recognised as a flaw in his approach the fact that if, as he believed, offspring were essentially a mix of the characteristics of two parents, new traits would be rapidly diluted in subsequent generations, leaving little chance of their being selected through natural selection. The solution to this puzzle was provided by Mendel, who proposed that there were units of heredity which were passed on from one generation to another.

   The science of genetics blossomed in the twentieth century on the basis of rediscovery of Mendel's idea. To many, this was confirmation of the views of biochemist, Jacques Loeb, whose influential work *The Mechanistic Conception of Life* (1912) proposed that the properties and functions of living organisms would ultimately be explained through physical and chemical concepts. With the discovery of the structure of DNA and an improved understanding of the way in which genes lead to the production of enzymes, the view advanced by Loeb seemed to gain momentum. The new discipline of molecular biology became the area in which much 'frontier' work was done. Yoxen (1983, 33–5) refers to the view of biotechnologists as a 'Meccano view of nature' in which life is reduced to the processing and transmission of genetic information.

In the early 1970s, the first successful experiments transforming the genome of living organisms were performed. Yet despite these efforts, which focused on ever smaller units of analysis within living organisms, very little progress has been made in understanding other important questions:

> you could say all the genetic and molecular biological work of the last sixty years could be considered as a long interlude ... Now that the program has been completed, we have come full circle – back to the problems ... left behind unsolved. How does a wounded organism regenerate to exactly the same structure it had before? How does the egg form the organism? (Francis Crick, cited in Capra 1983, 117)

The view held by some biologists that individual, or in some cases a few genes, are responsible for particular traits makes light of the interaction between genes, or epistatic effects. Such effects account for the fact that the same gene in different genomes can perform quite different functions.

Currently, a vast effort is being devoted to sequencing genomes of various organisms, a task akin to the 'puzzle solving' to which Kuhn suggested scientists revert in periods of normal science, their solutions being guaranteed by the potency of the paradigm.[37] Much of this work is undertaken with the aim of discovering genes responsible for particular traits. In plant breeding too, the prevalence of reductionist views is reflected in the emphasis on single gene traits, or qualitative inheritance, at the expense of multigenic traits, or quantitative interactions, particularly with respect to improved performance in terms of pathogen resistance (Robinson 1996).

2. The adoption of the *positivist view that judgements of aesthetic, political, personal, or moral value are deemed irrelevant to scientific enquiry* makes obtaining a complete view of technology impossible. In particular, if, as mentioned in Chapter 2, the central concerns of sustainability are environmental quality and issues of equity, issues of sustainability will be excluded from this positivist view. Yet positivism is certainly a cherished value of many natural (and indeed social) scientists. The instrumentalist view of science and technology holds that science deals only with facts, and technology deals with the means to meeting particular ends. Normative values, including the uses to which technology are put, are deemed completely separate from scientific and technological endeavour.
3. *Reductionism in the form of specialisation within science* prevents anything but a narrow segment of an already narrow perspective on reality being examined. This makes it extremely difficult to capture all the relevant dimensions of a particular phenomenon.[38] The possibility for interdisciplinary research to bridge the gap still exists, but it must overcome the division between 'white-collar' and 'blue-collar' discipline, and the mutual contempt in which the one holds the other.[39]

Expertise has become virtually synonymous with specialisation, both in the practice of science and technology as well as in the wider social world. Furthermore, the shifting nature of much modern knowledge, when combined with specialisation, allows one to choose one's own experts. Any damaging 'side-effects' of technology stem not from a conscious attitude of scientists predisposed to 'dominate nature' in the Baconian sense. The practice of science is carried out in environments which dissect and control in the quest for knowledge, and those who carry out the dissection do so because they believe (they have been taught) that this is how science 'gets done'. Yet the results are deemed to be applicable to a world of comparatively bewildering complexity. As a result, far from being dominated as Bacon envisaged, nature has tended to strike back under the pseudonyms of 'side-effects' and 'unintended consequences', generating what economists have labelled (not inappropriately) 'externalities'.[40]

The likelihood of such 'side-effects' occurring has been reinforced by the socio-economic institutions which surround technological development. Beck's disturbing analysis is worth quoting at length:

> Technological innovations increase the individual and collective well-being. The negative effects (deskilling, risks of unemployment

> or transfer, threats to health and natural destruction) have always found justification in these rises of the standard of living. Even dissent over the 'social consequences' does not hinder the accomplishment of techno-economic innovation. That process remains in essence removed from political legitimation, particularly by comparison to democratic-administrative procedures and the long periods needed for implementation; indeed, it possesses a power of enforcement virtually immune to criticism. *Progress replaces voting*. Furthermore: progress becomes a substitute for questions, a type of consent in advance for goals and consequences that go unnamed and unknown. (1992, 183–4)[41]

In short, for all the sophistication of the science employed in developing technology, technological change is encouraged and desired though no one knows quite how it will affect the lives of humans and other species with whom we share this planet. Despite the obvious limitations of such an approach, the institutional framework within which scientific and technological practice are situated presupposes the utility of outcomes, and tends to downplay the side-effects.[42]

Negative externalities inevitably proliferate in such an environment. As new technologies emerge, so they introduce new interactions with the existing world. Some of these will be foreseeable, others less so. As such interactions come more clearly into view, so they achieve the status of externality, the now-perceived downside of activities whose potential consequences were knowingly underexplored.

The dominant mode in formal agricultural research

Much formal agricultural research has evolved within, and helped shape, a breeding-chemical-mechanisation (BCM) techno-economic mode, in which genetics-based breeding activity is integrated with the development and use of synthetic chemical fertilisers and pesticides, and new machinery. HEIA is the technique corresponding to the BCM mode. Its main characteristics, admittedly characatured in what follows are:

1. Goals are set within the research organisation, often in advance of discovering, or in isolation from, what farmers want, or need (Pretty and Chambers 1993, 10–11). Expertise is held to reside with the researcher, with the farmer as passive recipient of new technology (Richards 1985, 138). The principal clients may not be farmers, and where they are, they may be a specific sub-set of the wider farming community;

2. Knowledge/data is acquired through controlled experimentation. A complex world is thereby simplified so as to examine cause-effect relationships affecting, usually, a single variable;
3. Related to 2, resources are concentrated on individual crops or even varieties of a crop. Dealing with intercropping patterns introduces unwanted complexity.[43] Furthermore, working with individual varieties allows for pronouncements on which variety is 'best';
4. Again related to 2, increased control over the natural environment so as to ensure replicability, and therefore, validity of experimental results. A single reality, what Busch *et al.* (1991, 43) call 'ontological monism', is assumed, despite multiple social and ecological realities;
5. Investigators may work alone, or in single disciplinary groups. Maxwell (1984, 37) notes that the dominant organisational structure is what he calls a role culture model;
6. Mechanisation of functions performed either on the farm, or off it (in handling, or in processing the product) is actively sought, and where 'bottlenecks' to mechanisation exist, research is oriented to removing these;[44]
7. Knowledge is transmitted to the field through extension agents. Knowledge flows in one way only (*from* the research organisation); and
8. There is an institutionalised inability to perceive the possible negative consequences of new technologies. Norgaard (1992, 78) attributes this to 'unforeseen' changes in social and environmental systems which are apt to occur due to science's belief in universals, and its production in controlled, replicable situations.

This mode reflects the characteristics of modern science and technological development outlined above. Since, as Rouse (1987, 118) points out, 'science sometimes "works" only if we change the world to suit it', over time, the field has been made to look more like the laboratory. As far as possible, nothing is present that should not be, and nothing happens that cannot be accounted for by agricultural scientists (Murdoch and Clark 1993, 6).

Just as the development of science and technology both shapes and is shaped by social change, so it is with the BCM mode. The increasing integration of agricultural producers into markets of growing geographical extent, and the tendency towards vertical integration in the food and agricultural raw material industries shapes, and is shaped by, the evolution of the techno-economic trajectory. The emergence of global agro-industrial complexes centred around specific commodities has been made possible by, and now supports the diffusion of, technologies developed in the BCM mode. Most authors who have commented on the emergence of these global empires speak, echoing Marx's view expressed in Chapter 2, of technological developments designed specifically to 'reduce the importance of nature in rural production' (Goodman *et al.* 1987, 3). Again following Marx (1959, 645), Marglin (1974, 1990), Noble (1984), Braverman (1974) and Marcuse (1964) amongst others have sought to show that:

> technological innovation responds to the exigencies of control as much as, if not more than, to the exigencies of efficiency ... capitalist production is shaped by the concerns of capitalists to create, and then to maintain, a place for themselves in the production process. (Marglin 1992, 14)

Thus, a drive to control both the production and labour processes in agriculture, whilst at the same time minimising the exposure of industrial capital to the residual climate- and other ecological-related risks associated with farming, has characterised the development of the BCM mode.

In the context of agriculture, Pretty and Chambers (1993, 8) criticise science not as a body of knowledge, principles and methods, but on the basis of the beliefs, behaviour and attitudes that accompany it. Marglin (1992, 32) goes further, suggesting that modern science, because it portrays itself as the totality of knowledge, cannot peacefully co-exist with the tacit knowledge of farmers. Yet in the opinion of Long and Villareal (1993, 163), Marglin's view 'founders ... upon the rocks of dichotimisation'. For Long and his followers, the encounters between scientists/administrators and farmers must be perceived as knowledge interfaces, and their actor-oriented perspective emphasises the social construction of knowledge (Long and Long (eds) 1992; Arce 1993; Long and Villarreal 1993; Arce and Long 1994).

Common to these perspectives is an appreciation, however varied, of the interplay between power and knowledge. The two are intimately bound and indeed one can postulate a dialectical interplay between the two. For Howes and Chambers (1979, 7), scientists' need to lean on scientific knowledge to legitimise their superior status is at the root of a bias against polyculture systems (see also Belshaw 1979, 24–5).

Crop ideotypes and technological interrelatedness

Genetics, and its application to breeding, changed the way in which plants were perceived by breeders:

> Species and varieties *which in themselves have little or no intrinsic value* become of first importance if they possess certain desirable characters which may be transmitted through breeding. In other words, *plant exploration is to a large extent becoming a search for genes.* (Ryerson 1933, 123–4, my emphasis)

Plant breeders began a quest for ideal plants:

> Now the breeder tends rather to formulate an ideal in his mind and actually create something that meets it as nearly as possible by combining the genes from two or more organisms. (Hambridge and Bressman 1936, 130)

Historically, a problem with the ideal plants has lain in the relationship between farmer and scientist, and field and laboratory, or field station. The scientists' ideal plants have not always been ideal from the farmers' point of view, their performance in the field being quite different from that in the laboratory or field station.

Many authors writing in the wake of the Green Revolution talked about high yielding varieties in the context of a *package* being promoted for use by farmers. Some suggested that the term highly responsive variety was a more apt description than high yielding variety since considerable use of synthetic chemical inputs, as well as available water, was required for their most profitable use.[45] That this should have been the case was part consequence of the mode of research described above, in which yield maximisation was pursued in controlled environments. Plant breeders often employ ideal plant concepts. Donald (1968; see also Reitz (1968)) used the term 'ideotype' to describe an ideal which would maximise biomass and partitioning of grain for wheat, and similar ideas were elaborated at the International Rice Research Institute (IRRI) in the Philippines.[46] In a widely-read and oft-cited text, Simmonds (1979, 63) states: '[T]he breeder will need to have some sort of ideotype in view, and this should largely reflect the breeder's appreciation of what is desired by markets.' As to how breeders should actually decide how to set priorities:

> in determining objectives, the plant breeder will no doubt listen to anyone qualified to comment but must resign himself to receiving nearly as many different bits of advice as he has advisers. Recognition of crucial elements is never simple but the breeder's own judgement on this is probably better than that of most others. (Simmonds 1979, 64)

This statement is important for two reasons. Firstly, there is no appreciation of possible disjunctures between breeders' and farmers' objectives, and secondly, it reflects the increasing number of interests which influence the work of breeders. These interest groups exert their influence precisely because plant breeders can and do reflect them.

The most in-depth studies of the emergence of a particular ideal plant concept have involved the International Rice Research Institute in Manila.[47] Oasa's (1981) account, based on a stay of several years at IRRI during which extensive interviews were carried out, is particularly thorough. What happened at IRRI exemplifies the transfer of the BCM mode from its birthplace in the United States to tropical Asia. In noting the emergence of the ideal plant concept, Oasa shows how alternative conceptions for dealing with particular problems were dismissed, even laughed at by IRRI some scientists. Interestingly, he notes that IRRI's varietal improvement staff had virtually no experience in tropical conditions. Despite this, the goals of the varietal improvement programme were established very early on the basis of procedures carried out in America (Oasa 1981, 183).

Beachell and Jennings (1965, 30) outlined the characteristics of their ideal plant at a 1964 IRRI Symposium, emphasising that most of the characteristics they deemed to be most desirable were scarcely found in the tropical rices grown at the time. Controlled conditions were essential:

> Controlled conditions made attainable a 'big jump' in yield. In terms of the potential success in research, they made possible the development of the ideal plant type ... Because only chemicals sufficed to eliminate weeds at the early growth stages of a short-statured variety, costly herbicides were indispensable to the ideal plant type's success. In the particular case of weed control, *the Institute was essentially locked into investigating a complementary technology* that could yield quick and visible results. (Oasa 1981, 205, 217, my emphasis)

All departments were affected by the ideal plant concept:

> The architecture of the rice plant established the parameters within which research was conducted and the types of question raised. In a word, the purpose of other departments was to locate "bottlenecks" and eliminate them ... Elimination of constraints meant elimination in the quickest and most visible way. (Oasa 1981, 232)

Note that the concepts of what an ideal type of plant would be were constituted without consultation with farmers. IRRI's Office of Communication conducted a one-way information transfer extolling the virtues of Institute technology. The notion of scientists as experts, and farmers as passive recipients of new technology, was regarded as unproblematic even in environments of which the scientists had no experience, and farmers themselves were experts, at least in relative terms. What could be achieved in controlled laboratory conditions was assumed to be possible in farmers' fields. It is important to state that at each and every turn, there were some who foresaw the 'unforeseen' side-effects, or second-generation problems of this technological package. Yet the 'success' of technologies in controlled conditions was apparently used to legitimate the view that worse than expected performance in the field was the fault of farmers themselves, or a result of 'constraints' which, it was implied, were no concern of the scientists themselves.

Oasa highlights the 'normal' professionalism which has been internationalised steadily in the post-Second World War years, particularly through the CGIAR centres. Goals are set in accordance with the prevailing techno-economic mode, and pursued using the box of tools that have shaped, and been shaped by, it. The idea that farmers might be experts in their local ecological conditions was not entertained. Yet as Richards (1985, 157) puts it, 'Why should a farmer expect to "hand over the controls" to an adviser who, in all probability, has never before piloted a farm "for real"?'

A selection pressure model of technological change

The aim of this section is to draw the threads of this chapter together. In the technical sphere, the BCM mode is dependent principally on the interplay of innovations in the physical, mechanical, chemical and biological sciences.[48] Each of these three contributing 'disciplines' shapes the heuristics which guides research in the others. At the institutional level, professional bodies, career structures, and education and training programmes anchor the mode. The innovations emerging as a result of these heuristics aim either:

- to create new wants in consumers (as Schumpeter (1912, 11) well knew, not all technologies seek to meet the expressed desires of would-be consumers. Many technologies aim to create new 'wants'); or

- to meet an 'obvious and compelling need', removing bottlenecks in the sequence of technological change;[49] or
- to exploit new opportunities arising as a result of the use of new technologies.

An economy becomes, 'a web of transformations of products and services among economic agents. Over time, "technological evolution" generates new products and services which must mesh together "coherently" to jointly fulfill a set of "needed" tasks' (Kauffman 1988, 126). The mesh becomes finer and the web increases in complexity as economic interests grow and diversify, as institutions are transformed, and as industries mature, increasing the degree of specialisation in production.

The heuristics shaped by the key disciplines of the BCM mode are the source of adaptations which increase the 'fitness' of either a machine, a seed variety, or a chemical input within the economic web. The organisation and development of science, and especially, its splintering into disciplines, have favoured a reductionist view of agriculture. Furthermore, the chronological order in which scientific disciplines have matured has focused attention first on pieces of the physical, then the biological world.

A new technology, or an adaptation of an old one would deform the fitness landscape to which other technologies are coupled. Thus, the ability to breed for uniformity alters the fitness of mechanical innovations which function best where plants are uniform. Evidently, not all technologies are coupled to all others. The way in which the coupling process itself evolves is a characteristic of the trajectory (of the BCM mode). Thus, the economic web encompasses arrays of coupled, co-evolving entities, the path of whose development is traced out as combinations drive the evolution of the web through deforming the fitness landscape, and challenging coupled technologies to evolve accordingly. In the BCM mode, agricultural input supply, engineering, and food processing industries have co-evolved with the ideal plant concept. The demands of these industries increasingly shape, and are shaped by, the appearance of the ideal plant. Institutional changes also deform the fitness landscape through altering incentives for investment in particular activities. Agricultural policies, and property rights regimes, especially regarding germplasm, are important in this regard.

The fitness of some erstwhile 'unfit' technologies can be improved by new technologies. In such instances, it might be more appropriate to speak of co-mutation than co-evolution, as the very existence of one technology in the web is predicated upon the existence of the other. The tomato harvester and tomatoes with skins able to withstand that process were developed in tandem. Similarly, some technologies which could improve the fitness of technologies used in other techno-economic modes never emerge, being rejected by the dominant mode.

The heuristics applied in each of the disciplines mentioned may vary to some extent over time as the state of scientific knowledge changes, or as new demands, or potential markets, are recognised. However, such changes are confined within boundaries. It is important to understand that the nature and extent of this confinement changes with time. In the early development of a techno-economic mode, the confinement may be relatively loose owing to the fact that little coupling of technologies exists. A range of possibilities may co-exist (interpretative flexibility). The web is immature and the shape of the fitness landscape is independent of inter-related technologies owing to their absence. As regards the ways of meeting a given end or purpose, the fitness landscape might be represented by a relatively stable topography with large numbers of low peaks, or small mounds. Each technology used is part of the way up one of the peaks, though there are also peaks which signify relatively under-used technologies.

As technologies mature in market economies, so the web increases in complexity. Specialisation of tasks, and the new opportunities opened up by the use of new technologies, causes the web to increase in the number of coupled technologies. These processes – the emergence of interrelated technologies, the evolution of institutions, the effects of learning processes, financial inducements, political decison-making, historical contingencies, and others – exert their influence in this model of technological change through deforming the fitness landscape, and selecting out one, or sometimes more than one, way of doing something, and effectively deselecting alternatives. These events and the emergence of one or other technology represent a co-evolutionary process where increasing returns operate. The greater the support for one or other technology, the more powerful are the interests which effectively lobby in favour of its continuing. On the fitness landscape, the peaks become fewer, farther apart, and more sharply defined. In these circumstances, it becomes likely that adaptive hill-climbing on one of the dispersed peaks will be the easiest path towards improvement, rather than seeking another of the widely dispersed peaks. Institutions support the development of this peak, lowering others in the process.

It is important to recognise that the landscape's topography is shaped, in part, by the way people see the world. If one were to remove this influence upon the landscape, the topography might well be substantially altered by virtue of the fact that the number of 'ways of doing something' exceed the number which we conceive of at any one time, a reflection of the fact that our thinking is path-dependent. Kauffman (1988,142–3) writes:

> The actual web is a subset of the possible; and the actual evolution of economic webs must be a path through the ever expanding set of next possible economic webs ... Since the actual evolution of the web is caught on the edge of addition and disappearance, under endogenous non-economic and economic forces, an eventual dynamical theory of web evolution must meld all these.

Similarly, a theory of lock-in seeks to explain why what disappears is rarely allowed to re-appear, and why these disapperances influence what can and cannot be added in the future. Options that are available may disappear due to seemingly insignificant events triggering the building of a new web on one peak as opposed to another, as the work of Arthur and David suggests (see above). Since, as the web increases in complexity, each new good or service generates possibilities for other new innovations which are coherent with the web structure, so the web's evolution takes on something close to an autocatalytic character.

Note also that it is not impossible for more than one technology to develop in fulfillment of a particular role. The way in which the fitness landscape is deformed does not preclude the simultaneous development of two or more technologies on different peaks which effectively compete in the marketplace as long as they both remain competitive. The rate of incremental innovation, the institutional framework (especially issues concerning the structuring of markets) as well as political factors, will affect the way in which the market is shared. Where changes occur as a result of changes which appear only indirectly related to goings on in the area under examination, it is useful to talk in terms of historical contingency. This term captures some of the historical events described by Arthur and David which may be 'small' and appear, at first glance, of limited relevance to the systems whose subsequent evolution they may determine (although they can equally be 'big', like war).

Whilst this idea has some general implications, it has some quite particular ones in the context of agriculture. As farming has become more commercialised and more specialised, partly due to technological changes 'outside' agriculture, so the fitness of one or other innovation has come to be determined by quite different selection pressures to those that might have operated previously. Indeed, because of the nature of the BCM mode itself, in particular, its desire to control the production process, the selection pressures themselves are increasingly derived from anthropogenic institutions, in particular, markets, rather than from 'natural' forces. Some of the risks associated with the 'natural' characteristics of agriculture are reduced, only to be replaced by risks emanating from exchange rate and interest rate policies, and commodity price fluctuations.

The fitness landscape representation allows one to visualise the historical process by which the BCM mode of doing agriculture has evolved. As will be shown in the case study material, the evolution of a way of doing agriculture based on genetic uniformity has proceeded in the context of evolving ways of interpreting the world, changing institutions, new opportunities for making money, and political decisions concerning the role of the state in agricultural research. In addition, as Chapter 4 will show, genetic uniformity has enabled the sorts of specialisation of agricultural tasks, in terms of mechanisation of harvesting and processing, that have enabled an increasingly complex technological web to emerge. Moreover, it necessitates the use of synthetic pesticides (sooner or later) and as such, is both the glue which both binds the BCM mode, and the heart of the process of locking-in.

Conclusion

Through elaboration and use of various concepts, a way of understanding technological change in agriculture within an evolutionary framework has been proposed. Having hinted that technological change generally is possessed of an evolutionary character, and that certain paradigmatic forms for such change exist, I have suggested that technological change in agriculture has taken place within a mode in which mechanisation, agrochemicals, and breeding are closely integrated, a more or less tangible illustration of which has been the elaboration of an ideal plant concept whose architecture enables and demands such integration. Lastly, I have proposed a way of viewing technological change in terms of the fitness landscape concept first used by Sewall Wright in his work on evolutionary biology.

The fitness landscape conceptualisation allows for a visual representation of the 'vestedness' of vested interests. It also makes clear how, as the techno-economic trajectory is carved out, and as an increasingly complex web arises to support it, it becomes increasingly difficult to do things in ways other than the way they are being done. Unless the mode reaches such a stage of maturity that exhaustion of the possibilities for adaptive uphill walks on an increasingly tightly defined landscape seems to be close at hand, then the only changes which are likely to be encouraged are those which would be selected by the prevailing selection environment, or something very close to it, in which case, the extent to which such a change really constitutes a change at all is likely to be subject to some debate. In the absence of some external intervention which forces a change in mode, the situation of lock-in is likely to persist.

In the next chapter, I will examine more closely the influences shaping technological change in agriculture. In addition, I will show how the formal agricultural research system has become increasingly locked into breeding for genetic uniformity, notwithstanding the possibilities for using alternative approaches requiring less pesticide use, and being less vulnerable to crop loss.

Notes

1. For an excellent exegesis, see Hodgson (1993). See also Clark and Juma (1987, 1988).
2. This implies that if economies can be described by evolutionary mechanisms, the future is uncertain, and one must abandon hope of predicting the future in a completely deterministic fashion.
3. It is important to note, as few tend to, that natural selection acts on the total phenotype, including behaviour. Since behaviour may not be genetically determined, it would be incorrect to state without qualification that natural selection has only a genetic basis (Rindos 1984, 37).
4. See Chapter 2. Arrow (1962b, 155) seems to be defending economists' dearth of this type of work in the first half of the century by assuring us that few would deny the importance of technological change in promoting economic growth. Why, then, were they, as Arrow himself points out, apparently happy to use, as a proxy in their theories (if indeed it was perceived as such), time?
5. '[I]n the economic analysis of technological innovation, everything is included that might be expected to influence innovation, except any discussion of the technology itself' (Pinch and Bijker 1987, 21).
6. See Rothschild (1988, 92–3).
7. Heertje (1988) points out that the emphasis shifted from the totally new to the partially new between the first and second editions of the German version (1912 and 1926 respectively).
8. This is a distinctly neo-classical interpretation of 'innovation', the irony of which will become clear below (see Mirowski (1989; 1993; 1994) for a discussion of 'neo-classical's' meaning).

9. Schumpeter held the view that entrepreneurial behaviour was by nature irrational since it implied non-routinised behaviour. Schumpeter's challenge to neo-classical economics is to ask how a world of perfectly rational beings could generate innovations. He overstates the case due to a failure to appreciate the cost of acquiring, and limits to the acquisition of, information, and the uncertainty inherent in the process of innovation.
10. Contrast, for example, the opinions of Rosenberg (1994, 48–50) and Hodgson (1993, 140–5). For Rosenberg, Walras *provoked a radical rethink* by Schumpeter, whilst for Hodgson, Walras was an *inspiration* to Schumpeter, who then misguidedly (in Hodgson's view) sought to imbue the Walrasian system with greater dynamism.
11. See Hodgson (1993, 145–50).
12. The concept of stasis is not wholly absent from biological systems. Mature ecological systems, having evolved through successionary periods, reach what ecologists call a climax state, which is more or less stable.
13. Neither the term 'evolution' nor 'survival of the fittest' appear in early editions of Darwin's *The Origin of Species*, perhaps because at the time of his writing, the theory of evolution (a term he did not invent) was much discredited (Burrow 1968, 27).
14. See Burrow (1968, 42–6), Rindos (1984, 59–69) and Hodgson (1993, chs. 4 and 6).
15. Giddens criticises what he calls 'unfolding models of change' (1979, 222–25. See also Giddens 1990, 4–6).
16. Hodgson (1988, ch. 1) appears to share this view.
17. There is a debate as to whether selection can operate on what does not already exist. Nelson and Winter (1982, 142) state 'selection works on what exists, not on the full set of what is feasible.' Perhaps the word 'selection' is problematic, but it is surely valid to ask *why* that which is feasible does not exist. Indeed, if something has acquired the status of 'being feasible', that implies the potential to exist. Selection may occur at the level of ideas, perhaps even prior to an idea being considered feasible. Ideas which occur to us form a set from which a 'feasible' subset is derived. But the set itself must be a subset of a larger realm of ideas, some of which will not occur to us, yet some of which would almost certainly be feasible. They are, to use an awkward phrase, ideas with the potential to be actualised. Yet the ideas which we have are shaped by our way of looking at the world, the roots of which are themselves to be found in our personal and collective histories. Knowledge, its creation, and its assimilation, both for the individual and the collective, are path-dependent processes. Researchers carry out their work in an environment shaped by the history of how it is that they have come to be who, what and where they are.
18. For further discussion, see Hodgson (1993, ch. 13).
19. In their landmark work, Nelson and Winter (1982) used the concept of organisational routines, likening these to genes which are selected by the market.
20. See also Dosi and Orsenigo (1988), Dosi (1988) and Orsenigo (1989).
21. Kuhn's ideas began a debate between his followers and those of Popper, with Lakatos trying to bridge the gap between them. Lakatos suggested that a negative heuristic was employed by scientists in the sense that

certain 'hard core' concepts were beyond questioning. Dosi recognises this (1982, fn.14), and we might comment that his drawing of an analogy between scientific and technological paradigms is selective. For an attempt to tease out the differences between the two, see Clark (1987).

22. On long-waves, see Mensch (1979), Freeman *et al.* (1982) and Coombs *et al.* (1987, ch. 7), and for a sceptical view, Rosenberg (1994, ch. 4) (first published in 1984, well before much of the work we are about to discuss was written).
23. See also Freeman (1992, chs. 4 and 6).
24. Elsewhere, van den Belt and Rip (1987, 141) note that 'the assumption of a selection environment that is truly independent of a particular technological trajectory is hard to justify'.
25. See Chambers (1993, 2–3) on the difference between developmental and scientific paradigms.
26. Ideas of equilibrium and diminishing returns were destined to be held dear to a discipline accused of 'physics envy' (Clark 1990, 36 – see also Mirowski 1989) as long as equivalent concepts were reified by physics itself. But autocatalytic systems shatter the physicists' notion of systems ceaselessly increasing their degree of disorder (entropy), and the world slowly approaching 'heat death' (see Prigogine and Stengers 1985). The study of 'complexity' embraces the study of systems of all sorts which appear to demonstrate 'emergent' properties (see Mitchell Waldrop 1992; Lewin 1993a).
27. For an attempt to model the importance of 'non-rational' behaviour, see Allen and McGlade (1987) and Allen (1988).
28. Agliardi (1991, 52–71) has modified Arthur's model to show how early adoption helps to eliminate uncertainties in respect of relative performance of competing technologies. This can itself act as a source of lock-in.
29. See Hawthorn (1991) for a thoughtful exploration of this question.
30. See Faucheux (1997) and Kemp (1993; 1997, chs. 10 and 11) for rare discussions of lock-in in the environmental context.
31. See Juma (1989, ch. 1); Shiva (1990, ch. 2); Busch *et al.* (1991, ch. 2) and eco-feminists such as Merchant (1980), Easlea (1981), Keller (1985), Harding (1986) and Mies and Shiva (eds) (1993). For a critical view, see Nanda (1991).
32. Positivism, officially founded by Comte and then adopted by the Vienna Circle, has many faces. Here we are concerned with its denying cognitive value to value judgements and normative statements (from which springs the so-called 'fact-value' debate), and its belief that there is an essential unity of scientific method. The positivist view has come under heavy fire from sociologists of science and knowledge. Feyerabend (1978, 39–40) in particular has proposed that there is no preferred methodology in science and that essentially, as he puts it, 'anything goes'. Many sociologists advance a social constructivist view of knowledge, and some appear to support a position of extreme relativism and irrationalism in science, a position criticised by Murphy (1994), Oldroyd (1986, ch. 9) and Nanda (1991).
33. The development of science and technology in their social context has spawned a massive literature, much of it either of a distinctly Marxian hue,

or from the school of critical theory. For an interesting review, see Feenberg (1991).

34. Comte's *Course of Positive Philosophy* was published between 1830 and 1842 in six volumes and established the science of sociology. Comte called it 'social physics' as he sought to establish sociology on the same principles as the physical sciences. On this, and the development of technocratic rationality, see Fischer (1990, ch. 3).
35. Habermas (1968, 104) refers to the end of the nineteenth century as the period of the 'scientization of technology' during which 'technological development entered into a feedback relationship with the progress of modern science.' Rosenberg places great emphasis on the development of industrial research laboratories, 'one of the most important institutional innovations of the twentieth century', in reinforcing the influence of technology upon science (Rosenberg 1994, 20; also Rosenberg 1982, ch. 7).
36. See Wallerstein (1991, ch. 18). Disciplinary fragmentation, and the resulting structures of knowledge production, are path-dependent processes.
37. Popper countered Kuhn's thesis by suggesting that 'The normal scientist, as described by Kuhn, has been badly taught. He has been taught in a dogmatic spirit.' I think this is partly what Kuhn was trying to say, and indeed, Popper continued 'I admit that this kind of attitude exists ... [But] I see a very great danger in it and ... the possibility of its becoming normal ... a danger to science and, indeed, to our civilisation' (Popper 1970, 53).
38. The disturbing existential implications of this are captured by Giddens (1991, 124) who suggests that we are all non-experts in most walks of life, and forced to make life-shaping decisions daily on the basis of conflicting evidence.
39. See Bunders and Broerse (1991, ch. 6) and Maxwell (1984).
40. As Winner (1977, 97) points out, these are not 'not intended' consequences. They are the outcome of what Beck (1992, 60) calls 'the economic Cyclopia of techno-scientific rationality'.
41. See Beck (1992, ch. 8) for more on this. Winner (1977, 98) calls negative side effects, 'necessary evils that we are obligated to endure'.
42. Beck (1992) refers to this as 'the confusion of the centuries'. The side-effects of modern industrial technologies are often more significant than the benefits they bestow on humanity, a reversal of the situation prevailing in earlier times.
43. Tripp (1989, 19) argues that the problem with commodity-oriented research is not that orientation *per se*, but that it does not take account of (some) farmers' views. I would argue that the two explanations are inseparable since both the failure to consult farmers and the concentration on single commodities can be traced to the mode in which research is conducted.
44. These functions may change as the trajectory is carved out within the paradigm. Fast-maturing rice varieties required rapid land preparation between harvests, a characteristic which led engineers to focus on mechanisation of these operations (Oasa 1981, 224).
45. This view has lost some, though by no means all of its validity as scientists in (especially international) research organisations responded to criticisms that such input-intensive varieties were, to use Lipton and Longhurst's phrase, 'hard on poorer farmers' (1989, 20).

46. On the ideal rice plant, see IRRI (1962, 12), IRRI (1963, 23), Jennings (1964), Beachell and Jennings (1965), Chandler (1968, 1972). Beachell and Jennings were plant breeders at IRRI, and prior to his posting, Beachell had been studying the yield performance of shorter strawed rice varieties in the United States (Oasa 1981, 178). Chandler was IRRI's Director and, by his own admission, had never seen a rice plant before becoming IRRI's Director.
47. See the works by Oasa (1981, ch. 5), Levy (1982) and Anderson *et al.* (1991).
48. Information technology has revolutionised a variety of activities in agriculture, principally in farm management and mechanization, but I will not consider it closely here. This extends the degree of apparent control over the farm system through precision farming (RAFI 1997).
49. See Rosenberg (1969).

4

Beyond Orthodoxy: Locking in to Genetic Uniformity

Introduction

The theory of induced innovation discussed in Chapter 2 has received a good deal of criticism, much of it based on the fact that the mechanism by which change occurs is heavily market-oriented. Markets do not of themselves develop new techniques or technologies. Nor do they create organisations. They merely provide more or less good signposts, which are not always followed, as to the path which changes might follow if the aim is to guide change along the most *financially efficient*, or profitable, road given the available options. Commenting on induced innovation and its structuralist offspring, Biggs (1982, 209) comments:

> While agreeing with the structuralists about the need to analyse the influence of special interest groups on technology generation, neither theory looks inside and analyses the decision-making processes of institutions and people who generate new technologies and new R&D institutions and research methods.

This is characteristic of the 'black box' approach to the development of technology and techniques that has been adopted by much of the economics profession until recently, as discussed in the last chapter.

Much attention in induced innovation theory is paid to the concepts of supply and demand. The use of these terms is best restricted to markets for goods or services which already exist rather than innovation. Markets are repeatedly restructured by institutional changes which are both prompted by, and which promote, the development of new technologies and techniques. There are difficulties in supposing that markets themselves propel institutional change, when often what is sought are institutional changes which will enable new markets to be developed. One could, in discussing technological change, try to maintain a 'supply' and 'demand' framework, thus 'extending' induced innovation theory, but one would need to entertain such broad definitions that the terms would become all but meaningless. Whilst Burmeister (1995, 52) has pointed out that retention of a supply and demand framework leads to identification problems, it would also convey an impression that the emergence of new techniques, technologies and institutions can be explained wholly in terms of market phenomena.

One of the issues which is ignored by analysts of change is that of the character of technologies, which, in turn relates to the path which technological change follows. Although there has been discussion in economics literature concerning the level of investment in research and development (R&D) and its relation to the appropriability of the innovation in question, this discussion is usually divorced from considerations of the development of appropriability regimes themselves, and the technological options available for addressing any given problem. If the appropriable nature of these innovations supposedly corrects for market failure in terms of R&D resource allocation, it is reasonable to ask whether appropriability effectively distorts decisions made concerning what gets researched, especially if private sector involvement in research effectively pushes the public sector into a more subservient position. Below, it is claimed that aspects of appropriability regimes currently in place lead to the ecological mis-shaping of agriculture.

The first part of this chapter discusses lock-in in the context of the many influences on the work of formal agricultural research organisations and the movement of their innovations into use in agricultural production. A stylised chronological picture of the innovation process (see Figure 4.1) will be used. Admittedly, this is not always followed, since innovation can, indeed should, be a recursive process. However, this sequence allows one to draw out some of the ways in which research can become locked in to a particular trajectory.

I use the terms 'formal' and 'informal' agricultural research in the same way as Biggs and Clay (1981, 323–7), who include public- and private-sector research organisations in their definition of formal research. They suggest that formal systems differ from informal ones in that:

(i) The generators of technology ... and the users of technology are no longer one and the same. Farmers have to wait for scientists to diagnose their problems before the resources of the formal systems are applied to solving them.

(ii) The formal R & D programme has an 'institutional memory', linkages to information and technology totally outside the local environment of the informal system and a capacity for looking into, and planning for, the future to a far greater degree than the informal system.

(iii) Government policy can have a major influence on the types of new agricultural technologies generated and the methods used.

(iv) Formal communication between scientists and between the scientific community and farmers (their clients) takes on a major significant role (Biggs and Clay 1981, 327).[1]

It should be stressed at this point that, notwithstanding the implied criticism of formal agricultural research organisations (FAROs) within this thesis, such organisations undoubtedly have a role to play in the future of agriculture. Without going into details as to what this might be, it will suffice here to highlight, again following Biggs and Clay (1981, 326) the limitations of informal research; possibilities are constrained by the locally available genepool, supplemented by new materials; environmental change and unforeseen consequences of technology transfer may have severe consequences; and an inability to undertake long-term, forward-looking research. The most substantive criticism of agricultural research is that there are few well-developed mechanisms for integrating formal and informal processes.

The second part of the chapter concentrates on explanations for the continuing dominance of the BCM paradigm through addressing processes outside the research organisation which can lead to lock-in. The focus is on the factors that mitigate against the re-introduction of diversity within the BCM mode.

Factors influencing the work of formal agricultural research organisations

Organisational framework

Analysis of the organisational framework (see Figure 4.1) for research concerns the number, size and nature of interaction between FAROs, including the relative importance of the private and public sectors. It goes without saying that FAROs will not be established unless somebody perceives that they are likely to fulfil a useful, or profitable end. The principal clients need not necessarily be farmers, and even where they are, this does not imply that all farmers are served equally by the organisation. It would be difficult to generalise concerning the effect of

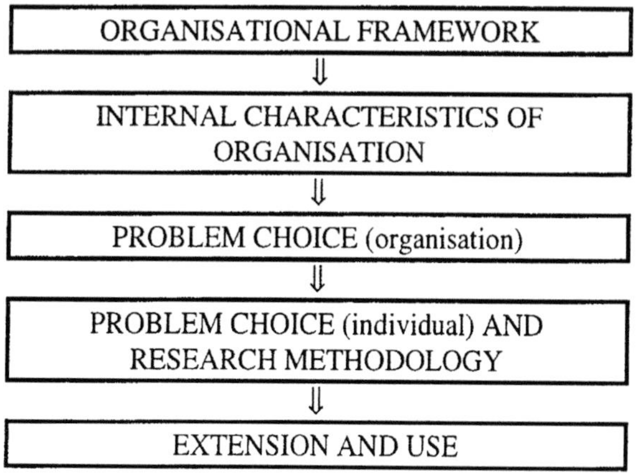

Figure 4.1 Stylised chronological sequence of events determining technological outcomes

various groups and their demands on the organisational framework for research. Nonetheless, it seems fair to say that there will be both quantitative and qualitative influences on the number and nature of FAROs in existence.[2]

A general desire for research organisations is insufficient for them to become established. A source of funding is necessary. The actions of those demanding such organisations, whether farmers or entrepreneurs, usually entails a search for funds. Whether or not this can be found determines whether an organisation will or will not be established. The source of funding is likely to reflect the extent to which innovations can be made privately profitable, reflecting their degree of appropriability through intellectual property rights (IPR) or other means.[3]

In most countries where a formal research system has evolved, the public sector has taken the lead in biological innovations, primarily due to their non-appropriable nature. However, the passing of the R&D baton from public to private sector, a defining feature of agricultural research in industrial market economies, is an important factor influencing the character of agricultural technologies. Over time, institutional changes have allowed private companies to appropriate streams of income relating to R&D activity based on these techniques. Today, chemical pesticides, and in some countries, plant varieties and seeds, are subject to IPR legislation. As Goodman *et al.* (1987, 2–4) suggest, different, sometimes competing industry interests have intervened piecemeal either to transform different aspects of the production process into industrial processes, or alternatively, to substitute for the commodity altogether. Kloppenburg's (1988) work on the commodification of the seed is a classic study of the way in which barriers to private enterprise becoming involved in seed production and sales are surmounted through political decisions affecting science and the technological sphere.

As is clear from the reservations expressed by many concerning the potential of biotechnology to serve agriculture in a positive sense, as critical as the question 'Which group is research done for?' are the questions, 'Where are these groups located, and who is research done by?'[4] The source of funding of research either determines, or is in the position of being able to influence the setting of, the goals and priorities of the research organisation. As Paarlberg (1978, 135) has pointed out, 'control of the research agenda is the central issue, for control of the agenda connotes the ability to keep items *off* the agenda, that most potent of all powers, the one the public seldom sees'. Again reflecting the way the source of funding influences research priorities, Ashford and Biggs (1992, 355–7) distinguish between public sector, private sector and non-government organisations on the basis of their conforming to different social welfare functions.

The crucial modification that many authors make to Hayami and Ruttan's theory reflects the belief that technology is shaped by social structures and that it often performs a specific social function which reflects the interests of particular sectors of society. This critique explicitly takes into account the unequal distribution of political resources in society and the implications that this has for the development of agricultural technologies. Piñeiro *et al.* (1979, 172) argue that the non-appropriable nature of much agricultural research makes it more, not less important to understand decisions made by the state in allocating resources to publicly funded agricultural research.[5] The relative political power of different social groups, and its concentration in certain hands increase the likelihood that state funded research will be congruent with the most powerful:

> Technical change is not only an instrument in the generation of an economic surplus but also an object and an instrument of social conflicts. Technical change conditions the social control of the means of production, the organisation of the labor process, the social division of labour, and the social appropriation of the surplus. As such it is a powerful instrument of social change or stasis. (de Janvry and LeVeen 1983, 27)

Social conflicts affect the rate and direction of technical change at several different levels. In a direct attack on the failure of induced innovation theory to incorporate the 'structural and economic determinants of the inducement of innovations', de Janvry and LeVeen (1983, 28) comment (revealing the path-dependent nature of the process under consideration):

> Land ownership is not established as a relation between people and land, but as a relationship of exclusion among individuals. The long history of the formation of the land tenure system must be understood as the struggle for appropriation of land and water with its associated dose of legal battles, bribery, violence and violation of the law.

Piñeiro *et al.* (1979, 176) and Biggs and Clay (1988, 38) make a similar point in noting that asset prices reflect their distribution rather than national factor endowments. States should recognise the significance of distributional issues in allocating research resources.

Internal characteristics of organisations

In referring to the internal characteristics of FAROs, I am concerned with staffing issues, organisational structure and hierarchy, the way the organisation works, and its culture. It is difficult to dissociate these from the decision to establish an organisation in the first place. With funding come demands from funders. Those who sought funding will also usually have a say in setting research goals, choosing personnel, and establishing the way in which the organisation goes about its work.

Many FAROs have mission statements which determine priorities. Who sets these, how and when they are set (that is, how early in the development of the organisation), and how narrowly they are specified is affected by, and will affect, the choice of personnel. Such latitude as exists in terms of these goals and how they are to be met is greatest when the organisation is first established since subsequently, the possibilities for radical changes without changes in personnel and working culture are greatly reduced.[6] Each step towards a tighter specification of the aims of the organisation constitutes a de-selection of other approaches to research. As the specification of problems is narrowed, so the related question as to how problems will be investigated, and more significantly, which personnel will be chosen to carry out the research, grows in importance. The choice and the approach are probably closely interrelated through the type of training that the individual or group has been exposed to. Thus, even where the underspecification factor is considerable regarding actual problem choice, the approach may be specified quite tightly through the selection of personnel in an organisation.

The history of a research organisation shapes its future, and its early history may have a disproportionate influence on its subsequent activity. For this reason the use of one organisation as a blueprint for another can be particularly significant. Such transfer can foster the emergence of a culture not just within a given organisation, but across organisations. Richards (1994b) makes the point that 'Organizations come and go, but change in institutional culture tends to take place more slowly.' In terms of the landscape model of Chapter 3, the fitness landscape is progressively deformed in such a way as to de-select possible alternative aims and ways of doing things. The more organisations that choose to follow the BCM mode, the more complex the economic web supporting it becomes, and the more its claims to knowledge and expertise are reinforced. Thus, McCalla (1978, 89) notes: 'the research establishment has a tremendous tendency to maintain the status quo and to continue its lumbering, disjointed movement along well-trodden paths'. The BCM mode becomes, in a Kuhnian sense, the 'normal' way of doing things, and those who constitute it become practitioners of 'normal professionalism.'

An accepted way of doing agricultural research corresponds, and indeed, is derived from, a particular way of training would-be researchers. Curricula in Colleges and Universities, as well as in-house training, are inevitably closely aligned with the normal professionalism. Once these curricula are in place, and as the BCM mode matures, so those interested in agricultural research increasingly see that their career paths, whether in private or public service, will be smoother if they seek advancement within the accepted way of doing things.[7] Peer review and career prospects impart a certain logic to the continued survival of the BCM mode. Biggs (1990, 1489) refers to these two influences, respectively, as 'research as academic pursuit', and 'research as administrative behaviour.' In addition, as Oasa (1981) showed, the political context in which researchers' work is carried out, as well as the work they do, determines which researchers are listened to. Biggs (1990, 1491), Oasa (1981) and Anderson *et al.* (1991) all point to the fact that the IRRI reward system, and the policy and political context in which it functioned, effectively by-passed scientists with skills for breeding for a wider range of environments than the lab-like conditions on which IRRI focused.

Historically, as mentioned in Chapter 3, the process of knowledge production has fragmented along disciplinary lines. Associated with this process is a perceived hierarchy of disciplines, and of particular relevance to agriculture is the fact that this hierarchy favours the scientific 'hardness' of the discipline concerned (Maxwell 1984). Unfortunately, the theoretical rigour of researchers' work, though it may enhance the status of agricultural science among peers in the world of science, does not guarantee relevance in the field, and tends to concentrate the domain of applicability of research to controlled environments. The ranking of disciplines along lines of scientific 'hardness', and the general favouring of reductionism over systemic, or holistic approaches mentioned in Chapter 3 does not work in favour of ecologically adapted research.

Key issues that shape the organisation's structure are:

1. How, and on what basis, research resources are to be apportioned, often reflected in, or reflecting, the lines along which the organisation is divided (commodity, discipline, farming system, region etc.)[8] and its remit;
2. How these divisions interact;
3. How the work of each division is performed (if this is specified at this level – see below);
4. How, and with whom, the organisation interacts with those outside its physical boundaries; and
5. The sources of funding and the extent to which these can be relied upon in the future.

Each of the issues 1 to 4, and sometimes 5, influence, and are influenced by, the organisation's staff. Motivations for change in the above-mentioned characteristics may be mediated by clients, who may feel that the organisation is ill-designed to cope with their needs. It may also come from funders, who may feel that the organisation is unable to fulfil its mandate, or that it's mandate should change, or that a particular line of research is not worth pursuing further.[9] Alternatively, funders may be reducing or withdrawing support. Lastly, demand for change may come from within the organisation itself. This could reflect frustration felt by staff, or the emergence of new knowledge or technologies which appear to justify change. It might also reflect new appointments, or internal struggles aimed at re-ordering a perceived hierarchy in the organisation. Alternatively, it might occur in the wake of an appreciation of the fact that the organisation's output is not meeting an applied set of criteria and that change is necessary in order, for example, to sustain funding levels or stay in business.

It goes without saying that, in keeping with the hypothesised inertia, a publicly stated change does not always confirm change within an organisation. Lipton (1994, 609) comments that the CGIAR presents the current (and past) way of doing things as sustainable rather than interpreting demands for sustainability as requiring something new. He feels that new institutions are needed to tackle new (actually, rather old) problems. This view is echoed by RAFI and GRAIN (1995), who comment that the CGIAR has no track record in incorporating poverty alleviation or environmental issues into their work.[10] They add that 'There is a tendency to recycle trustees from board to board and to resurrect past trustees on review panels or special committees'.

Problem choice (organisation)

There are two clear levels at which problem choice operates, that of the organisation, and that of the individual. Depending on the structure of the organisation, other levels may be more or less relevant. Successful research organisations create outputs which match the needs of clients and they choose research problems accordingly. Yet the principal clients vary across organisations. This raises questions as to whether an organisation's mandate specifies with sufficient precision who clients *should* be, or whether there appears to be some leeway in effectively selecting clients. For many organisations, a mandate is outlined in terms of goals rather than client orientation. This gives considerable leeway as to chosen client orientation since it can be argued that a given aim can be achieved through a variety of approaches. Indeed, as regards international (public) research organisations and their clients, few have any problem with their mandates, noble as they are, as stated (Ravnborg 1992, 6–7). Where debate has been more heated is in the interpretation of that mandate, and the approach to achieving the aims stated therein.[11]

There has been a tendency for FAROs to listen to those who have the loudest voice (Biggs 1990, 1490). FAROs, whether publicly or privately funded, have tended, when they work closely with farmers, to work most closely with those who are members of organised lobbying groups or producer associations. Indeed, the interests of farmers themselves are not always the most relevant to an organisation. Agribusiness and other sections of the business community, as well as consumers, have all played key roles in the shaping of agricultural research (Piñeiro *et al.*, 1979, 173).[12] As McCalla (1978, 77) puts it: 'The "agricultural research establishment" is a morass of loosely related, sometimes complementary, sometimes competitive, organisations, so intertwined with the "publics" they serve that it defies simple definition.' Donors, too, are often able to specify the projects for which a particular tranche of funding is to be used.[13]

The problem of selective listening affects not only private FAROs, who might be expected to listen selectively in the interests of maximising returns on research investments. Public FAROs have often felt the need to give their work commercial or political appeal to secure continued flows of funds. They are least likely to achieve this through appeal to the value of helping less commercial producers increase production whilst employing minimal external inputs. Thus, neither the private sector, nor, under certain circumstances, the public sector, are likely to throw resources into such work. As a consequence, FAROs tend to supplant informal ways of doing things, the more so since the two have failed to reach a working partnership. As long as this remains the case, informal approaches to agricultural technological development will tend to square up to formal approaches rather than each accommodating the other.

It might be supposed that questions of problem choice are less relevant in considering the private sector's work than the public sector's, since it is expected that the private sector simply responds to market opportunities. Yet because FAROs, certainly in the case of germplasm, are born principally through public funding, to the extent that the private sector exists at all in a certain area of business, it does so by dint of institutional changes designed to open up an area of research to private enterprise (see above). This is related not only to technical and technological changes, but to shifts in the institutional division of labour (between private and public sectors). The type of research carried out by private and public FAROs has to be different to justify their co-existence, so that the institutional framework, through its influence on the public/private division of labour, has an impact on problem choice, and therefore, on whose demands should be met.[14]

Appropriability regimes, or appropriability characteristics (of technology) influence not only this division, but in the most commercially oriented of the FAROs (including some publicly funded ones), they actually have a bearing on which projects should or should not be pursued. To the extent that the political economy of the public/private division of labour is closely associated with the prevailing economic philosophy and/or budgetary constraints, this bears on whether the public sector completely vacates a given area of research, or whether it remains in the same area, but carries on work in that area that would not otherwise be undertaken by private research organisations.

Where the public sector remains active, it is likely to do one or more of three things; concentrate on so-called basic research; aim to produce outputs which are patentable and can be licensed to third parties; or concentrate on the same problems as the private sector but through alternative, non-appropriable approaches. It is notable how, in the developed countries, the first two absorb more resources than the third (Kenney 1986, 230–4; Kloppenburg 1988, 234–41). This is important in the context of the genetic uniformity since, as we shall see below, IPR as currently constituted encourage uniformity. If the public sector vacates an area of plant breeding on the basis that it has become profitable to private industry to carry out such research, it becomes inevitable that approaches to plant breeding which might have encouraged the use of genetic diversity, but which, by the very nature of appropriability regimes are not attractive to the private sector, will be dropped.

Patents are usually discussed in terms of their capacity to promote innovation, but they also have restrictive effects depending on the institutions governing patentability.[15] The breadth of a patent's coverage is all important in this respect (RAFI 1994a). Furthermore, if process innovations are patented, research and development, particularly in smaller or less-well financed organisations, might be hampered by prohibitive licensing fees (Sederoff and Meagher 1995; Day 1995; Lacy and Busch 1991). Patents also affect problem choice. The value of patents to innovating companies has led to a situation in which some company lawyers have as much input into the make up of the project portfolio as the researchers themselves (Hobbelink 1991, 108). Lastly, the blueprint for patent systems tends to be the formal research process, so that by denying the work of informal researchers a status equivalent to that of FAROs, the institutional arrangement itself prevents informal work from acquiring commercial value. It is clear from these considerations that institutions play an important role in deciding which innovations should be in use in agriculture.

Problem choice (individual) and methodology

The tendency to listen selectively can be modified by the manner in which the organisation seeks to listen. The organisation can make it easier or more difficult for different groups to make their demands known, either through appropriate staff selection, or through identifying clients and making appropriate changes in methodological approach, which probably amounts to the same thing unless the change is a cosmetic one. Researchers', and the organisation's perceptions of what constitutes, and who is likely to possess, useful knowledge influences problem choice through the impact on client choice. It is important to point out that choosing a particular problem to be researched does not mean that the problem will be solved. At one level, this may be considered to be a consequence of insufficient advance in scientific knowledge, but it also reflects both the importance of methodological, as well as problem, choice, and the fact that research is an inherently uncertain process. Yet the chances of success are not purely a matter of luck. The skill of researchers is an important factor, as indeed is their perception of what constitutes a success (it is difficult to separate the idea of researchers' 'skill' from that which they are supposed to be skilled at). Hence, the innovations which emerge from research organisations are closely bound up with the skill sets of those who staff (or are chosen to staff) them.

Individual researchers may simply choose to work on a problem because they perceive it to be a problem worth researching.[16] The incomplete specification of exactly what, and how research should be done leaves individual researchers some room for manoeuvre in determining how to go about what they are being asked, or advised to do.[17] This room for manoeuvre may be quite substantial, and even where it is not, individuals may find ways to make it so. Not all researchers feel bound by the policies and mission statements of the organisation for which they work. Thus, whatever the extent of the latitude that exists, the choice of problem may ultimately come down to that of individual researchers. Narrowly specified, the choice of problem need not imply a choice of methodology. In practice, however, it may. Both in the matter of problem choice, and in the choice of approach to a given problem, the freedom to choose may actually be more apparent than real since the menu over which choice is expressed reflects researchers' backgrounds, as well as that of the organisation itself. Peer recognition can impart a certain focus on a subset of choices and reward structures may persuade researchers to narrow that focus further (McCalla 1978, 87).

Co-evolutionary views discussed in Chapter 3 suggest we should perceive researchers as actors in a social world perceived by them principally through the institutions which they constitute, and through which they, as individuals, are constituted. Choices are in general bound up with individuals' education, culture, race, and gender. In other words, their choices reflect who they are, and their past, present, and even future (through, for example, perceived career prospects) in the social world. This gives rise to a paradox. On the one hand, the space for autonomous action is constrained both by institutions, and the resulting structures within which the individual functions. On the other hand, the institutions, and the structures within which individuals function, and which they help constitute, change only by virtue of the actions of people like them. They co-evolve.

There has indeed been a degree of evolution in FAROs' approach to their work, particularly with respect to developing country agriculture, where past failures of research to reach large numbers of farmers working in complex, diverse, risk-prone environments have been understood in a changing light. Chambers' (1993, 67) stylised chronology of the explanations offered for farmers' non-adoption of technology holds that in the 1950s and 1960s, this was understood as the result of ignorance on the part of farmers, and the prescription was extension. In the 1970s and 1980s, with a more positive view of farmers' knowledge emerging, the focus shifted towards farm-level constraints, leading to efforts to understand these and the emergence of farming systems research (FSR). However, in the 1990s, there has been a growing willingness to accept that non-adoption is an outcome of the fact that the technologies being made available simply do not fit the needs of the would-be adopters. This has led to greater focus on the research process itself, and on the need to change the behaviour and attitudes of those involved.

The issue of information flow reflects more than a simple current of demands impinging on the researcher. It is central to the methodological approach adopted. The changing nature of the explanations for non-adoption reflect past failures to understand the significance of farmers' knowledge and needs in technology generation, not just in terms of the problems farmers face, but also in terms of which technologies will actually work in the farming system or systems involved. This was institutionalised through researchers' biases in terms of where, and with whom they felt expertise resided (themselves), and where they carried out their work (in controlled environments such as laboratories and experiment stations). The technologies thus generated were, as a result, most suited for fields where such control, especially in terms of water, could be exercised. It was also reflected in inherently centralist tendencies fostered by an emphasis on transfer of technology. Organisation's were not designed to be flexible, nor were they designed to learn from feedback coming from the fields, not least because such feedback was not actively sought.[18] The limitations of such a strategy became even more stark in its encounters with ecological diversity (Biggs and Clay 1988, 24–6; Anderson *et al.*, 1991).

Since the mid-1970s, when Farming Systems Research (FSR) began to attract attention from researchers, a whole range of research methodologies have been developed. Some authors (for example, Tripp *et al.*, 1991) use FSR as an umbrella term for these methodologies whereas others (for example, Scoones and Thompson 1993; Pretty and Chambers 1993) view FSR as but one of a number of methodologies which vary in the nature and extent of participation of farmers in the research process.[19] The FSR approach they regard as somewhat limited, being 'extractive' in its approach to information exchange between researcher and farmer (Howes 1979, 13). However, each of these methodologies seeks to step out of the linear transfer of technology approach to technology generation and delivery so as to give greater priority to the needs and/or knowledge of rural people.[20]

The extent to which participatory methodologies have been employed in FAROs has been mixed.[21] It has been argued that such approaches can be time-consuming and costly in terms of the use of research resources. Even in public sector research organisations, there has been a reluctance to devote substantial resources to this type of research, a tendency which has strengthened with the growth in interest in trying to measure the rate of return to investments in agricultural research (Anderson 1994, 11; Scobie 1984). The problem with measuring rates of return to research is that financial and social rates of return are often at variance, and efforts to adjust the one to account for the other are open to widespread abuse (Winkelmann 1994, 89–93; Lipton 1994, 601–4). Furthermore, the highest financial rate of return is achieved where one variety is planted over a large area.[22] This is exactly what makes the crop concerned more vulnerable to pathogens, and is likely to require use of crop protection chemicals for that very reason. Few studies on rates of return account for the 'empty wells' discovered by researchers. None can properly consider benefits that might be foregone from wells which no one looks into.

Whilst the choice of a particular problem and methodology effectively may close off other options, it is equally possible for useful innovations to be produced, but to go unrecognised as such. That innovations can go unrecognised results from the way in which they are tested, and the people involved in that test. If the criteria to be applied in determining whether an innovation is useful or not, and therefore, whether or not it will be supplied, fails to reflect the real-world circumstances in which an innovation could be used, it becomes almost inevitable that some potentially useful innovations slip through the net, so to speak. There are several known cases where those working in research organisations have absconded with seed varieties whose properties were not well appreciated (Maurya 1989). Such examples have been used to support arguments in favour of on-farm research, and to underscore the value of farmer experimentation (Maurya *et al.* 1988).

Similarly, it can be argued that innovations which should not have been used at all have gained widespread use as a result of a system of legislation/supervision of technological development that seems reluctant to stem the tide of anything that is new. Some pesticides have been withdrawn from use in some countries owing to their suspected carcinogenic properties, yet they are exported for use in other countries (Clapp 1994, 15–16). Whether or not they are admitted for use in the first place depends on somewhat bogus science which claims to be able to judge the toxicity of chemicals to humans by testing them on animals (Conway and Pretty 1991; Beck 1992; Cantley 1995). The question of technological risk assessment is a 'scientific' hot potato which has less to do with objectivity and truth, and more to do with subjectivity, profits and politics.[23]

Extension and use of innovation

A market-oriented view suggests that the use of technologies depends on the extent to which supply meets an unmet need or demand. This somewhat heroically assumes the availability to farmers of information concerning innovations from which they can the pick and choose, and the availability of the innovations themselves. Use actually depends firstly upon a product being brought to the attention of potential users, secondly, on the availability of information concerning the product which is available, and thirdly, on the existence of known alternatives.

The way information reaches farmers is related to the methodology implied in technological development. If research is participatory, some farmers are immediately made aware of the product, the design of which they have some control over. If research takes place in the laboratory, or on an experiment station, or both, the innovation has to be taken out of that context and effectively 'marketed.' The principal 'marketing' tools have been field demonstrations, extension personnel (public and private sector agronomists), the media, billboards, trade journals, trade fairs and seed catalogues. Farmers' 'demand' will be affected by the information which they receive concerning innovations (is it relevant to them? Is it in a language or form which they can understand?), the channels through which they receive it (do they trust the provider?), and whether or not the innovation is appropriate to their needs. The notion of 'demand', therefore, masks the various factors determining end use, the concept not always reflecting the desires of end-users themselves (a point well understood by Schumpeter 1983, 11). The means through which the level of adoption can be manipulated range from simple inducements, for example, through subsidies, or free gifts, to the downright coercive.[24] Alternatively, extension personnel can, through inappropriate recommendations, whether they make them knowingly or in ignorance, encourage a level of use above or below what might be appropriate.[25]

Three other issues affect demand. The first is complementarity of innovations, the demand for one product being conditional on the presence of its complement. The second is the way in which innovations spread. Varietal use may spread by means of farmer-to-farmer communication, by 'learning by watching', or through informal exchange. The freedom with which these processes may occur will increasingly be conditioned by IPR systems which may increase demand through formal channels at the expense of curtailing the diffusion of varieties through informal channels. The third factor is the adaptability of an innovation. Some innovations are more versatile than others, and indeed, some innovations may take the form of knowledge which can be applied across many farming systems, whilst some plant varieties may have limited ecological adaptability. Such ecological constraints inevitably affect not only final demand, but also, decisions as to what should be supplied in the first place.

It goes without saying that the effect of an innovation in use should feed back into decisions concerning problem choice. However, this may not always be easy to achieve, especially if such information is not actively sought. Furthermore, many organisations have promoted outreach programmes, but no inreach programmes (Biggs 1990, 1490). Thus, and for reasons mentioned above, research organisations may become locked-in to certain ways of doing things. The second half of this chapter explores other ways in which this might occur.

The BCM treadmill – causes of lock-in and lock-out

A crucial element in the trend towards uniformity has been the path of development of agricultural research in its social context. The view of the Committee on the Genetic Vulnerability of Major Crops (NRC 1972, 15, 286–9), that breeders are bombarded by demands from society for uniformity which they were simply obliged to fulfil, is lopsided because it is ahistorical. These demands certainly exist, but their existence depends in part on the approach which breeders have adopted as a matter of choice in the past. The demands and the approach have co-evolved. If it were true today that breeders are seeking to 'switch to diversity', then if they could not, owing to the catalogue of demands for uniformity placed upon them, we would be well-placed to appreciate the path-dependent nature of the problem, and our predicament of being locked-in to uniformity. The following account addresses some of the factors originating outside the research organisation which combine to deepen the lock-in process.

The treadmill

The treadmill metaphor is one that has been applied widely in the literature on all sorts of change that have taken place in agriculture (Ward 1993). In the economic sense, it was first used by Willard Cochrane (1957) who used the metaphor in an attempt to explain the implications of the logistical curve thought to describe the process of diffusion of most agricultural technologies, and its implications. Cochrane argued that early adopters of new technologies were usually the more technologically literate and entrepreneurial farmers. Successful innovations, in Cochrane's model, lead to a lowering of unit production costs, but in the early stages of the adoption process, insignificant lowering in the market prices of the commodity being produced. This phenomenon enables early adopters to profit from rents accruing from their cost reduction.[26] As more and more people adopt the innovation, market prices adjust to reflect the fall in costs. This in turn forces more people to adopt the innovation since failure to do so leads to economic failure on the farm as a result of a cost–price squeeze. Further innovations may promote new rounds of adoption.

In Cochrane's model, market forces keep the wheels turning on a technological treadmill which farmers step off at their peril. Models of diffusion such as Cochrane's are, up to a point (mathematically speaking, the point of inflexion on the logistical curve), models of increasing returns. As such, they are apt to exhibit path-dependence, possible multiple equilibria, possible inefficiency, and lock-in (Arthur 1988a, 10). Environmentalists have not been slow to adapt Cochrane's model for their purposes (Clunies-Ross and Hildyard 1992, ch. 2). In doing so, they typically complement a Cochrane-type economically propelled treadmill with an ecological one. It is the combination of the two forces that keeps not only farmers, but also researchers on a treadmill, forcing them to continue adopting, and creating, respectively, new technologies with increasing intensity owing to the nature of the interaction between the technology used and the ecological setting of the farm.

Because over time, farmers in the BCM mode come to rely on external sources of technology and information, to step off the treadmill is to step into a world that is unrecognisable from the one from which they first stepped on board, with all the risks and insecurities that this entails. The treadmill is of little comfort to farmers employing technologies which they know less and less about. Unease within the farming community stems partly from their being told that the levels of chemical input which they once used are no longer deemed safe (Erwin 1993). Circumstances such as these lead to a breakdown in the trust upon which properly functioning regulatory institutions depend (Giddens 1991). When stepping off the treadmill, farmers also find that the farm itself has been changed by the adoption of a production process which, increasingly, they do not understand. An example is provided by the changes wrought by fertiliser use (Clunies-Ross and Hildyard 1992, 61–2). The ecological impacts of fertiliser use and misuse are increasingly well-researched, though there are still areas of disagreement as to exactly what these are.

The use of fertiliser has been predicated on soil chemistry which has shown that three elements, nitrogen, phosphorous, and potassium, promote plant growth. The history of fertilisers is largely one of devising new delivery techniques for compounds comprising these three elements. Yet one of the subjects that fascinated Darwin more than a hundred years ago, the biological activity in the soil, remains vastly under-researched. A focus on soil chemistry has occurred at the expense of attention to the impact of synthetic fertilisers on biological activity in the soil.

There is some agreement that the continual use of artificial fertiliser is detrimental to natural soil fertility and structure, thus reducing its water retaining capacity. The availability of trace elements, such as copper, magnesium and zinc, also appears to be negatively affected, whilst phosphate fertilisers are a source of cadmium in the soil (Bonnieux and Rainelli 1997). Humus content declines, earthworms become more scarce, and the reduced aeration of the soil affects living organisms in the soil. Soils tend to become more acidic. The poorer structure and falling index of organic matter requires that, in order for yields to be maintained, fertiliser applications have to be stepped up each year. The fall in pH necessitates applications of lime. These processes are not completely irreversible, but some of them are in the short-term. Soils cannot be rejuvenated in a single season, but with time, soil building can alter soil properties. Given the economic pressures under which many farmers operate, desisting in the use of synthetic fertilisers is often not a viable option in the absence of external support or other sources of income. As a result, therefore, of a combination of the effects of competition and their impact on soil quality, fertilisers can lock producers in to their use.

As the BCM mode replaces other ways of doing agriculture, many useful cultural practices are forgotten, and since knowledge of many of these are maintained effectively through their use, they may be irreversibly lost. At best, the development of such knowledge systems is curtailed (Bell 1979, 47).[27] Where introduced practices are adopted, but subsequently rejected, such practices may produce ecological impacts which, in the short-term, are irreversible, and which have an impact on the effectiveness of the cultural practices. Such irreversibilities characterise what I term the ecological aspects of lock-in, those in which a given practice produces changes which are reversible only in the medium to long term, and which make it difficult to switch instantaneously between techniques. In Marglin's (1992) terms, *episteme* confronts, and defeats, *techne*, and the cultural equivalent of genetic erosion, the process of knowledge erosion, sets in (see Chapter 5).

Institutions governing transactions

In all economies, the institutional framework governing transactions has a bearing on the products which enter into commerce, broadly understood.[28] Paradoxically, a tendency to genetic uniformity in the field seems to be fostered both by competitive market economies and by systems of state planning. This reflects the fact that ultimately, decisions as to what, when, and how to plant are often taken out of the hands of farmers, in the former case owing to specialisation, the increasing prevalence of contract farming and the power of retailers, and in the latter, out of the state's desire to control inputs to, and outputs from the farm (Watts 1990; 1994; Clunies-Ross and Hildyard 1992, 68–70; Lang and Raven 1994; Fowler and Mooney 1990, x–xi). The difference in the two cases lies in the basis of the decision-making on the part of the respective decision-takers.

In market economies, even where farmers act with greater autonomy, a purely commercial outlook would appear to favour specialisation in line with climatic and soil conditions (both by, and within, species). The degree to which this specialisation is liable to occur depends to a certain degree on developments in the transportation and marketing infrastructure serving the industry. The more sophisticated these become, the more feasible it becomes for a farmer to plant a single crop on her or his land, not only because the whole crop can be sold, but also because the farmer too has access to food products grown in distant places. Neighbouring farmers are likely to follow similar strategies, increasing vulnerability to pathogenic attack (Clunies-Ross and Hildyard 1992, 15; Norgaard 1988a). Also encouraging greater specialisation, and in many cases underpinning the BCM mode, are agricultural policies. Many of these, reflecting the political effectiveness of commodity organisations, have developed along commodity lines (Hardin 1978), and the way in which some are designed actually constitutes a disincentive for farmers to diversify outputs (NRC 1989). Whether specialisation necessarily takes the extra stride and embraces uniformity within the crop species depends principally on whether the crop is harvested mechanically, the nature of the available seed, and relatedly, institutions governing seed commerce (see below).

Decisions made by the state concerning the desired direction for agriculture can affect the demand for innovations virtually by dictat. Indeed, in post-Second World War Britain, the government established County Agricultural Committees, which were invested with powers to evict farmers variously described as 'recalcitrant' and 'incompetent', and impose upon them practices of 'good husbandry' (Holderness 1985, ch. 2; Pretty and Howes 1993, 1–2). Such recommendations as were made advised the use of fertilisers, new seeds, and new uses of land. Similar activities on the part of the state, explicitly coercing farmers into using specific technologies, are documented in other countries.[29] One author has suggested that, over time, the technological structure of agriculture, once one where demand determined supply, has acquired it own momentum: 'It is through the supply of technologies that farmers' behaviour, including the required demand for technology, is defined' (Van der Ploeg 1992, 26).

Explicit or implicit subsidies or taxation also affect demand. Price policies in the European Community (now including the nations of what was the European Free Trade Area) and Japan certainly affect the use of inputs, raising demand above levels which would prevail in their absence (Harvey 1991), whilst those of the US probably do so too, despite attempts to claim that direct support does not affect farm output. Where price policies work in conjunction with desires on the part of the state to encourage the adoption of specific technologies, Burmeister's (1987) use of the term 'directed innovation' seems particularly appropriate. The path-dependent aspects of the processes of adoption, ecological change, and knowledge erosion suggest that, in some circumstances, though subsidies appear undesirable, the short-term impact of their removal may do more harm than good (Mearns 1987, 20; Lele *et al.*, 1989, 49).

Breeding for mechanisation

The increased use of mechanical power as opposed to animal traction has had a number of impacts on the structure of the farm. Not least has been the reduced need to feed animals on the farm, though of course, the feeding of animals has itself become a specialised industry. However, animals were not simply suppliers of power. Their manure was a vital source of fertility and organic matter for the soil, helping maintain soil structure too. This fertility gap was first met through increased use of organic manures such as guano, but increasingly, it has been met through the application of chemical fertilisers, derived from natural gas and limited deposits of phosphate which vary in their associated cadmium concentrations (ERM 1997).

This is primarily a post-Second World War development, after which the application of synthetic nitrate to the soil became economically viable. The industrial plant used in the manufacture of explosives was put to use in the fertiliser industry, and indeed, many major chemical companies flourished on the basis of this function (Kloppenburg 1988, 118). Plant breeders have increasingly sought to breed plants which can respond to heavy levels of chemical fertilisation. Typically, such plants must be short and stiff-stalked so that they do not suffer from lodging, which would otherwise be more likely since high levels of nitrate in the soil promotes sappy growth. In turn, their stiffness makes them more suitable for mechanical harvesting.

Mechanical harvesting is facilitated by a crop which matures at the same rate, and has uniform architecture, so encouraging uniformity in the field. Further pressures operating to promote uniformity derive from the end use of the products in question. Uniformity is often desired not so much from the perspective of harvesting machinery, but also from the standpoint of handling machinery, and the packaging and processing of outputs.[30]

Institutions governing seed use and trade

In the seed sector, there are effectively three types of institution affecting the production and use of seed directly; those governing which seeds can be traded; those governing ownership, of intellectual property; and, less often recognised, plant germplasm quarantine legislation. The last of these is not considered below, but its importance, obviously related to international germplasm transfer, is discussed in Plucknett and Smith (1988).

In most countries with a developed seed industry, certification systems are in place and to a greater or lesser extent, these determine which seeds can be marketed. These began to emerge in the mid-1920s in Europe (Vellvé 1992, 58). In 1973 in the UK, following accession to the European Community, it became illegal to sell seed which was not on the UK National List, or to sell it under a name different from that on the List (MAFF 1996, 2; Clunies-Ross 1995, 31). To maintain a variety on the list involves payment of a fee for testing and an annual fee for maintaining the variety on the List. In 1980, the European Community sought to rationalise the different lists of each country and drew up a Common Catalogue for the Community as a whole. In a move ostensibly designed to delete existing duplicates, having asked seed companies their view as to which were duplicates, a list of 1547 vegetable seeds were suggested for deletion. Yet the UK's Henry Doubleday Research Association (HDRA) found, upon closer inspection, that only 38 per cent of these were true duplicates, arousing suspicion that the seed industry was taking the opportunity to rid itself of unwanted competition. A thousand unique cultivars were thus written out of commercial existence (Vellvé 1992, 60).

The most common means of conferring IPR on plants has been through Plant Breeders' Rights (PBR), which are patent-like rights designed specifically for the plant breeding industry. PBR legislation requires that varieties, in order that they qualify, should meet the requirements of Distinctness (the variety is distinct from all others), Uniformity and Stability (the variety can be relied to pass its traits from one generation to the next), the so-called DUS test. Exactly the same criteria apply for varieties seeking a National Listing, and for most agronomic crops in the EU, VCU (value for cultivation and use) standards also apply (Thelwall and Clucas 1992, 26), so the two types of legislation tend to function in tandem.

A series of consequences follows from the marketing and PBR legislative framework:

1. Most obvious of all is that the range of varieties available in commerce is a highly restricted subset of what might otherwise be available. Not all seeds that could be traded would pass the DUS criteria necessary for a variety to obtain a National Listing. Without such a Listing, the 'variety' in question cannot be sold, presumably preventing varietal mixtures and multilines from being sold.[31] Furthermore,

if a variety which is on the List is not maintained there (someone allows it to 'fall off'), it is extremely difficult for it to be reinstated, so it is effectively banned from commerce.[32]

2. DUS requirements, most obviously that of uniformity, but also that of stability, which together entail that a variety be uniform with a high degree of homozygosity, clearly favour plant breeding strategies whose goal is to produce uniform varieties. Similar considerations arise in respect of patents (see Chapter 7).
3. It may take eight generations of crossing for a variety to be ready for DUS tests, and these may take a further two years. Only 40 per cent of varieties will make it on to the National List (Clunies-Ross 1995, 35), and once there, an annual fee must be paid to keep it there. The fees for obtaining PBRs are not trivial. In the UK, for cereals excluding maize, the fees for National Listing as laid down in the Seeds Regulations of 1995 (MAFF 1996) were;

Application	£245
Tests	£665 for each year in testing
Trial	£1085 for each year in testing
Award	£105
Renewal	£245 per annum.

Since it is rare for tests to last less than 2 years, the cost of having a variety listed and maintaining it for seven years would be just over £5500. If one adds to this the research expenditure involved in developing the variety prior to applying for a listing, one begins to appreciate that recouping such an investment would require a variety to be widely planted. Indeed, Clunies-Ross (1995, 35) suggests that in the UK, for a variety to be commercially successful, it almost certainly has to make it onto the Recommended List of the National Institute of Agricultural Botany, tests for which take a further three years. In short, the logic of the legal institutions surrounding the industry reinforces a tendency towards a smaller number of uniform varieties being planted over larger areas. They effectively outlaw the possibility of farmers in the UK experimenting with varieties with a view to selling the result (a practice which is not such a distant relic of the past as might be supposed). There is reason to believe that, ironically, this logic reduces the commercial lifetime of each variety since selection pressures work to increase susceptibility of the variety over time.

4. Some seed certification schemes (for example that of the UK) run on the basis that, for a new variety to be added to the list of certified

> varieties, it must be 'better' than a variety already on that list. This implies a decision as to what should be the relevant criteria for making such a judgement, though varieties are usually judged against reference varieties. Although a range of indicators may be used, including disease resistance, a critical indicator is yield (indeed, disease resistance is effectively a contributory factor in determining yield). Yet tests for yield cannot be de-linked from the circumstances in which those tests are carried out. Presumably different varieties perform better or worse than others depending on the farming practices used. Interestingly, yield tests are rarely performed in the absence of synthetic fertilisers. It is hardly surprising, therefore, that in the UK, 'varieties specifically bred for organic production are not available' (NIAB 1996, 123).

These forms of legislation seem to impose demands for uniformity on the seed industry. The industry itself has been largely responsible for the development of these institutions, and increasingly, seeks patent-like protection for its products (see Chapter 7).

Institutions governing use and trade of agrochemicals

If a trend towards increasing genetic uniformity has been supported by the seed industry over the years, this may not have been possible in the absence of the development of synthetic pesticides. Uniformity, encouraged by the institutions mentioned above, may not have been so attractive in the absence of the development of synthetic pesticides. The combination of uniformity and use of pesticides selects for matching pathotypes, as well as new pathogens emerging as a result of the use of chemicals. The more of a pesticide that is used in the field, the quicker resistance builds up.

By the late 1970s, twenty-four of the world's twenty-five major pests were secondary pests, which is to say, they emerged as a result of the loss of natural predators due to pesticide use. Some 440 insects and mites, and 70 species of fungus are now resistant to all major agrochemical formulations (Clunies-Ross and Hildyard 1992, 61–3). In addition, Le Baron (1991, 27) reported 107 herbicide resistant weeds world-wide. Farmers find themselves on a chemical treadmill owing to the ecological disturbances implied by pesticide use. Yet despite the dramatic increase in the use of pesticides, pre-harvest losses as a percentage of the field crop have increased over time. The use of herbicides discourages any form of intercropping (Edwards *et al.* 1990, 174). The rationale behind herbicide development is the elimination of all plants that are not the crop being grown. To plant more than one crop in a field, a practice common in large areas of many countries (see Chapter 1), is to render this approach useless. Many herbicides work on the basis of the elimination of monocotyledon weeds among dicotyledenous plant populations, or vice versa. Eliminating like from like presents more intractable problems. Thus, if grasses are cultivated continuously, grass weed problems become more and more intractable. Increasingly, because of the problems of grass weeds, break crops are used in rotation in arable farming in the UK. This allows for more effective elimination of grass weeds whilst a dicotyledonous plant is being cultivated.

In the absence of rotations, and even with them in place, varieties have a finite lifetime. This can be shortened in proportion to the level of pesticide use in a given region, necessitating replacement by varieties resistant to emerging pathotypes.[33] Research organisations that operate in the BCM mode never finish their work since genetically uniform cultivars carry an obsolescence which is built-in to the breeding technique. Hence the need to lock up resources, which could be usefully employed elsewhere, in maintenance breeding activities. The situation is accurately captured by Lowe *et al.* (1990,5), who suggest that 'successive cycles of innovation, needed to maintain capital accumulation, are repeatedly justified as providing the solutions to problems which have arisen in part from previous cycles'. Over time, attempts have been made to improve the specificity of action of pesticides to ensure that non-target organisms are not affected by designing the chemical such that it interferes with biological processes specific to the target organism. However, the costs of doing this are substantial and growing. Already, due to the fact that the time taken for, and stringency of, environmental and safety tests, the costs of bringing a new pesticide to market are of the order of $35 million (this does not include the design of production plants). Ironically, these escalating costs, through making it more difficult to obtain an adequate return on an investment, tend to mitigate against designing pesticides with specific targets since they are unlikely to reach as wide a market as desired.[34]

The history of pesticides has been a controversial one and continues to be so (MacIntyre 1987). Not only are their affects in terms of activity somewhat perverse, but they have caused considerable ecological damage, and many are toxic to humans (Pimentel *et al.* 1993; Beaumont 1993; Conway and Pretty 1991, chs. 2 and 3; Dudley 1987). Given their crucial role in the BCM mode, and given the fact that their effects on human health and the environment continue to arouse considerable concern, it seems pertinent to ask how pesticides were sanctioned for use in the first place. The science of risk assessment as it has been developed and applied for use in regulating the pesticide industry (see Conway and Pretty 1991, 4–8) is a bogus one, relying as it does on what amounts to guessing, *ex ante*, what an 'acceptable level' of a particular (toxic) chemical may be.[35] Quite apart from the fact that those who seek to elicit this level cannot possibly know what they purport to know, Beck (1992, 65) makes the telling point:

> Acceptable levels in this sense are the retreat lines of a civilization supplying itself in surplus with pollutants and toxic substances. The really rather obvious demand for non-poisoning is rejected as *utopian*.

This situation led MacIntyre (1987) to postulate the existence of a pesticide sub-government in the United States.[36]

All of these policies relating to the use of technology must be placed in context. What Beck (1992, 60) calls the 'economic Cyclopia of techno-scientific rationality' finds its fuel, in the late twentieth century, in the quest for 'competitiveness.' If in the past, 'progress' legitimated the free development of new technologies, today, it is in the interests of the one-eyed pursuit of competitive advantage that society is asked (or required) to ignore the possible 'side-effects' of innovations whose impacts cannot be known in advance.

Malthus again – more and better food

The last factor that should be mentioned with regard to locking in to uniformity is one which I have sought to show is contentious. This is the argument that genetic uniformity is a necessary component of any strategy aimed at feeding the world's growing population (NRC 1972, 83; de Greef 1996). To accept this logic is to accept the use of pesticides as a necessity, and a position in which, 'The devil of hunger is fought with the Beelzebub of multiplying risks' (Beck 1992, 43).

Perhaps more difficult to deal with are the impacts that diversity and the reduced use of synthetic pesticides have for the appearance of food. Consumers tend to purchase unblemished produce whose shape fits their perception of the ideal product, the cost of producing which is genetic uniformity and high levels of chemical pesticide use. As markets have expanded, the relatively fixed costs of transport have dictated that only premium quality (in the sense of being unblemished, visibly) produce and that of maximum durability is marketed. Ironically, given the arguments raised concerning food supply, one report suggests that as much as half of organic produce being sold to supermarkets is deemed inedible through not meeting cosmetic standards (Clunies-Ross and Hildyard 1992, 89; see also Rosenblum 1994). It is difficult to separate the issue of consumers' perception from that of food marketing, and it is not clear whether the desirability of such produce was first proclaimed by consumers or the food industry itself, which has considerable power to shape such issues. It has been suggested that in some countries where consumers are acutely aware of problems associated with pesticide use, they deliberately purchase blemished produce.

The irony of this situation is that, whilst food companies equate appearance with quality, the nutritional content of food produced organically has, in some studies undertaken, been demonstrably superior (Maga 1983; Widdowson 1987, 148–9). The exact reasons for this are unknown. Possible reasons include the effect of reduced soil organic matter on the activity of mycorrhizal fungi in the soil which are important in facilitating uptake of these nutrients. When one adds that many non-organic foods are contaminated with pesticide residues, sometimes at levels above even that deemed acceptable by the bogus scientific procedures mentioned above, the external costs of cosmetically superior produce begin to appear appreciable (Clunies-Ross and Hildyard 1992, 29–31). Finally, with reduced diversity, many varieties which were quite different in taste, and which had specific culinary uses, are no longer available to consumers owing to institutions such as the National Lists mentioned above.

Conclusion

This chapter has sought to illustrate how research organisations can become locked-in to a particular way of doing things, and how various factors make it more likely that this will be a technique based on the production of genetically uniform crops. Ecological factors introduce irreversibilities which enhance the lock-in effect, at least where short-term considerations predominate.

It is striking how the institutions which I have identified as entrenching lock-in are so well matched to the mode which they have sought to encourage. The reason for this is to be found in the co-evolutionary nature of their development in parallel with the commercialisation and industrialisation of agriculture. Lock-in is the direct manifestation of the process which Marx foresaw when he observed that the tendency of capital is to transform what was once superfluous into a necessity. As industry has become more deeply involved in the production process, so institutions have developed which make it increasingly difficult for things to be done in ways other than that with which industry is involved. In particular, the legal institutions surrounding the seed are designed on the basis of work done by breeders on genetically uniform ideal plants. Even though it is technically possible to carry out research in ways which do not engender genetic uniformity in the field, it is increasingly impossible to do this on a commercial basis.

It is possible that as institutions and a breeding approach have co-evolved, so the key driving forces towards uniformity have changed. Initially, research scientists and entrepreneurs would have promoted uniformity. Increasingly, the institutions which have been developed, and which they have lobbied for, have acquired a force of their own, making it increasingly difficult for anyone to 'change their mind' concerning the desirability of the ongoing process. Ecologically speaking, this is what makes lock-in such a dark force. Sociologically, it becomes difficult to observe, since orthodox ways of looking at the world prevent reconsideration of problems in radical ways. Thus, for example, in recent discussions within the Department of the Environment concerning how to minimise pesticide use, no one seriously posed the question as to why pesticides are used in the first place.[37] The desirability of genetic uniformity was never questioned, and indeed, genetics were only considered through discussions on the potential of genetic engineering to improve matters regarding the ecological impact of agriculture (see Chapter 7).

Notes

1. I will argue below that farmers are not always the clients of agricultural research.
2. The work of Engel (1993) on national systems of innovation is an attempt to create a typology of the types of framework that evolve.
3. Arrow (1962a) sought to show that knowledge was a pure public good and that firms would not undertake 'basic', or 'fundamental' research, because of its inappropriable nature. This view has been rigorously questioned by Rosenberg (1982, ch. 7; 1990; 1994, chs.1 and 8; Mowery and Rosenberg 1989, ch. 1), who highlights the importance to private firms of doing their own basic research. Notwithstanding Rosenberg's important points, that appropriability provides an important incentive for private sector investment in *any* R&D remains valid.

4. See for example Kenney (1986), UNCTC (1988), Hobbelink (1991), Hindmarsh (1991, 1992), Kloppenburg (1988, 1993), Junne (1992), Jefferson (1993), and Nana-Sinkam *et al.* (1992).
5. For Hayami and Ruttan the critical questions are not how or why, but whether state funded agricultural research became established. They imply that where publicly funded research did not come into being, it was because farmers lacked the political clout to command resources from the government. They miss the point that politically powerful interests, even landowners, may perceive that lobbying for agricultural research organisations runs counter to their own interests (see Piñeiro *et al.*, 1983).
6. See Maxwell (1984) on the problems that this has presented social scientists who join established research organisations.
7. A recent contribution from Dietrich (1997) addresses professionalism through the lens of path dependence, and career choice becomes a matter of strategic lock-in.
8. See McCalla (1994) for a discussion of this issue in the context of the CGIAR.
9. Concerning the influence of donors, and changes made within the CGIAR, see Ravnborg (1992) and McCalla (1994). See also Oasa (1987) concerning the CGIAR's response to criticisms of the Green Revolution.
10. RAFI and GRAIN (1995) suggest some in the CGIAR system 'believe that high external-input agriculture should be called "sustainable agriculture" until it collapses.'
11. Most of this debate concerns the continuing attention given by CGIAR centres to more-favoured environments, and the related ecological impacts of their work, which they defend by hypothesising a consequent reduction in encroachment on marginal areas due to the increased output from the more-favoured zones (Graham-Tomasi 1991; Winkelmann 1991, 5–6). Where the two overlap, it is in the appreciation of the fact that greater emphasis given to research on complex, diverse environments would of necessity be a location-bound activity, raising the chances that outputs would be ecologically adapted to the agroecosystem in question. For discussion, see Oasa (1987), Koppel and Oasa (1987), Anderson *et al.* (1991); Ravnborg (1992), Winkelmann (1994) Lipton (1994).
12. In a 1992 report, the OECD (1992, 138) commented 'the final food sector is increasingly replacing the farmer as the seed industry's privileged partner'. Several authors, Madden (1987), Berry (1986) and Hightower (1973) among them, have criticised the accommodation of agribusiness interests by public research organisations in the US. As a consequence of Hightower's book, representatives of the US public agricultural research system were called before the Senate. One, John Caldwell of Carolina, when asked whether he thought that small farmers could afford the machinery being engineered by the colleges, replied that they could not, but that 'Then they are free to go out of business' (in Cleaver 1975, 156–7). The concept of client-driven research is a theme of the work of Busch *et al.* (1992, 49), who note that; 'science is conducted in response to client demands – demands expressed not through the market but through negotiation, persuasion, and coercion'.
13. Within the CGIAR system, in a move that appears to reflect donor dissatisfaction with the way in which funds are used, more and more of the

system's funding has been channeled into non-core, or project-specific funding. Furthermore, the centres within the system are increasingly advised (and under pressure, given stagnant funding in real terms) to seek contracts from outside their traditional list of donors (Ravnborg 1989, 16–18; personal communication with Sir Ralph Riley of the CGIAR's Technical Assistance Committee 1996; Marozzi 1997).

14. In virtually all developing countries where the private sector has taken a stake in the seed industry, it is in hybrid maize (and vegetable) seed production that this occurs first owing to the appropriable nature of hybrids, and the fact that farmers cannot save seed for planting the following year (for a discussion, see Dalrymple and Srivastava (1991) and Echeverría (1990)). What happens to the public sector under such circumstances varies. Pray *et al.* (1991) note that a flourishing private sector hybrid millet and sorghum breeding industry exists in India. Like others, they advocate, to my mind foolishly, the public sector simply moving upstream to conduct basic research for the benefit of the private sector, a course of action which would leave all eggs in the privately owned basket.
15. There is a separate debate here concerning the secrecy engendered by the increasing use of IPR in the biological sciences, and the effects this has on the rate at which knowledge is generated (Kenney 1986, 1995; George 1990, ch. 6; OTA 1988).
16. A comprehensive study of what influences problem choice in agricultural research is that of Busch and Lacy (1983) in the context of the American Land-Grant College system.
17. This is what Anderson *et al.*, (1991, 13–15) term the Underspecification Thesis.
18. Lipton (1994, 613), speaking of the need for formal and informal agricultural research, makes the point rather starkly; 'Today, they seldom speak the same language – often literally not ' (see also Marglin 1992, 32–4). A reading of Anderson (1994, 9) suggests formal and informal systems are still regarded as quite separate.
19. See Tripp (1989) for a critical view of this ranking approach. Farrington (1997) suggests the term participation is being devalued. He suggests it needs to be situated in its agro-ecological and socio-economic context.
20. Chambers argues that without deep-seated changes in the attitudes of all those involved in technology generation and diffusion, agricultural research will continue to fail to meet the needs of 'the third agriculture', that which is 'complex in its farming systems, diverse in its environments, and risk-prone'. (Chambers *et al.*, 1989, xviii). His 'Rural Development: Putting the Last First' (Chambers 1983) followed the publication by the Institute of Development Studies of an issue of its bulletin devoted to the question, 'Whose knowledge counts'?(IDS 1979), based on a workshop held in 1978). He calls for a new professionalism, an assault at the level of ideology and a change in reward systems in agricultural research which challenges the established professions and their hitherto accepted orthodoxy (Howes and Chambers 1979, 7, 9–10; Chambers 1993, ch. 1). Chambers sees participatory methodologies in terms of a new paradigm for agricultural research (e.g. Howes and Chambers 1979, 5; Chambers 1993, 2–3, 60).

21. RAFI and GRAIN (1995) refer to 'the continuing absence of real farmer participation' as 'an enormous barrier to genuine progress'.
22. Typically, CIMMYT (Centro Internacional de Mejoramiento de Maiz y Trigo) measures its success in terms of area planted to germplasm based on its research (CIMMYT 1992c, 22–4). Winkelmann (1994, 88) notes that transnational corporations organise some of their research along international lines hoping that their work will fit a given environment in a number of countries.
23. A good example is provided by the recent debate on genetically manipulated maize, and genetically engineered soya in the European Union. A combination of concern for consumers' right-to-know in terms of product labelling, and worries over the use of antibiotic resistant marker genes have led to major Europe-wide campaigns to demand the labelling of genetically manipulated soya products and have Ciba's (now Novartis') maize banned on health grounds (ENDS 1996; Greenpeace Business 1996/7; Europe Environment 1996a,b; Barber 1996; Maitland 1996; Hogg 1997).
24. Government officials in Indonesia have been known to burn down traditional rice varieties even though traditional varieties were economically competitive (MacDougall and Hall 1994, 7). Credit and loan facilities often favour the cultivation of uniform varieties.
25. Ward *et al.* (1993) suggest that advisors are extremely important in telling farmers what, when and how much pesticide to spray. A survey by the NFU in 1984 found that although the Government advisory service, ADAS, was well respected, 'merchants' reps are more likely to be used for advice on day-to-day and other matters because they are often on farms anyway whereas the ADAS man would have to be called out' (Crossley 1984). More recent information from Hearn (1997) suggests that very few UK farmers make their own decisions regarding pesticide use. Low *et al.* (1991) show what difficulties extension officers, trained to disseminate 'correct' techniques, encounter in varying recommendations according to farmers' needs.
26. Labelling farmers as leaders or laggards can have the effect of entrenching discrimination among extension agents (Pearse 1980, 16).
27. Jiggins (1986, 45) points out that gender biases in the work of research organisations can result in the loss of (often) both women's seed, and stock selection abilities.
28. The importance of the institutional framework for promoting innovation has been the subject of much research, some of it extolling, vis a vis other organisational forms, the virtues of capitalism as an 'engine of progress' (Nelson 1990; Rosenberg 1994, chs. 5 and 6). A key factor is deemed to be the way in which risk-taking by private enterprises is encouraged by capitalism.
29. For brief accounts of the situation in Britain, France, Spain and the US, see Clunies-Ross and Hildyard (1992; ch. 2). On the UK situation, see Lowe *et al.* (1986); Bowers and Cheshire (1983). On the US, see Berry (1987).
30. The NIAB Recommended List of cereals gives an indication of the specification for barley varieties used by maltsters. These include nitrogen content, varietal purity, germinative energy and grain size. Wolfe cites this an obstacle for uptake of varietal mixtures and multilines (ABN Bulletin

1992). Wheat varieties are assessed on the basis of Hagberg Falling Numbers, protein levels, specific weights, and Chopin Alveograph tests for baking quality, in which a disc of dough is inflated until it bursts to determine the dough's strength and extensibility (NIAB 1996, 114–5).

31. Texas changed its seed laws such that they emphasised the grain, not the variety, opening the way for use of mixtures (Clunies-Ross 1995, 55).
32. Farmers' World Network have been trying to revive old varieties in the UK, but have been told by MAFF that it would be illegal for them to engage in trading non-listed varieties (Clunies Ross 1995, 27).
33. There is some evidence, however, to show that *lowering* pesticide dose rates can quicken the build up of resistance through enabling resistant tails of a population to flourish.
34. In the UK, a levy on pesticide sales (aimed at the producers) is used not only to fund parts of the approvals and monitoring processes, but also to subsidise the development of pesticides with small markets such as those used in crops such as watercress. This is because the costs of research and development cannot be recouped fully through sales..
35. The tests used are the Lethal $Dose_{50}$ (LD_{50}) and Effective $Dose_{50}$ (ED_{50}) tests which seeks to discover the dose of a toxin required to kill or affect, respectively, half the exposed population. The populations are usually small mammals, yet these differ enormously in their responses to the same doses. Furthermore, it is rarely clear how these results should be interpreted in order to make them applicable to humans, amongst which there may also be considerable variation.
36. The pesticide industry itself has been described in the language of techno-economic modes and trajectories (Achilladelis *et al.*, 1987; also MacIntyre 1987). Different periods have been dominated by searches for new active agents, and once a useful agent has been established and patented, competitors seek to innovate around it to shorten its life (and a competitor's monopoly).
37. The author was privy to these discussions through work with ECOTEC Research and Consulting Ltd.

Introduction to Case-Studies

The view that science is not a positivistic sphere of enquiry where the enquirer can stand apart from what is being studied is now widely accepted (though not always within science itself). Science is shaped by social, political and cultural forces, and as such, it is infused with subjective and normative judgements on the part of those involved in its shaping. Thus, as Biggs (1990, 1489–90) notes in discussing the 'second generation problems' of the Green Revolution, closer examination, 'often reveals that there were also "non-center" scientists who had, at the time of the original decision, correctly predicted the outcomes of the proposed actions of the center'. The significance of lock-in lies precisely in the existence of 'non-centre scientists', doing things in 'non-centre ways'. Typically, there exist alternative ways of understanding and approaching a given problem, and these throw open the plausibility of counterfactual worlds. I have sought to highlight in the following case studies the two questions of greatest importance; first, the obvious one, why was it that things were done as they were done?; and second, the often ignored one, why, and how, was it that alternative approaches were marginalised or ignored? The answers to these two questions inevitably overlap, but they are not one and the same.

Regarding seeds, even if alternative approaches to breeding existed at the time at which one or other gained supremacy, the reasons for the alternatives not receiving greater attention will not necessarily be the same today as they were at the time the choice was effectively made. Certainly, evidence may arise in the context of the use of a given technology that make alternatives appear more attractive. This is not exactly the same as saying that we might have made a different decision with the benefit of hindsight since even the sharpness of our vision as we look back in time is shaped by the very decision (and the information flowing from it) that led initially to lock-in. Hindsight does give us a slightly better perspective, and allows us to reconsider whether, as is so often implied, the only counterfactual worthy of comparison with a situation prevailing in the wake of change is a projection of the *status quo ante*.

In these case studies, I look both at why things happened as they did, and why alternatives were ignored. I review these in the light of ongoing assessments of the relative merits of the approaches under examination. In the final case, that of biotechnology, we are not in a position to employ hindsight *vis-à-vis* the technology itself. However, I employ the framework developed in Chapter 3 to suggest an explanation for what is currently occurring in the development and use of agricultural biotechnologies.

5

Hybrid Corn in the United States, 1900–35

Introduction

The development of hybrid maize caused dramatic changes both on and off the farm. It both assisted, and took place against a background of, the professionalisation of plant breeding. Breeding was taken out of the hands of farmers, and became the exclusive domain of new specialists, whose expertise was founded on the application of new insights into the mechanism of heredity. Despite the euphoria with which hybrid maize was greeted, especially following its widespread adoption, change was a long time in arriving. Whilst the key theoretical work had been all but completed by the end of the first decade of the twentieth century, it would be twenty years before hybrids would be planted on significant acreages in the Corn Belt. Once this point was reached, however, the change was swift, though the rapidity of this change was not due solely to the alleged superiority of hybrids in the field.

I begin this case study by setting the scene in the relevant theatres of action. Through understanding the prevailing situation, one can better appreciate the significance of some of the decisions taken along the road which led to the virtually universal adoption of hybrid maize in the United States by the time the Second World War began. These decisions effectively locked out alternative approaches to breeding which allowed for genetic exchange within the field. The use of hybrids entailed uniformity, a fact that led to catastrophic losses in 1970 (see Chapter 7).

Hybrid maize in context

The state of US agriculture

As a result of the Civil War, the price index for agricultural products rose from a low of 93 in 1857 to 200 in 1865. Following the war, however, many young men started up farms in the newly broken prairie lands. Wheat production soared in the late 1860s and prices began a downward drift. The period marked the beginning of the US's dependence on export outlets for a growing proportion of farm income. Large volumes were exported to Europe where Malthus' and Ricardo's warnings haunted a continent for which, though industrialisation proceeded apace, food self-sufficiency was increasingly problematic.

The late nineteenth century was, for the United States, one of the century's most depressed periods of economic activity. The Populist party was formed in 1890 in response to rising discontent, especially in the rural areas, where the activities of railroad owners were regarded with increasing suspicion. Despite the growth in farm exports, the pace of railroad development outstripped the growth in railroad freight. To compensate for the lack of traffic, freight rates were high and discriminatory. Farmers, faced with falling prices and growing problems with pest and insect epidemics, united and mobilised around the issue.

Once economic conditions began to improve, protest gave way to hard work (Benedict 1953, 52–3, 70, 84–5). Cochrane (1979, 100) notes that farm prices rose steadily in every year from 1897 to 1910, and the terms of trade shifted in favour of agriculture. This was the Golden Age of American agriculture: 'Farm product prices were high and stable. The terms of trade were strongly in favor of farmers. The country was settled. The world was at peace. Hard work, thrift, and "right thinking" had indeed paid off for farmers; the good life was a reality' (Cochrane 1979, 100). After the outbreak of the First World War, farm prices again began to rise. Between 1916 and 1920, farm prices more than doubled, making cropland a valuable commodity only 20 years or so after the frontier had been reached. But commodity prices plummeted after 1920, and recovered only mildly in the mid-1920s, leaving thousands of farmers bankrupt as they defaulted on loans used to purchase land during the boom period. This agricultural depression was compounded by the Stock Market Crash of 1929, and the onset of The Great Depression.

It was already clear that the solution to agricultural problems was less simple than blithe advocacy of modernisation and increases in productive efficiency. With this in mind, the United States Department of Agriculture's (USDA's) Bureau of Agricultural Economics was established in 1922. Other social sciences also flourished as the complexity of the farm problem, which most now agreed was not *one* problem, revealed itself. Political response to the Depression was characterised by an interventionist policy designed to correct disequilibria in agricultural markets. Roosevelt's New Deal policies included the Agricultural Adjustment Act of 1933 which in large degree has shaped US agricultural policy to the present day.

The state of the seed industry

As early as 1819, US consulates abroad were urged to send home seeds for introduction to the US. A Division of Agriculture was established in the Patent Office in 1836, and in 1838, the First Patent Commissioner of the US, Henry Elsworth, asked for money to be used for seed collection and distribution, and the collection of agricultural statistics. Elsworth's distribution of germplasm was not welcomed by all. Many agricultural journals objected to the distribution of free seed, not least because some of them used free seed packages themselves to solicit subscriptions. That germplasm was distributed freely reflected in part the fact that farmers were best placed to determine whether new germplasm was of any use. By 1878, the growing USDA was spending a third of its budget on the collection and distribution of germplasm (Ryerson 1933, 116–8).

As Fowler's (1994) account of the seed industry suggests, its development cannot be understood in isolation from other changes affecting agriculture. As late as 1820, in the developed north-eastern regions, less than a quarter of agricultural produce was being marketed. Only in the 1870s did the number of people employed in agriculture fall below that in other occupations so that, from the 1880s, although they were a dominant political group, they no longer constituted a majority (Benedict 1953, 87).

The picture changed dramatically as a consequence of new technologies which, at first glance, had little to do with farming. Growth in industrial production led to the emergence of large cities, whilst railroads assisted in the settling of, and marketing of produce from, the west. This promoted greater specialisation amongst farmers, especially in the east as competition from the newly settled west intensified. The process was encouraged by the railroads, many of whom were given land grants by the government, and encouraged settlement through land sales. They acted as evangelists for a market-oriented agriculture so as to increase freight traffic. The postal service, again not directly related to farming, brought the seedsmen to every door, and made seed catalogues a crucial element in companies' marketing strategies (Pieters 1900, 558–9; Fowler 1994, 37–40). The industry was given further impetus by agricultural fairs and the farm press, the former acting to convince the farmer as to the form of the ideal crop, the latter being a valuable advertising forum. Last but not least, as seems so often the case in changes in agriculture, war intervened. The American Civil War encouraged the growth of seed production for local markets, and limited availability and high prices promoted specialisation in seed production (Pieters 1900, 558–9).

As the seed industry began to grow, the first signs of dissent as regards federal largesse in the distribution of free seeds began to be heard. At the end of the twentieth century, J. Sterling Morton, then Secretary for Agriculture, tried to put a stop to the free seed programme, arguing that this was crowding out private enterprise. That distribution was so quickly resumed probably reflected, as Kloppenburg (1988, 63) notes, less a concern to maintain this public function than a wish to continue a means by which congressmen gained favour with their electorate by sending accompanying letters with packets of seed.

If the emerging seed industry appeared to be taking breeding out of the hands (and eyes) of farmers, other developments reinforced the trend. First, the land grant colleges, then the state agricultural experiment stations became important avenues for distribution of new plant materials by the USDA. Increasingly, especially from the turn of the century, exotic germplasm was distributed not to farmers as had been the case in the past, but to scientists, so preventing farmers from testing new materials in their fields (Ryerson, 118–20; Fowler 1994, 55–6).

Corn seed in the nineteenth century

The nature of cereal crops provided many obstacles to private industry's attempts to commodify cereal seed. Of considerable importance was the fact that farmers could simply save seed from one harvest for planting the next crop. Additionally, with few exceptions, the early work on corn improvement was based on mass selection in which farmers selected desirable ears from the corn crib or from the field so as to produce next year's crop.

The corn ear is the result of pollination of many (female) silks by (male) pollen grains from any number of parent plants. Selection of ears, therefore, implied selection on the female side of the reproductive process only. In the 1870s, W. J. Beal began talking to farmer organisations about the importance of also selecting for the pollen parent. He performed the first controlled crosses whose express purpose was increasing yields, and by the 1880s, most midwestern experiment stations were using varietal crossing as a means to improve corn varieties.[1] Even so, most varieties that achieved notoriety in the Corn Belt in the late nineteenth and early twentieth centuries were developed by farmers, among the most famous being Reid corn, and Krug corn (Wallace and Brown 1956, 81–4).

Science and the farmer, 1840–1930

The early history of agricultural research in the United States owes much to the agricultural experiment stations of Europe, especially Germany. Many scientists from the United States were trained in German stations and British universities and brought back with them not just knowledge, but a commitment to the value of agricultural science (Rosenberg 1976, ch. 8).

As mentioned above, well into the twentieth century, farmers' corn varieties were the best, and they were also the major innovators in machinery. Innovations were usually brought to the attention of other farmers through agricultural newspapers and societies, of which there were more than 900 on the eve of the Civil War. Field trials at state farms and plots owned by agricultural societies enabled farmers to be both innovators and adjudicators of their own innovations. The emergence of agricultural science must be understood within this context. If there were experts in farming, it was farmers themselves, and many perceived that academics and laboratory scientists could do little for them. After Michigan, Maryland and Pennsylvania had already established their own agricultural colleges, President Lincoln signed the Morrill Act in 1862 which allowed for the donation of public land to the states for colleges of agriculture and the mechanic arts (Ross 1938; Benedict 1953, 83–4).[2] These land grant colleges received amounts of land assessed on the taxable base of the state in which they were located.

It was in soil chemistry that scientists first attempted to ascend the ladder of legitimacy in the eyes of farmers, but soil chemists made claims for soil-testing which could not be sustained and the fad quickly passed from view. Farmers were, however, keen to expose fraudulent fertiliser manufacturers and were enthused by the prospect of chemists analysing fertiliser samples. It was this function more than any other that furthered the cause of agricultural science (Rossiter 1975, 109–24, 149–71; Marcus 1985, 27–58). Many states provided for a state chemist in the 1870s, and their time was devoted almost entirely to fertiliser analysis.

The first experiment station to widen its investigations was established in New York in 1882. The Hatch Act of 1887 aimed to replicate the New York experience on a nation-wide basis. Each state and territory was awarded $15 000 annually to fund an experiment station devoted exclusively to agricultural experimentation and investigation. However, unlike in the New York case, and as a result of astute lobbying of Congress, the agricultural colleges managed to ensure that the stations were attached to the colleges in states where experiment stations had not been set up already. Their manoeuvres had forced farm lobbyists, who were hardly receptive to the idea of the colleges having such an important role, to accept the Hatch Act as it was, or nothing at all.[3]

Scientists found themselves closely tied up in politics throughout this period and beyond. They sought to gain funding for the practice of research, the quality of which they wanted judged by the standards of academia. Yet to secure funding, the enterprise had to give itself some economic, if not explicitly commercial, appeal. The question remained as to which interests should be courted. This period saw the emergence of what Rosenberg (1976, 159–65) calls research entrepreneurs, men who saw the logic in trading off some independence in their intellectual activity if serving clients would lead to increased funding from federal and state legislatures. They directed their activities in such a way as to benefit farmers, usually the better educated and capitalised ones, in an attempt to provide economic justification for their work. But as the number of agricultural scientists grew, many grew disenchanted with compromises made by these research entrepreneurs. Demands from farmers were preventing scientists from engaging in pure research.

It became common to assert that research was essentially of two types. Basic research, the scientists asserted, sought to establish scientific principles which could help explain what happened in farmers' fields. This was best conducted in the laboratory, where phenomena could be controlled and examined in detail. Applied research entailed application of these principles to practical farming problems. Without basic research, it was argued, no applications would be forthcoming. The scientists' pleas were acknowledged with the passage of the Adams Act in 1906, which doubled the federal funding for each experiment station, stipulating that the new funding be spent on 'original investigation'.[4]

The farming community was far from heterogeneous and saw different ways of dealing with rural discontent. For Marcus (1985, 1987), as for Rome (1982), the transition in the way in which American farmers saw themselves and their occupation was already underway by the 1870s. Farming had lost its rationale as a way of life, was a commercial enterprise, and most farmers wanted to update their production methods with the help of science. Marcus (1985, 7–26), in dividing farmers' into systematic farmers and the scientific agriculturists, claims that both were aware of major problems in farming and rural life. Systematic farmers viewed farming primarily as a business, and they sought solutions to agriculture's problems in learning, through book-keeping, which methods worked best. For scientific agriculturists, agriculture's salvation lay in the use of observation as a means to uncover the underlying principles of farming. Danbom (1979, 17–19) on the other hand, seeks to show that much of the farming population was quite content, and resented intrusion into its affairs.

Probably all these elements were present in a farming population which constituted a diminishing percentage of the total, and which was under pressure from outside to rationalise, specialise, and increase the efficiency of its operations. It was fashionable to associate moral and material progress as two sides of the same coin, so those who perceived rural life as disorderly and ill-disciplined sought solutions in the rationalisation of agriculture through the application of science. As Danbom (1979, 23–6) notes, it was principally the influence of off-farm interests – railroad men, bankers, and through these, the scientists themselves – that led to clamour for a transformation of rural life through education, and a refashioning of agriculture through the application of science.[5] Rosenberg (1977, 417–8) makes the point that since increasing agricultural production was viewed as unambiguously good, any side-effects were regarded as acceptable trade-offs. Increased production clearly hurt some farmers, but agricultural scientists invented a future infused with social Darwinism. The means to improve efficiency was through science, and inefficient farmers would fall by the wayside in an increasingly competitive world. Ironically, therefore, although the scientists' rhetoric dwelt on their ability to assist small farmers, and thus, to maintain a degree of social order through preserving a Jeffersonian ideal, as time passed, the scientists found themselves working increasingly with established producer associations, large and highly capitalised farmers, and producers of fertilisers, seeds, and livestock. This was most inevitable where state support for the experimental work was underwritten by support from those with significant agricultural interests in the state legislature.

Scientists also succeeded in convincing scientific agriculturists that they were guilty of confusing two separate professions, that of science and that of agriculture. In achieving the division of labour they had sought, scientists achieved authority in agricultural matters and farmers became dependent on their expertise.[6] As importantly, scientists claimed that since they dealt with generally applicable scientific truths, they should be exempted from accusations of failure that arose from specific contexts of their application. These specifics were to be left to farmers to deal with, not scientists (Marcus 1985, 27–32). By the 1920s, following the establishment of the co-operative extension service by the Smith Lever Act of 1914, scientists were completely free from obligations to carry out field work, and were able to concentrate on basic research.[7]

The rise of mendelism and genetics: the professionalisation of plant breeding

The rediscovery of Mendel's work on inheritance

In 1900, Webber and Bessey (1900, 486) wrote that corn 'has been greatly modified and improved by hybridisation, but no improvement stands out as marking a distinct epoch'. The possibility that a new epoch might soon emerge was given a boost by the rediscovery of Mendel's work, apparently independently by de Vries, Correns and Tschermak, in 1900. That Mendelism was so quickly adopted by many scientists in the United States was principally due to recognition that its insights might be applicable to the growing number of studies of crossing not just corn, but many other crop species too. As Kimmelman (1983, 166–7) notes, these scientists did not see the study of the fundamentals of breeding, and their application to agriculture, as separate enterprises. Indeed, as the above account suggests, the strategy by which scientists sought to professionalise agricultural science was predicated on the need for basic research to underpin new applications.

Fitzgerald (1990), Pallodino (1994) and Kimmelman (1983) all appear to attribute significant weight to social factors in the development of Mendelian theory and ultimately, in the rise of new 'scientific' plant breeding methods. Yet although one cannot dismiss social constructivist elements, neither can one dismiss the significance of agreed upon observations in verifying or refuting a particular theory. Fitzgerald (1990, 23) talks, in Kuhnian terms, of a change in paradigm. Yet paradigm changes occur where accepted wisdom clashes with a new, and usually competing, school. Prior to Mendel, the state of inheritance studies could be summed up in the expression 'like breeds like,' which was more a reason to study inheritance than an explanation of it. Furthermore, the emerging science of genetics did not so much reject the dictum 'like breeds like' as explain it, and show why it did not always hold true.

Mendelian theory and the science of genetics that blossomed in its discovery, paved the way for the emergence of two competing approaches to plant breeding, misleadingly referred to as Mendelism and biometrics. Both were compatible with Mendelian theory, as had been illustrated by Emerson and East (1913). Yet in plant breeding, it is generally accepted that the Mendelian approach has held sway over biometrical approaches (Robinson 1996; Fitzgerald 1990), and it would appear that the competition is played out in the socio-economic rather than the scientific sphere. The latter is in essence, a form of selection, but to equate it with the practices of farmers at the turn of the century would be to belittle advances made in genetics and related statistical techniques.

The key difference between the two emerging approaches as regards plant breeding was that Mendelians effectively sought individual genes which were responsible for certain traits or characteristics. In their interpretation, finding these would enable control over the final form of the plant. Biometricians, on the other hand, recognised that a given characteristic or trait may be controlled by a number of genes at different loci. In order to maximise the expression of a given trait, statistical techniques were necessarily the best. For Mendelians, the natural route was to seek genes for combination into a single variety. For biometricians, on the other hand, the end point might not be a specific variety, but an improved population. Statistical-based selection methods could tolerate diversity, whereas Mendelian approaches implied uniformity.

A key issue in this chapter is why the crossing of inbreds to produce hybrid corn was so vigorously promoted at the expense of selection methods, even as corn breeders were simultaneously improving methods of selection, partly on the basis of new statistical techniques. The channelling of resources towards crossing inbreds was influenced by many factors. Fitzgerald (1990, 28) suggests, quite rightly I believe, that these shifts, 'had less to do with the scientific superiority or inferiority of one or other theory and more to do with the economic, social, and professional interests of the hybridisers.' For plant breeders in particular, the potential rewards from this somewhat speculative exercise were a rationalisation of plant breeding, and concentration of expertise in the hands of researchers as opposed to farmers. But this would also have been the case with the newer methods of recurrent selection. However, for reasons that will become clear below, recurrent selection was no match for crossing inbreds as a method of producing commercial seed. The tight-knit nexus of science, politics and mammon effectively locked-out alternative approaches to the hybrid route.

The work of East and Shull

In 1896 at the University of Illinois, Cyril G. Hopkins pioneered ear-to-row selection in which seeds from a given ear of corn were planted in a row, enabling a certain degree of control to be established over the pedigree of a particular cross (Wallace and Bressman 1949, 24). Hopkins sought to select corn varieties for differing protein content in order to provide better quality feed for livestock breeders. However, although he could manipulate the protein content of corn varieties, the more carefully he selected for these, the less these plants seemed to yield.

Archie Shamel, Edward East and H. H. Love were appointed by Dean Davenport to work on corn at Illinois. East and Love were hired explicitly to work for Hopkins. In 1904, East took over the plant breeding work at Illinois. When Hopkins decided to terminate the breeding work at Illinois, East was offered a new post at Connecticut by Edward H Jenkins. Love, who had befriended East at Illinois, sent East the inbreds which Hopkins had developed at Illinois (Crabb 1947, 20–36). East's early work at Connecticut mainly involved promoting varietal crossing to farmers as a means of improving corn. But few farmers had time to engage in anything other than selection to improve their seed stock. East began experiments involving the crossing of inbred materials in 1907. The results astounded him, both top crosses and single cross hybrids giving significant increases in yield over their parent material. In 1908, East attended an American Breeders' Association conference where he heard a paper given by Dr George Shull of the Carnegie Institution's Station for Experimental Evolution. Shull's paper helped explain some of what East had observed, in particular, that inbreeding constituted a mechanism whereby the different biotypes which constituted the corn plant could be separated out (Crabb 1947, 45–6). Shull was more interested in the theoretical aspects of the effects of self- and cross-fertilisation than in corn *per se*. He suggested that open pollinated varieties of corn were essentially 'a mass of very complex hybrids' so that selfing served to separate and purify the strains, and generally led to a loss of vigour.[8] By crossing inbreds he noted that the lost vigour could be restored (Wallace and Brown 1956, 106; Sprague 1983, 49). This effect, the restoration of vigour through crossing, is still known as heterosis, the name given to it by Shull.

Shull's theoretical bias led him to overstress the potential economic significance of single cross hybrids to the farming community.[9] By contrast, East appreciated that not only would farmers find this method too complex, but that even if the work were to be carried out on experiment stations, the fact that the seed would be borne on one of the weak inbred parents would make it prohibitively expensive as a method of seed production. East believed the method offered some hope of a more rational and scientific approach to corn improvement, but as it stood, it was all but useless to farmers (Wallace and Brown 1956, 109–12).

It is not a simple matter to discover exactly what impact, in terms of research funding, the experiments of East and Shull had on corn breeding at the time. However, the written accounts of those involved universally support the view that during the following years, substantial resources were channelled towards methods involving the crossing of inbreds.[10] Many researchers, seeking a more scientific approach to breeding, followed the work of East and Shull with interest and not a little hope. Sprague (1983, 49) speaks for the breeders when he states that 'Mass selection, as a technique for increasing yield, was generally deemed ineffective and inbreeding and hybridization appeared to offer an alternative route for improving yields.' Yet twenty-five years would elapse before suitable hybrids were available for much of the United States despite a burgeoning research effort. Furthermore, if corn yields showed little sign of improvement at the time, this had less to do with selection as a method, and more to do with the criteria, as laid down by corn shows (see below), by which selection proceeded. In addition, improved understanding of the statistical basis for selection was enhancing the potential utility of alternative techniques.

Donald Jones and the double-cross

In 1910, East moved from Connecticut to Harvard's Bussey Institute. His place was taken by Herbert Hayes who worked under East's direction whilst at Connecticut before moving to the University of Minnesota in 1915. Hayes' place was then filled by Donald Jones. Jones had studied under East at Harvard while at Syracuse University. Within 3 years Jones had proposed the double cross hybrid, a cross between two single crosses, reasoning that if seed could be borne on one of the highly productive single crosses, hybrid corn might become commercially viable. The success of Jones' experiments owed much to chance:

> only a rare and occasional combination of single crosses will effectively blend into a double-cross hybrid. It easily might not have happened again with single crosses, untested for combining ability, in a hundred, even a thousand trials ... The flipped coin had come down and stood on its thin edge. (Crabb 1947, 86)

With Jones' discovery, there now existed an economically viable method of improving corn which was, to some degree, under the control of the breeder, yet too complex and time-consuming for most farmers. Most importantly, hybrid corn did not breed true in the field. The implication of this was that farmers who sought to save seed would suffer reductions in yield the following year. Breeders had long been concerned with the competition to their trade represented by federal seed distribution, and they were also aware that it was easy for farmers to set aside a portion of last year's harvest for seed the following year. Hybrids potentially altered this relationship. None of this was lost on Donald Jones:

> It is not a method that will interest most farmers, but it is something that may easily be taken up by seedsmen; in fact, it is the first time in agricultural history that a seedsman is enabled to gain the full benefit from a desirable origination of his own or something that he has purchased. The man who originates devices to open our boxes of shoe polish or to autograph our camera negatives, is able to patent his product and gain the full reward for his inventiveness. The man who originates a new plant which may be of incalculable benefit to the whole country gets nothing – not even fame – for his pains, as the plants can be propagated by anyone. There is correspondingly less incentive for the production of improved types. The utilization of first generation hybrids enables the originator to keep the parental types and give out only the crossed seeds, which are less valuable for continued propagation. (East and Jones 1919, 224)

Since the exact nature of the inbreds used to create the hybrid could be kept secret, and since using hybrids implied seed saving was useless, a biological patent (Buttel 1990, 169) was effectively conferred on hybrid seed.

Adaption and adoption of hybrid corn by scientists and farmers

To read many accounts of the history of hybrid corn, one could be forgiven for thinking that following Jones' discovery, its development was untroubled, and an inevitable consequence of the lack of available options. Neither is the case. The applicability of the method across the US was far from assured and indeed, the process took many years. Even then, the problem of adoption would remain, the specific characteristics of hybrids offering barriers to simple replacement of open-pollinated varieties. Furthermore, as hybrids were being developed, other routes were being explored with quite different characteristics. These did not seem, and neither do they now (with the benefit of hindsight), any less good in terms of performance than hybrids. In this section, I seek to understand why hybrid corn was adopted so vigorously as the method of choice by public research organisations, and why other options were ignored.

Support for hybrid corn and its regional adaptation

Jones' work aroused great optimism, but although the hybrids were being used on Connecticut farms by 1920, it would take more time to convince many researchers of the possibilities of hybrid corn. In areas where no viable hybrids had been developed, scepticism was understandably prevalent. Over time, research resources began to pour more freely into the hybrid route to corn improvement, and if East and Shull started the flow, Jones' work unlocked the flood gates (Fitzgerald 1990):

> The United States Department of Agriculture, private breeders and experiment stations launched a vast programme of developing new inbred lines and new combinations of double cross hybrids (Wallace and Brown 1956, 113). Beginning shortly after 1920, there was a very rapid expansion in the number of inbreeding projects. (Jenkins 1936, 472)[11]

How the USDA came to launch its programme on hybrids provides an interesting and critical insight into how hybrid corn actually came to see the light of day instead of ending up as a failed idea.[12] Shortly before his father, Henry C. Wallace, was publicly declared the new Secretary for Agriculture under Warren Harding, Henry A. Wallace visited the USDA at his father's suggestion to form an opinion of those conducting research on corn. Henry A. Wallace had conducted experiments of his own with inbreds, and was aware of the work of East, Shull and Jones.[13] He felt that the best work was being done by Frederick Richey, but that the Principal Agronomist in charge of corn investigations at the Office of Cereal Investigations in the USDA's Bureau of Plant Investigations (BPI), C. P. Hartley, was hindering Richey's work with inbreds. Hartley was 'definitely opposed to hybrid corn', and convinced that selection was the most promising way forward in corn improvement (Wallace and Brown 1956, 109). Over a period of time, Hartley's views came to be marginalised and in 1922, he was replaced by Richey (Crabb 1947, 98–9; Hayes 1963, 30–1; Kloppenburg 1988, 102–3).

With Richey at the helm, the USDA's Office of Cereal Investigations took up the hybrid corn work. The 1920s were devoted to seeking inbreds which would combine effectively so that good hybrids could be developed (Crabb 1947, 100). Only a limited amount of material was suited for use in making hybrids. Hence, the search for these new inbreds, and for suitably adapted combinations of them, was a mammoth task.[14] The quest was given a great boost in 1925 with the passage of the Purnell Act. Kloppenburg (1988, 103) notes that the funding which the Act made available:

> provided Richey with both the institutional authority and the financial clout to organize an unprecedented venture in directed scientific investigation ... [and] ... allowed Richey to support hybrid development while isolating or bypassing those recalcitrant departments that resisted the new direction taken by research.

Richey appointed a committee to formulate a Co-operative Programme of Corn Improvement whose work:

> resulted in a plan for a group attack which functioned formally until 1932, and this had an important influence in coordinating the research, in developing a very free interchange of breeding material, and in promoting far more rapid progress than otherwise would have been possible. (Jenkins 1936, 472)

What Kloppenburg (1988, 104) calls 'agriculture's Manhattan Project' required Herculean effort from researchers over a decade and more. By 1935, public agencies had developed hybrids which yielded 10–15 per cent more than their open-pollinated counterparts. So, between 1933 and 1943, the percentage of Corn Belt lands sown with hybrids increased from less than 1 per cent to 78 per cent (Wallace and Bressman 1949, 27).

Why was the hybrid route pursued so vigorously? Was it, as Simmonds(1979, 159) suggests, a result of some objective assessment of the prospects of this method of improvement on the part of the scientists? Or was the decision made through appealing to the commercial significance of the innovation (Berlan and Lewontin 1986), an episode in the inevitable march of commodification of seeds and genetic resources (Kloppenburg 1988)? Or is Fitzgerald (1990, 72) right to argue that it was the result of breeders' and plant geneticists' obsession with Mendelism, and their desire to foster the emergence of a new scientific breeder? In fact, all these factors and others were present, and all were interrelated:

- Jones' and subsequent discoveries indicated to breeders that they could begin to direct evolution more precisely than one could through mass selection methods. The resultant increase in the degree of control over breeding opened possibilities for achieving whatever was required in the field. Combining inbreds with specific characteristics allowed for a rationalisation of plant breeding, took breeding out of the hands of farmers, and invested authority in the researchers.
- That the necessary resources were made available at all was in no small part due to the appeals made by researchers to industrial interests concerning the value of their work,[15] this despite the fact that as agriculture entered a depression, research increasingly came under fire, especially since its primary goal seemed to be increasing production. The changes sought, if not wrought by agricultural science were viewed by some farmers as the key reason for their precarious predicament (Paarlberg 1964, 16; Rasmussen 1987, 893; Kloppenburg 1988, 84–7). Yet under Richey, a mission-type mentality embraced those engaged in the project of developing hybrids adapted to Corn Belt conditions, leaving alternatives to the hybrid route largely unexplored.
- Hybrids, double-crosses to a lesser extent than single-crosses, were extremely uniform in the field. This characteristic was desirable beyond the corn show (see below). It is common in historical, and autobiographical accounts, to read of scientists' (and farmers') pleasure at seeing uniformity in terms of plant height, shank, and timing of maturation. There may have been cultural, sociological, or aesthetic reasons for this.[16] But there were more important economic

reasons. Hybrids came to be designed for machinery as much as for the field itself (Wallace and Bressman 1949, 28; Wallace and Brown 1956, 128; Steele 1978, 31). An important characteristic in this respect was resistance to lodging (that is, toppling over), allowing heavier fertilisation of the crop. Stiffer shanks also facilitated harvesting by machine.[17] As late as 1938, only 15 per cent of US corn acreage was machine harvested. Between 1930 and 1950, the number of mechanical corn pickers and combines with corn heads underwent a nine fold increase (Cochrane 1979, 198). In addition, the uniformity of single-cross hybrids fitted the needs of those growing sweetcorn for canning (Richey 1933, 187). The emerging ideal plant was designed with mechanisation clearly in mind. Kloppenburg (1988, 299) notes that not only did breeders' efforts facilitate machine harvesting, they actually made hand harvesting more difficult.
- The appropriable nature of hybrid corn conferred upon it by its biological characteristics, and the fact that it made saving seed on the farm unviable, excited the fledgling seed industry. One can see why hybrid and selection-based methods were incompatible from its point of view. The latter was a path which, if it could be closed off, would force farmers to purchase new seed for each crop.

If the above issues appear to have made the choice of hybrids all but inevitable, consideration of alternative routes suggests otherwise.

Alternatives to hybrids

I have already indicated Jones' awareness of the commercial potential of hybrid corn. He was also, it appears, alive to the threat to that potential posed by synthetics, which allow farmers to plant farm-saved seed without appreciable (if any) loss of yield. East, and Jones himself, had experience with synthetics. Jones saw synthetics only as a means of developing inbreds, but not because he felt they would be inferior in their performance. In a letter written to Henry A. Wallace in 1925 by Jones' commercial associate, George S. Carter (the first farmer to grow hybrid corn), he states that synthetic varieties and the possibilities they presented 'would spoil the prospects of anyone thinking of producing the seed commercially' (cited in Kloppenburg 1988, 102).[18]

It is also surprising how quickly the promise of synthetics seems to have evaporated from the thoughts of Frederick Richey. In 1922, he wrote tentatively on the prospects for hybrids, suggesting that they could outyield others, but that 'the number of experiments in which crosses have been tested under strictly comparable conditions seems too small so far to warrant definite conclusions as to the practical possibilities of this method as a means of obtaining larger yields of corn' (Richey 1922, 13). However, the degree of control which the process allowed, as well as Jones' theory of linked dominant growth factors (which purported to explain heterosis) provided, for Richey, sufficient rationale for concentrating on hybrids: 'There seems to be little reason for hesitating between methods ... [pure-line approaches provide] the only sound basis for real improvement in corn' (Richey 1922, 14).

In 1927, having dismissed mass selection and older methods, he noted that single- and double-crosses, *as well as synthetics*, increased yield. He added: 'As yet, however, there is no evidence as to which of these methods will prove most *profitable*' (Richey 1927, 249, my emphasis). By 1933, in an article extolling the virtues of hybrids (and in which he felt the need to deny that he was writing for propaganda purposes), all mention of synthetics had disappeared (Richey 1933).

All too often, the history of technological change tracks only the success stories. Questions concerning alternative routes and possible counterfactual worlds are thereby ignored, diminishing our understanding of the processes involved. Many authorities on plant breeding have implied that any difficulties which existed with selection as a breeding method were more apparent than real. The problem lay not with the principle, but in the practice (Kloppenburg 1988, 101–5, 111; Simmonds 1979, 142). Though the attitude of breeders of the day was to condemn selection methods to irrelevance, it actually became clear relatively early on that maize responded quite well to selection for yield improvement and other traits. Ironically, the refinement of selection methods occurred in response to the need to isolate new inbreds for the hybrid programme. In any case, selection was not as ineffective as the breeders would have liked to believe. A pressing problem for corn growers was the new wave of pathogens, particularly the corn borer, that were attacking corn (Collier 1928; Gray and Merrill 1930). Ground was already being gained in tackling these diseases through education regarding selection, and by encouraging farmers to test their seed in portable germinators (Fitzgerald 1990, 70, 117–9).

Heterosis, an agreed explanation of which continues to elude biologists, apparently occurs not only between inbreds, but also between populations of diverse origin. Thus, the phenomenon which attracted so many breeders' attention following Jones' discovery did not exclusively apply to inbred lines, so that it need not have implied a resort to genetically uniform crossed inbreds.[19] As early as 1919, Hayes and

Garber wrote a paper in which, after observing that crossing inbreds 'deserves further trial' (313), they examined the production of synthetics.[20] Synthetics are open-pollinated varieties produced by intercrossing selected plants, these often having been selfed once, and maintained through routine mass selection. Early studies of synthetics made no evaluation of the combining ability of lines to be included, so that yields tended to be relatively low (Lonnquist 1961, 4).

Synthetics and their production can be used as the basis for a method of corn improvement known as recurrent selection. As outlined by Jenkins (1940), this involved selfing lines for one generation, crossing them with open-pollinated varieties, intercrossing the best yielding progeny, growing the synthetic for one or more generations, and then repeating this procedure. Inbreds could be added to the synthetic as was deemed desirable. In a later paper with Sprague (Sprague and Jenkins 1943), Jenkins suggested also that synthetics could be used as sources of germplasm for use in developing new inbred material. Different methods of recurrent selection have evolved, and they can be divided into two groups, intra- and inter-population improvement (Sprague 1983, 62).

When corn yields appeared to have plateaued in the 1940s, a reappraisal of corn breeding methods took place. Hull (1945) gave details of recurrent selection with regard to specific combining ability of inbreds. Some breeders felt that genetic diversity had been exhausted and that further yield increases would be difficult to obtain, but others felt that breeding methods themselves were at fault. In this regard, Hayes (1963, 33, 35) wrote of the need in developing hybrids to maintain a considerable degree of heterozygosity without too rapid approach to homozygosis. Selection could then be conducted over a somewhat longer period than is usually the case when continued self-pollination is followed by selecting and producing inbred lines (so that homozygosis is approached more rapidly).

Hull (1945) had suggested that earlier experiences with selection in corn belt varieties had dissipated what he referred to as additive genetic variance and that this was the cause of the failure of mass selection and ear-to-row selection to increase yields of these varieties. However, Lonnquist disputed this, suggesting the problem lay in the accuracy of selection. In support of his position, he concludes his 1961 study (he began working with synthetics in 1943):

> The average increase in productivity of three varieties in comparison with the parents after one cycle of selection for general combining ability was 13 per cent. The improvement in yield shown by the

> syn-2 (F2) synthetics over their parental varieties was maintained in advanced generations of synthesis in isolated blocks *through normal visual selection procedures*. (Lonnquist 1961, 61, my emphasis)

He goes on to mention that subsequent rounds of selection improved yield further and that the synthetics out-yielded a hybrid used as a check in the experiments.

Gardner (1978, 207) is another who laments that selection methods were so readily dismissed by those interested in crossing inbreds:

> If population improvement through the use of well-designed cyclical selection and recombination procedures had been practiced in corn over the past half century along with inbred line and hybrid development, there is reason to believe yields might be substantially higher than they are today.

He goes on to note that recurrent selection is a central component of most hybrid corn-breeding programmes today, and gives substantial evidence on performance in terms of yield and other traits (Gardner 1978, 212–23).

For some time now, therefore, it has been clear that varieties could be developed which were more genetically diverse than hybrids, whose seed could be saved and replanted by farmers, and whose performance would not have been inferior. However, though this route appeared promising, and would potentially be cheaper for farmers, the methods of improving open-pollinated varieties only found use in areas where environmental conditions mitigated against the use of hybrids, and as part of the quest for superior inbreds. It is striking that advocates of population improvement in the US (including Lonnquist and Gardner) rarely extend their advocacy beyond recommending the method as a way of improving populations from which to select inbreds for hybrid combination. Only in the wake of the outbreak of southern corn leaf blight in 1970 did the Committee on Genetic Vulnerability of Major Crops consider population improvement as a viable way forward for breeders, partly on the basis of its promise in terms of augmenting diversity in the field (NRC 1972, 115). It noted that the best interpopulation hybrids were equal in yield to standard double cross hybrids. Such evidence leads Simmonds (1979, 162) to ask:

> Are HYB [hybrid] varieties really necessary, even in outbreeders? In trying to answer this question, the undoubted practical success of

> HYB maize is irrelevant; a huge effort has gone into it over a period of nearly sixty years; population improvement ... is evidently capable of rates of advance which are at least comparable and may well be achieved more cheaply.

Only in cases where genetic uniformity is desirable, and where primacy is given (for whatever reason) to stimulating a private seed industry, does it seem logical to have followed the hybrid route. At the time of its development, genetic uniformity appears not to have troubled breeders, indeed, probably the reverse is true. Furthermore, key players in the development of hybrids were, as I have shown, acutely aware of the commercial possibilities presented by hybrid corn.

From corn shows to yield tests

One of the principal aims of farmers' selection practices at the turn of the century was winning prizes in the increasingly popular corn shows:

> As corn gained in importance in the middle west, increasing attention was paid to the shape of the ear and its kernels. There came to be fairly general agreement about what an ear of corn should look like, and thousands of farmers endeavoured to produce this ideal out of the flint-dent mixture. Gradually a fairly definite type took shape, varying somewhat with the preferences of the region and of the individual farmer. (Wallace and Bressman 1949, 20–1)

The popularity of corn shows, established by 1890, peaked around 1910. Although early shows emphasised the importance of yield, as time passed, emphasis came to be placed on appearance, and uniformity (Wallace and Brown 1956, 99). The emphasis on uniformity and on selection for particular characteristics was detrimental to yield as diversity narrowed and characters became fixed: 'Continued selection for ear type without doubt led to a form of inbreeding with consequent reduction in yielding ability' (Hayes 1963, 26). Those who bred corn for yield were often greeted with derision by their counterparts who bred for the corn show scorecard. If hybrid corn was to achieve widespread adoption, farmers' focus on appearance rather than yield had to be changed. Hybrid corn was exceptional in the eyes of scientists because it performed so well on their key criterion, yield, and scientists, true to positivist ideals, were unimpressed by aesthetic judgements.[21]

The yield depressing feature of the corn shows was soon recognised as such through a variety of experiments. In 1903, P. G. Holden had the idea of using County Poor Farms as places to test comparative yields of corn from different farmers. In 1904, a yield test was carried out under the supervision of a student at Iowa State University, Martin Mosher, but the results were never published (Wallace and Brown 1956, 101). However, farmers from Woodford County, Illinois, heard about Mosher's continuing work in Clinton, Iowa, and hired him to find the best yielder in the County on the basis of a three year test. The winner of the 1919–21 test was not a corn show star, yet it was through this work that Krug corn was brought to the notice of the rest of the Corn Belt (Krug was from Woodford). Prof H.D. Hughes also did similar work, showing that corn-show winnings were unrelated to yield (Wallace and Brown 1956, 101; Wallace and Bressman 1949, 23).

Both Fitzgerald (1990, 65–9) and Kloppenburg (1988, 104–5) make the point that Henry A. Wallace was an important figure in persuading, or coercing, the BPI to support the work of Mosher in Iowa. Fitzgerald's view is that Henry C. Wallace stepped in on the side of his son to persuade Richey of the value of these tests. Henry A. Wallace saw their value not just in promoting the adoption of hybrids through focusing farmers' attention on yield, but also in uncovering the farmers' best varieties for use in the search for quality inbreds. In the 1919 Christmas issue of Wallace's Farmer, he wrote, 'If it is impossible to tell much about the yield of corn by looking at it, why shouldn't the corn show branch out into ... a yield test under controlled conditions' (Wallace 1919, 2509). Later, he would reflect: 'Without the Iowa corn yield-test it is doubtful that hybrid corn would have swept away corn shows and corn judges so abruptly' (Wallace and Brown 1956, 102). Wallace's own Copper Cross hybrid won the gold medal in the Iowa yield test in 1924. He contracted with the Iowa Seed Company to sell the variety (developed from inbreds produced in public institutions) to farmers in 1924, and promptly set up his own seed company, Hi-Bred Corn Company, the first to engage in commercial production of hybrid varieties.[22]

In Illinois, another major corn producing state, it was another seed industry man, Eugene Funk, who first suggested that there should be performance tests in 1916 (Crabb 1947, 116–7). The 10-Acre Corn Yield test was inaugurated in 1930 (Fitzgerald 1990, 127–8). These changing attitudes paved the way for hybrid corn. The role of hybrid evangelists such as Wallace and Funk in this process should not be underestimated.

Farmers' adoption of hybrid corn

Hybrid corn spread rapidly through the Corn Belt in the 1930s. It might be supposed that this can be attributed to the promise of higher yields which yield tests appeared to substantiate. But the evidence suggests that these promises were not fulfilled. There were problems with adoption of hybrid corn (Fitzgerald 1990, 189–97): 'Every experiment station that has distributed hybrid corn has been impressed with the difficulty of convincing growers that seed should not be saved from hybrid plants for future planting' (Jenkins 1936, 468). Furthermore, hybrids were best suited to the best soils.[23] As a result of these two facts, some were disappointed with the performance of these new seeds, which they had been led to believe were superior in every respect. Subsequently, in the wake of a good deal of hype, farmers were cautioned against rushing into the market for hybrid seed, further confusing the issue.

The number of hybrids being marketed in these early years is no measure of the variety of the product on offer. The lack of private investment into research on inbreds ensured that virtually all useful hybrids were combinations of a small number of elite inbreds developed publicly. Seed companies were reluctant to divulge the pedigree of their new miracle seeds (Jenkins 1936, 479; Kloppenburg 1988, 107), exacerbating problems of adaptation of the seed planted. Hybrids needed adequate supplies of water and fertility to perform at their best. In the absence of an increased supply of plant nutrients, their extensive root systems would have hastened the ongoing decline in soil fertility (Bray and Watkins 1964, 760–2).[24] We are left, therefore, with a complicated picture as to what exactly was the impact of hybrids on corn yields in the late 1930s and early 1940s.

Notwithstanding farmers increased concern with yield as opposed to appearance, and although inbreeding resulting from corn show scorecards might have been less an issue than in the 1910's, yields fell between 1925 and 1937. By 1940, they were still slightly below yields achieved in 1870, even though the acreage in use was 'better' land (Bray and Watkins 1964, 752) and hybrids were in widespread use. Thus:

> Failure to take capital depletion of natural fertility into account means that estimates of factor productivity increases have overstated real economic gains. Secondly, without the introduction in the early 1940's of artificially manufactured cheap nitrogen – in other words, without the introduction of a restorative technique, the yield potential inherent in hybrids could not have been realized. They did, in fact, by accelerating fertility removal, make more urgent the need for finding a solution to the fundamental problem of fertility maintenance. (Bray and Watkins 1964, 761)

The solution, in the form of cheaper nitrate fertilisers, was forthcoming as a result of war, and the manufacture of explosives. This was not lost on breeders, whose strategies were oriented towards inbreds which could resist lodging under hefty applications of nitrogen.[25] In 1940, Corn Belt fertiliser consumption was much the same as in 1930. But the following fifteen years saw a seven-fold increase (by weight) (Bray and Watkins 1964, 762).[26]

Given all these issues, it may seem surprising that farmers adopted hybrids as quickly as they did. Working in favour of adoption were droughts in 1934 and 1936 that constrained yields, and reduced the amount of good seed available for planting. With many farmers forced to purchase seed anyway, it seems likely that a good number would have been tempted to give hybrids a trial, though the issue of whether or not to purchase would have been a difficult one in the depressed economic climate of the day. This decision would have been made easier for farmers choosing to participate in the government corn-hog programmes, the first of which commenced in 1933. These offered guaranteed prices to farmers, making investment decisions, of which buying hybrid corn seed was one, less risky. The various acts aimed at controlling the corn market sought to reduce supplies to a glutted corn market where prices had fallen dramatically. Participation in the programme (well summarised in Paarlberg 1964, ch. 23) was always voluntary, and even those who participated by idling land might have sought to intensify production on remaining land. If the adoption of hybrids is indicative of such intensification, it is clearly what occurred (though with initially limited success).

It seems untrue, therefore, that it was high-yielding hybrids that were responsible for the failure of the Act to bring supply in line with projected demand,[27] and indeed, there was general concern expressed in advance of widespread adoption of hybrid corn (Bean 1935). It is true that the corn-hog programme achieved little in terms of alleviating the problem of chronic over-production, but this was not solely attributable to the adoption of hybrids. War then came to the rescue.

Henry A. Wallace, by then Roosevelt's Secretary for Agriculture as well as owner of Hi-Bred Corn Company, was forced to weather charges of conflict of interest during this period.[28] This did not stop him from promoting productivity-enhancing technologies even as he recognised the problem of surpluses:

> Agriculture needs not less science in its production, but more science in its economic life ... Farmers cannot have too much productivity or

> production power, provided they keep it under control ... It is half-science that turns research into a Frankenstein, and leads to demands for a halt in technical progress. (Wallace 1934)

Elsewhere, in the 1933 Yearbook of the USDA, a Department which Wallace did so much to politicise, it was boldly stated that: 'Scientific methods have to do with the cost of production, and do not determine the volume of production' (USDA 1933, 22) (a claim which seems, to say the least, difficult to sustain given the motivations of scientists outlined above, and the claims subsequently made for science and its ability to stave off hunger). To this end, scientists, who lobbied hard for further increases in appropriations for research following the passage of the Adams and Smith–Lever Acts, and who had been rewarded in 1925 with the passage of the Purnell Act, were further rewarded in the passing of the Bankhead-Jones Act of 1935.[29]

Implications for diversity

That hybrids led to the loss of genetic diversity in US corn is not seriously questioned. Of the accounts of those involved, Wallace was the one who most clearly recognised the significance of what was occurring. Wallace and Brown (1956) note:

> So far as the corn farmers of the central United States are concerned, the corn shows caused the elimination of many hundreds of varieties.
> ... From 1920 to 1955, the process of hybrid cornbreeding, based on inbreds, completed the job of eliminating practically everything which did not produce *a single well-developed ear.* (121)
> Fortunately, a few farmers never took any stock in Holden or Reid corn ... we can also give thanks for the skeptics and the 'forgotten corns'. (97–8)

Hayes (1963, 33) notes that it was L. J. Stadler who first emphasised the importance of retaining a broad source of germplasm, leading to a combined effort to retain material of the more valuable open-pollinated varieties in each state and region. He goes on:

> the writer joins with Henry Wallace in emphasizing the importance of the work of Dr. Paul Mangelsdorf, who incidentally was a student of Dr. East, in leading in the arrangements, under the auspices of

> the Rockefeller Foundation and the National Research Council, to maintain a collection of different corn strains of Mexican and South American origin for future use of students of corn breeding and of corn genetics.

Although variation was to provide breeders with the materials necessary for them to improve hybrids and the inbreds used to make them, the real limitations of uniformity were revealed much later through an adaptation of hybrids which made them cheaper to produce. The female parent of the hybrid had to be manually detasselled to prevent fertilisation by its own pollen (Wallace and Bressman 1949, 40–1). The search for alternative ways of achieving this occupied Donald Jones between 1920 and 1944.

A 1930 paper by Singleton and Jones suggested that this was a possibility, alluding to segregation among plants for male sterility. Jones, however, did not carry the subject further, and it was only in 1944, when Jones met with Mangelsdorf, that Mangelsdorf informed Jones of the existence of a strain of cytoplasm which conferred male sterility on corn plants. In 1949, Jones proposed the use of a fertility restorer gene into male parents, and the source of cytoplasmic male sterility in the female parents, to produce fertile hybrids without the need for detasselling. The commercially aware Jones patented his method of producing hybrid corn, a move that led to seed industry attempts to discredit him, and to his censuring by the American Society of Agronomy. Though the method of producing hybrids spread quickly throughout the United States, so that by 1965, virtually all corn plants in the US had the same cytoplasmic DNA that conferred male sterility on the female plant, all companies were in breach of Jones' patent, granted in 1956 (Kloppenburg 1988, 115). Royalties were paid to Jones only after a lengthy defence of the patent found in his favour in 1969.

The popularity of the technique led to increased vulnerability of the crop owing to the uniformity of cytoplasmic DNA. In 1970, almost a quarter of the US corn crop was lost to southern corn leaf blight (see Chapter 1). The limitations of uniformity were revealed in dramatic fashion, as was the prescient nature of the words of Wallace and Brown (1956, 126–7):

> we have no way of knowing today what kind of corn will be needed for breeding purposes some twenty-five, fifty or one hundred years from now. Without this information, it seems wise to save *all* kinds of corn on the theory that some day some one or more of them may

> provide the element we are looking for ... it seems nonsensical to allow any corn, now available, to disappear. (126–7)

Conclusion

In achieving the division of labour which it had sought since the mid-1930s, the private seed industry effectively situated itself in a key position between the researchers and the farmers. Distribution came to be carried out largely by private sector seed companies. Indeed, given the requirements for seed resulting from the need to repurchase from one year to the next, it was probably inevitable that private sector companies would become involved. Within this industry, a small number of players quickly emerged as pre-eminent enterprises in a substantial market:

> By disengaging from its link to the commodity-form, public breeding not only ceased to discipline the market but also surrendered its autonomy. Ultimately, research has value only insofar as its fruits can be applied to production in some fashion. With seed companies alone producing commercial hybrids, private enterprise is interposed between public research and the consumer of seed. The products of public research can enter production, and thus have value, only if seed companies chose to use them. Public breeders are therefore structurally bound to set their research agendas in accordance with the goals of private enterprise ... To control the shape of the commodity-form in the market is effectively to control all upstream research. (Kloppenburg 1988, 110–11)

What Kloppenburg is talking about is a process of locking-in to the hybrid route.

Through gaining an early lead on other technologies, the reasons for which, as we have seen, lie less in the narrowly technological domain than in the realm of political economy, the networks which built up around, and in support of, hybrid seed effectively determined that no other route to developing maize seeds could come into use. It is incumbent upon us to reflect upon how a new technology or technique could displace one around which such a vast industry has been built. Thus although it seems that population improvement methods need not be inferior to the crossing of inbreds, particularly if it were the subject of the same federal largesse which hybrids have enjoyed, hybrids continue to dominate the maize seed market in the US.[30]

It is often noted in literature on agricultural research how public research has become subordinate to private capital accumulation. But I would argue that it has never been otherwise. I noted above some of the contradictions which faced scientists in the early years of agricultural research, and researchers today have inherited the same ambiguities that haunted their predecessors. Agricultural research has always been subservient to private capital, more so than to the farmers who are often assumed (frequently incorrectly) to be its constituency. Recent developments have simply made this relationship more visible, and the subservient nature of public research has become more obvious.

Jones, Funk, his de facto employee, Holbert – all were key men in the development of hybrid corn and their commercial outlook unites them as much as their involvement in the hybrid enterprise. The almost omnipresent Henry A. Wallace and the influence that he was able to bring to bear must also be held to be highly significant. Recall his role in having Frederick Richey instated as Agronomist in charge of Corn Investigations at the USDA. Less than sixty days after his instatement to this post, Richey issued what Crabb (1947, 247–48) calls the 'Magna Carta' by which cooperative work on corn was to proceed:

> All experimental evidence indicated that *the older methods of open-fertilized breeding of corn varieties are of little value ...the fundamental problems of corn improvement can be solved only through investigations based upon self-fertilized lines*. For these reasons, a program of corn-breeding investigations should concentrate principally on pure-line methods as a means of obtaining larger yields of corn and the following program is concerned chiefly with such experiments (my emphasis).

Richey's statement belies the promise, of which he was clearly aware, that new methods of breeding open pollinated varieties might have held for corn research. His decisions, possibly influenced by how he had come to be in the position he was in, were largely responsible for ensuring that the principle method of corn improvement was one which reified uniformity and which prevented farmers from saving seed from year to year. We are now better placed, though by no means ideally so, to speculate as to what might have happened if the resources devoted to hybrids had been channelled in the direction of recurrent selection methods. If nothing else, the economic calculus which persuaded farmers to adopt hybrids might have looked very different if population improvement techniques had been further advanced prior to the development of hybrids.

Notes

1. For details of Beal's work and the influence of Darwin's work upon him, see Wallace and Brown (1956, 64–79; Fitzgerald 1990, 13).
2. Quite how the Morrill Act and the elevation of the Patent Office Division of Agriculture to departmental status (as the USDA in 1862) were regarded by farmers is debated in the literature. The latter opened the door for more effective political representation of agricultural interests. Yet the first decades of the land-grant colleges were undistinguished. One eminent historian of agricultural research has written of them; 'bereft of students or appropriate curriculum, they displayed little prospect of attracting either. Would-be disseminators of agricultural wisdom met only discouragement in their efforts to attract students – or even to stimulate interest' (Rosenberg 1977, 403).
3. A comprehensive account of the passage of the Hatch Act is found in Marcus (1985).
4. Exactly what constituted 'original investigation' was determined largely by then Director of the Office of Experiment Stations Alfred True and his lieutenant E. W. Allen. See Rosenberg (1976, ch. 10).
5. Danbom (1979, viii) writes: 'In the attempt of urban America to industrialize agriculture the primary motivation was not to aid the farmer ... their principle purpose was to make the farmer a productive supplement to the increasingly dominant industrial sector of the nation'.
6. This developed into what Rossiter (1975, 241) calls 'an addiction ... Government science seemed a free and highly beneficial commodity, and farmers wanted more and more of it, even to the point of overproduction and agricultural depression in the early 1920s'.
7. Rosenberg (1976, ch. 8; 1977) makes much of the ambiguities which arose as a result of compromises made in the establishment of agricultural science. These, he believes, are at the root of the contradictions which have plagued the enterprise ever since (also Nicholson 1978).
8. Darwin noted, in the 1860s, that inbreds usually lost vigour (see Wallace and Brown 1956, 64).
9. Shull had made three-way cross and double-cross hybrids four or five years before Jones (see below), yet failed to understand their commercial significance (Wallace and Brown 1956, 108).
10. A Table in the 1936 USDA Yearbook of Agriculture shows a steady increase in the number of programmes on experiment stations involving crossing inbreds (USDA 1936, 499–503). However, data in financial terms is more difficult to come by. Further significant financial assistance was given in 1925 through the Purnell Act (see pp. 121–3 this volume).
11. See also, Hayes (1963, 36, 38) and USDA (1936, 499–503). Recall also that the Smith–Lever Act was passed in 1914 freeing up more resources for research.
12. See Fitzgerald (1990, 43–72) on this issue.
13. He was also editor of the influential *Wallace's Farmer*. He studied the effects of self-fertilization as early as 1913, and took part in some of the early meetings of corn breeders working on issues related to hybrid development (Hayes 1963, 30).
14. See also Hayes (1963, 32–3) on the difficulties facing breeders in the 1920s.

15. Regarding the linkages between agricultural research and industry, one of the most strident views I have come across in this period is the Presidential Address given to the American Society of Agronomy by Sidney B. Haskell (1923) in which he talks dispassionately about the way in which agricultural research has freed up a labour force for industry by effectively driving people off the land (see also Kloppenburg (1988, 77, 83–4)).
16. See Wallace and Bressman (1949, 28); and Wallace and Brown (1956, 102). Steele (1978, 34) explains the increased use of single-cross hybrids in the US since 1960 by appealing to 'uniformity, beauty, and performance'. Don Duvick, recently departed from DeKalb Seed Company, suggests that uniformity was idealised and idolised beyond the needs of the market place (Fowler 1994, 69).
17. As early as the 1920s, Wallace had offered to develop such a hybrid for a major manufacturer of harvesting equipment (Wallace and Brown 1956, 111).
18. As will be shown below, Wallace was hardly a disinterested spectator.
19. The issue of control, and its links with Mendelian thinking was probably important here. Two inbreds could be selected with particular traits, and if these were the result of the expression of single dominant genes, combining the inbreds would combine the traits (I am indebted to Stephen Brush (personal comm.) for this insight).
20. I am using the terminology employed by the US maize breeders themselves, particularly that of Sprague since he was involved in the development of these methods as applied to maize. Confusion can arise since whilst Sprague (1983, 62) talks of different types of population improvement as methods of recurrent selection, for Simmonds (1979), different selection methods constitute means of population improvement.
21. Note that taste was barely an issue. With the exception of sweetcorn, corn was already principally a feed crop.
22. Wallace's company, renamed Pioneer Hi-Bred a few years after its establishment (Steele 1978, 30), is today one of the world's largest seed companies.
23. Steele (1978, 32) reports that as late as 1944, hybrids were still unable to out-compete open-pollinated varieties in southern, western and north-western states.
24. There was growing concern at the state of soils in the US in this period, leading to the setting up of the Soil Erosion Service in 1933. A few years later, the Dust Bowl problem blew up (literally), an event which (quite apart from the prevailing conditions of over-supply) would lead to further questioning of the pursuit of high yields through what was essentially a soil mining plant (Walden 1966, 52–3; also Pfister, cited in Crabb 1947, 243).
25. Nitrogen tends to promote sappy growth in most plants, and increased levels of the nutrient are apt to lead to the plant collapsing under its own weight.
26. According to Griliches (1958, 594–5), this was a result not of falling prices of nutrients, but of a decline in their price relative to an index of farm prices, most of which occurred in the 1940–45 period. The cost of nutrients in absolute terms fell most between 1920 and 1935.
27. This claim is made by Kloppenburg (1988, 119) and Fitzgerald (1990, 197) amongst others, even though both, in support of other claims that they

make, downplay (correctly in my view) hybrids' impact on yield in pre-World War II years.

28. Mechanisation, which followed the spread of hybrid corn, was not without its critics too. See Hamilton (1939) for an excellent early critique.
29. Kloppenburg (1988, 85–6) notes the strong influence exerted by Wallace in having the Bankhead-Jones Act passed.
30. Furthermore, hybrid maize invariably provides private seed enterprises with their entry point into developing country maize markets.

6

The Road to Mexico's Green Revolution: Maize Research, 1940–55

Introduction

The term Green Revolution was first coined in the 1960s, and used as a political metaphor in a Southeast Asian context (Spitz 1987, 56). But the Green Revolution began in earnest in Mexico, and its roots are found in the United States (see last Chapter). The changes that took place in Mexico were no less devoid of a political dimension than the events that took place in Southeast Asia (Cleaver 1975, 54). The similarities in the cases do not begin and end there. In both cases, the so-called Green Revolution made its most dramatic impact in areas that were well-irrigated. Along with the environmental critique related to the increased use of chemical inputs occasioned by the use of new, uniform crop varieties, a regional bias and the focus on a narrow range of crops proved to be enduring criticisms of the Green Revolution as international agricultural research came under the occasionally one-eyed microscopes of social scientists.

In Mexico, a brief glance at the headway made in key grain crops suggests that the initial progress was made almost exclusively in wheat. Here, yields underwent a quantum leap the like of which they have not seen since. Wheat was a crop grown primarily in the irrigated lands of Mexico's north-west.[1] The staple food crop of most Mexicans is maize. Here, one saw slower yet more consistent progress in raising yields, partly reflecting the fact that maize tends to be grown in more diverse environments than wheat. Maize, therefore, is rarely talked of as a Green Revolution crop, yet interestingly, progress has probably been as successful over the longer term as with rice (CIMMYT 1990, 3).

In this chapter, I concentrate on the very early days of the Rockefeller Foundation (RF) sponsored research programme in Mexico, seeking to reveal which decisions effectively determined the path research would follow in subsequent years. I show that there were alternative paths open to the researchers and their sponsors, but they were largely ignored. There are clear links with the study on hybrid maize in terms of personnel involved, the approach to research, and the obvious attempt to establish research organisations in Mexico modelled on the US design. Where and when staff were not US nationals, they were trained in US organisations.

The genesis of the programme

The idea of an agricultural programme in Mexico had been mooted by Dr. John Ferrell of the RF's International Health Division in 1936, who suggested that a co-operative venture between the Foundation and the Mexican government be established. Ferrell's idea, outlined in a letter to Foundation President Raymond Fosdick, was not taken further at the time was due to the souring of US–Mexico relations as a result of the actions of Lazaro Cardenas' government. These included expropriation of the Rockefeller-owned Standard Oil in 1938.[2]

In 1940, Avila Comacho was elected to the Mexican Presidency. At this stage, the issue of the expropriation of Standard Oil had not been completely settled, and the company was seeking compensation for its losses. Since half of the RF's financial assets were in Standard Oil Company stock (Cleaver 1975, 305), there was a considerable overlap in the interests of both Foundation and corporation.[3] Cleaver (1975, 305–8) suggests that the Foundation and the company developed a two-pronged strategy, the company pushing for compensation, and the Foundation working for improved US–Mexico relations.

For both Jennings (1988, 47–8) and Stakman *et al.* (1967, 20–2), a key figure, as with hybrid corn, was Henry A. Wallace, who, following his time as Secretary for Agriculture, attended the inauguration of President Camacho in Mexico as the US's Vice President elect. He stayed in Mexico for a month as the guest of the US Ambassador to Mexico, who was impressed by work that the Rockefeller-funded General Education Board and the Rockefeller Sanitary Commission had done in the American South (see Cleaver 1975, chs. 2 and 3). In considering the possibility of an agricultural programme in Mexico, Fosdick sought the opinion of Henry A. Wallace. Wallace, fresh from his stay with Daniels, suggested that, 'Raising the acre yields of corn and beans ... would have an effect on the national life of Mexico greater than almost anything else that could be done' (Stakman *et al.*, 1967, 22). Fosdick then consulted with Warren Weaver of the Foundation's Division of Natural Sciences. Weaver held that the best way to find out what could be done would be to send some agricultural scientists to Mexico to study the situation (Stakman *et al.*, 1967, 22). Even at this early stage, the problem was being perceived as one with a purely scientific solution.

Stakman *et al.* (1967) recount that following this discussion, a staff meeting took place on February 18, 1941, where it was decided that if an agricultural programme was to be undertaken, it would be administered by the Division of Natural Sciences, whose acting head (whilst Warren Weaver was performing war duties) was Frank Hanson. A committee was set up to investigate the possibilities of the programme taking place, consisting of Hanson, Harry Miller Jr. (a colleague from the Natural Sciences division), and A. R. Mann, vice president of the Rockefeller-funded General Education Board, who had earlier been Dean of Agriculture at Cornell University and director for agriculture in the RF's International Education Board. This Committee chose three scientists to staff a Survey Commission to report on the Mexican situation; Stakman (plant protection); Mangelsdorf (plant genetics and breeding); and Bradfield (soils and agronomy).

The account of these three (Stakman *et al.*, 1967) suggests an unproblematic progression of the Mexican Agricultural Program (MAP). However, the problems and debates which occurred around this time have been explored by four authors, Cleaver (1975), Hewitt de Alcantara (1976), Jennings (1988), and Marglin (1992). Marglin's (1992) work is an extension of a line of enquiry in which he contrasts two kinds of knowledge, *episteme* and *techne*.[4] Though these should bear a symbiotic relation to each other, he suggests that the one, *episteme*, has sought to devalue the other, *techne*. To oversimplify, *episteme* is modern science, whilst *techne* represents a more tacit form of knowledge gained through experience. Marglin uses the case of the MAP to argue that formal agricultural research, as *episteme*, has promoted itself as complete knowledge independent of *techne*.

Jennings' (1988) addresses the attempts to shape the form of scientific disciplines through support of individuals and institutions that supported a particular viewpoint in a given field. He suggests that the Programme sought to systematically undermine the prevailing conception of the Mexican agricultural problem as it was understood by such Colleges of Agriculture as existed in Mexico at the time. In so doing, the Foundation succeeded in eliminating from orthodox discourse the view that agrarian problems in Mexico were in any way associated with the structure of land ownership and tenancy, as had been suggested in the Cardenas era.

Hewitt de Alcantara (1976) gives more regard than Marglin or Jennings to the history and politics of Mexican agriculture. Her view largely backs up that of Jennings, though her concerns are focused less on the influence of the foundation on scientific thought and more on the social and political impacts of their work. Her work overlaps with Jennings' in showing that the Mexican agricultural scientists such as they were in the early 1940s, had a different conception of the way in which agricultural problems could be solved, particularly as regards the technological means through which to increase maize production. Cleaver's (1975) study forms part of a Marxist thesis concerning the transformation of agriculture along capitalist lines. In this context, the Green Revolution constituted an important moment in altering the structure of agricultural production in the developing world and tying it to the market economy. For Cleaver, the ultimate goal of the Foundation was to supplant peasant agriculture with a commercially oriented agribusiness, from which process, various Rockefeller-owned enterprises would ultimately stand to gain. In the rest of this chapter, building on the work of these authors, I will analyse first of all why the RF became involved in Mexico at all. This sheds light on the second line of enquiry, why the MAP took the form that it did.

Why philanthropy, and why Mexico?

US–Mexico relations and malthusian arguments

Both the part that the RF might play in improving US–Mexico relations, and the resolution of the issue of the expropriation of US property were discussed at a meeting that took place on 3 February 1941 between Fosdick, Wallace and Ferrell in Washington (Jennings 1988, 48). At this meeting, Wallace opined that President Comacho 'wants peace with the United States and that a basis for the settlement of controversies about American property is in prospect' (Ferrell 1941, 2,8).

It seems as though Wallace was trying to persuade the RF that since it was essentially regarded as non-political, it was better placed than the US government to carry out the type of programme discussed. Wallace's arguments drew criticism form Carl Sauer (see below) in 1941. Wallace's spell as head of the USDA was notorious for a 1935 purge of the Department (Lowitt 1979), and Sauer (1941b, 1) criticised the 'aggressive political philosophy' the USDA had developed with regard to agriculture. This was as true of its outlook on Latin America as on domestic agriculture:

> What I do not see is that this aegis is likely to provide the means of inquiry into other modes of life except insofar as we can influence them to our national advantage. This ... lies in the field of politics, not in that of research ... If the Department of Agriculture now wishes to survey the whole of New World agriculture, it can, especially in terms of world emergency, easily throw funds ... into a Pan-American program. Let it do so if it wishes, rather than a foundation. (Sauer 1941b, 1–2)

Sauer's view reflected his belief that the political overtones of the programme made it an inappropriate project for philanthropy (though the implicit assumption, that philanthropy should be non-political, is suggested as utopian by Cleaver (1975, 156–7)).

At the same meeting, Wallace showed his appreciation of the problems which could emerge if food production failed to keep pace with population growth. The use of US agricultural scientific expertise to combat hunger, and thus stave off the threat of communism was becoming official strategy. Point Four of President Truman's foreign policy suggested 'a bold new program for making the benefits of our scientific advances and industrial progress available for the improvement and growth of underdeveloped areas' (see Johnson 1950, 178–9). Several articles, the most straightforward of which was King's (1953), appeared in the influential journal, *Foreign Affairs*, advocating increasing foreign assistance in agriculture as a means to show the benefits of capitalist America. Lastly, a report by the Survey Commission in 1951 read:

> Whether additional millions in Asia and elsewhere will become Communists will depend partly on whether the Communist world or the free world fulfils its promise. Hungry people are lured by promises, but they may be won by deeds. Communism makes attractive promises to underfed peoples; democracy must not only promise as much, but must deliver more. (Stakman *et al.*, 1951, 3–4)

It is extremely debatable that Mexico, at the time the MAP began, was in the grip of a Malthusian crisis. Hewitt de Alcantara (1976, 6) is at pains to point out that the *ejido* (or communal landholding) sector of production, which arose chiefly in the Cardenas era, was as, if not more, productive than the private sector by 1940. Unfortunately, the indicator she uses to support her view, value of production in pesos per hectare, suffers serious shortcomings when it comes to demonstrating increased output of maize (or indeed, any other commodity). The late 1930s were a period of notoriously unstable market conditions with the Second World War looming. Furthermore, the indicator makes no allowance for the cost of inputs and for the regional variation. It might also be that, if *ejidos* were more likely to be producing food products, then given the inelastic response of demand with respect to price, higher peso per hectare output in the *ejido* sector could even be construed as evidence that food products were in increasingly short supply.

However, her version of events does receive some support from Martinez Saldaña (1991, 321–2), whose account of Mexican agriculture in the 1930s shows matters improving considerably at the end of the decade due to increased production from the *ejido* sector following a period of crisis. Later in her work, Hewitt de Alcantara (1976, 110–13) cites statistics for corn and wheat production and trade from the Mexican Secretariat for Agriculture and Livestock (Table 6.1). For maize, yields in the 1930s were more than 10 per cent less than they had been in the second half of the 1920s. However, in the mid 1920s, imports were quite high, then fell in the early 1930s to practically zero, before increasing again in the late 1930s. Total production in the 1930s was never as high as in the late 1920s, and declined year on year from 1931 to 1936 due to a combination of a drop in the land devoted to its production and a fall in yield. One could hazard a guess that it was the best land that was being taken out of corn production and that this was being planted to different crops in response to weak domestic demand. This hypothesis gains some confirmation from the fact that during the mid-1930s, despite falling production and per capita consumption, Mexico began to export corn.

The late 1930s were characterised by a revival in demand both domestically and in international markets. With the ongoing process of land reform redistributing land from commercial farms oriented to export markets to communities, more land began to be devoted to corn production. In the late 1930s, Mexico was producing more corn. But with per capita consumption returning to the levels of the early 1930s and an influx of Mexicans expelled from the US, Mexico became a corn importer. With the exception of an awful year for maize in 1940, these trends continued up to 1942 with yields increasing too. The late 1930s and early 1940s, therefore, appear to show an improving situation as regards supply and consumption of maize. If supply was being outstripped by demand, this was due only in part to population growth. At lower levels of per capita consumption than prevailed in the 1940s, Mexico had been a corn exporter.

Table 6.1 Maize and wheat in Mexico: harvested area, yields, total production, foreign trade and consumption, 1925–70

Year	Maize							Wheat						
	Harvested area (has)	Yield (kgs/ha)	Total production (tons)	Foreign trade: Imports (tons)	Foreign trade: Exports (tons)	Consumption: National (tons)	Consumption: Per capita (kgs)	Harvested area (has)	Yield (kgs/ha)	Total production (tons)	Foreign trade: Imports (tons)	Foreign trade: Exports (tons)	Consumption: National (tons)	Consumption: Per capita (kgs)
1925	2 936 169	670	1 968 732	66 432	197	2 034 967	133.807	455 050	655	298 131	43 758		341 889	22.487
1926	3 137 289	680	2 134 842	109 300	62	2 244 080	145.079	517 987	646	334 365	84 795		419 160	27.099
1927	3 181 384	647	2 058 934	28 423	2	2 087 355	132.632	528 022	729	384 768	37 706		422 474	26.844
1928	3 112 274	698	2 172 845	9941	3	2 182 783	136.324	516 475	691	356 951	47 437		404 388	25.255
1929	2 865 119	513	1 468 805	7898	1	1 476 702	90.618	520 771	704	366 744	96 107		462 851	28.403
Average	3 046 447	644	1 960 832	44 399	53	2 005 178	127.367	507 661	686	348 192	61 961		410 153	26.052
1930	3 075 043	448	1 376 763	79 315	1	1 456 077	87.966	489 772	756	370 394	69 527		439 921	26.577
1931	3 377 538	633	2 138 677	18 731		2 157 408	127.839	604 224	869	525 071	30 091		555 162	32.897
1932	3 242 647	609	1 973 469	37	4	1 973 502	114.941	444 708	703	312 532	67		312 599	18.207
1933	3 193 494	602	1 923 865	117		1 923 982	110.132	472 327	830	392 249	1648		393 897	22.547
1934	3 266 766	580	1 723 477	16	71 019	1 652 414	92.956	492 900	719	354 324	220		354 544	19.945
Average	3 171 821	576	1 827 250	19 643	14 217	1 832 676	106.744	500 786	781	390 914	20 311		411 225	23.952
1935	2 965 633	565	1 674 566	19	81 015	1 593 570	88.093	460 162	753	346 630	46		346 676	19.165
1936	2 851 836	560	1 597 203	10	4452	1 592 761	86.518	508 410	864	439 464	95	2	439 557	23.877
1937	2 999 907	545	1 634 730	3663	1	1 638 392	87.442	484 207	707	342 259	4932	1	347 190	18.530
1938	3 093 878	547	1 692 666	22 062		1 714 728	89.912	500 790	771	386 349	89 684		476 033	24.961
1939	3 266 766	605	1 976 731	53 899	2	2 030 628	104.601	563 371	761	428 784	51 086		479 870	24.719
Average	3 035 604	565	1 715 179	15 931	17 094	1 714 016	91.443	503 388	772	388 697	29 169		417 866	22.293
1940	3 341 701	491	1 639 687	8271		1 647 958	83.850	600 645	772	463 908	1225		465 133	23.667
1941	3 491 968	608	2 124 085	318	2	2 124 401	105.126	582 759	745	434 293	124 117		558 410	27.633
1942	3 757 937	629	2 363 223	1014	1	2 364 236	114.453	60 161	815	489 144	119 646		608 790	29.472
1943	3 082 732	587	1 808 093	751	15	1 808 829	85.464	509 574	715	364 294	296 891		661 185	31.240
1944	3 354 933	690	2 316 186	163 658	2	2 479 842	112.781	527 223	710	374 421	438 845		813 266	36.986
Average	3 405 854	602	2 050 255	34 802	4	2 085 053	100.561	564 072	754	425 212	196 145		621 357	29.968
1945	3 450 889	634	2 186 194	48 586		2 234 780	98.989	468 491	740	346 757	311 873		658 630	29.174
1946	3 313 194	719	2 382 623	9745	914	2 391 454	103.154	415 435	819	340 441	259 655		600 096	15.885
1947	3 512 264	717	2 517 593	695	106	2 518 182	105.758	498 861	846	421 859	279 023		700 882	29.435
1948	3 721 770	761	2 831 937	305	273	2 831 969	115.773	576 950	827	477 156	286 965		764 121	31.238
1949	3 792 497	757	2 870 640	310	14 924	2 856 026	113.641	534 868	941	503 244	250 927	8696	745 475	29.662
Average	3 558 123	719	2 557 797	11 928	3243	2 566 482	107.688	498 921	838	417 891	277 869	1739	693 841	29.113

Table 6.1 continued

Year	Maize							Wheat						
	Harvested Area	Yield	Total Production	Foreign Trade		Consumption		Harvested Area	Yield	Total Production	Foreign Trade		Consumption	
				Imports	Exports	National	Per capita				Imports	Exports	National	Per capita
	(has)	(kgs/ha)	(tons)	(tons)	(tons)	(tons)	(kgs)	(has)	(kgs/ha)	(tons)	(tons)	(tons)	(tons)	(kgs)
1950	4 327 772	721	3 122 042	363		3 122 405	121.066	644 428	911	587 297	427 074		1 014 371	39.330
1951	4 427 696	773	3 424 122	50 735		3 474 857	130.910	672 768	877	589 898	378 247		968 145	36.474
1952	4 235 665	756	3 201 890	24 820		3 226 710	118.521	593 381	863	512 212	452 310		964 522	35.347
1953	4 856 700	766	3 721 835	376 788		4 098 623	146.085	657 347	1020	670 629	249 437		920 066	32.793
1954	5 252 779	854	4 487 637	146 716	2	4 634 351	160.617	764 867	1098	839 466	68 515		907 981	31.469
Average	4 620 112	777	3 591 505	119 884		3 711 389	135.917	666 558	960	639 900	315 117		955 017	34.974
1955	5 371 413	836	4 490 080	993	58 629	4 432 444	149.344	799 887	1063	849 988	9545	23	859 510	28.960
1956	5 459 588	803	4 381 776	119 011	534	4 500 253	147.365	936 944	1326	1 242 538	84 886	251	1 327 173	43.460
1957	5 391 800	835	4 499 998	819 084	6798	5 312 284	169.040	957 911	1437	1 376 502	19 058	55	1 395 505	44.406
1958	6 371 520	828	5 276 749	810 436		6 087 185	188.180	839 602	1592	1 336 759	431		1 337 190	41.338
1959	6 324 018	880	5 563 234	47 894		5 611 148	168.479	937 060	1351	1 265 526	566	12 386	1 266 092	38.015
Average	5 783 668	837	4 842 371	359 484	13 192	5 188 663	164.933	894 281	1358	1 214 263	22 897	2543	1 237 094	39.324
1960	5 558 429	975	5 419 782	28 484	457 450	4 990 816	142.909	839 814	1417	1 189 979	4363	125	1 190 016	34.075
1961	6 287 747	993	6 246 106	34 060	78	6 280 088	174.115	836 538	1676	1 401 910	7605	234	1 372 532	38.030
1962	6 371 704	995	6 337 359	17 902	3829	6 350 432	170.501	747 728	1946	1 455 256	27 127	1034	1 414 093	37.979
1963	6 963 077	987	6 870 201	475 833	411	7 345 623	190.927	819 210	2079	1 702 989	46 163	72 633	1 676 519	43.576
1964	7 460 627	1133	8 454 046	46 696	282 811	8 217 731	206.802	742 680	2056	1 526 613	62 411	576 343	1 012 681	25.486
Average	6 528 317	1021	6 665 499	120 555	148 916	6 637 138	177.985	797 194	1835	1 455 349	29 534	130 130	1 333 168	35.829
1965	7 718 371	1158	8 936 381	12 033	1 347 189	7 601 225	185.221	773 791	2144	1 658 673	12 535	684 947	986 261	24.033
1966	8 286 935	1119	9 271 485	4502	851 865	8 424 122	198.754	726 595	2218	1 611 947	1122	478 27	1 565 242	36.929
1967	7 610 932	1130	8 603 279	5080	1 253 963	7 354 396	168.005	750 913	2745	2 061 433	1172	279 053	1 783 552	40.744
1968	7 675 845	1181	9 061 823	5500	896 607	8 170 716	180.724	704 551	2527	1 780 057	1599	2978	1 778 678	39.342
1969	7 103 509	1184	8 410 894	8442	789 083	7 630 273	163.412	782 715	2447	1 915 445	762	252 875	1 663 332	35.622
Average	7 679 118	1153	8 856 772	7111	1 027 737	7 836 146	178.824	747 713	2415	1 805 511	3438	253 536	1 555 413	35.495
1970	7 439 684	1194	8 879 384	760 990	2065	9 638 309	199.860	762 558	2817	2 148 395	788	41 681	2 107 502	43.706

Source: Secretaria de Agricultura y Ganaderia, Consumos Aparentes and Centro de Investigaciones Agrarias and Comite Interamericano de Desarrollo Agricola (1970), Estructura Agraria y Desarrollo Agricola en Mexico. Mexico, DF: CIDA 1970 (Vol.1, p. 239) (in Hewitt de Alcantara 1976, 110–11).

Wheat yields were greater in the 1930s than the 1920s, but again, demand appears to have been weak in the early 1930s. As with maize, land in production was slightly less in the early 1930s than in the 1920s, but picked up in the late 1930s as demand increased. Total production, and land in production, increased year on year from 1937 to 1940 inclusive. Per capita consumption increased, but yields showed no obvious upward or downward trend. Though imports had begun to respond to greater demand in the late 1930s, they fell to almost zero in 1940 before exploding in 1941. In 1944, wheat imports actually exceeded production, and though yields began to increase in the second half of the 1940s, imports remained substantial.

It seems fair to say that the situation regarding consumption of both products in 1940 was better than it had been for a decade or more. This probably owed itself in part to improvements in global economic prospects, but also to the land redistribution efforts of the Cardenas administration and the reduction in rural unrest. It is worth remembering that Ferrell first raised the idea of the MAP in 1936, when matters were somewhat worse.

Wallace had emphasised the importance of raising yields of corn and beans. Yet the MAP's distribution in expenditure on research by crop led to wheat, despite its being grown on a much smaller area of Mexican agricultural land (see Table 6.1), absorbing a share of the MAP's budget roughly equal to that devoted to maize (see Table 6.2). Such a distribution hardly supported the Malthusian rationale given to the programme (see also Hewitt de Alcantara 1976, 24, 26). This suggests that the Malthusian rationale is only part of the explanation for the involvement of the RF in Mexico.

Philanthropy, investment promotion and social engineering

Cleaver (1975, 131–5) suggests several possible attractions of philanthropy. It could demonstrate how wealth could be socially responsible even amidst agrarian unrest in the late nineteenth century. At that time, farmers concentrated their anger on monopoly capitalists, of which Rockefeller, with interests in railroads, banks, oil, and agribusiness amongst others, was the largest. Philanthropy could also have served as a shelter for some of Rockefeller's wealth. Yet Cleaver (1975, 130–5) argues that philanthropy aimed principally to foster social stability and a more favourable climate for capitalist investment.

Jennings (1988, ch. 2) shows that Rockefeller philanthropy often focused on education and science as the means through which a particular social order could be achieved. Philanthropic funds were used to

Table 6.2 Research expenditures by crop, 1943–60 ('000 pesos)

Year	*Corn*	*Wheat*	*Beans*	*Potatoes*	*Horticulture*	*Soil*
1943	171	171				85
1944	347	347				173
1945	606	440				275
1946	1068	712				445
1947	1638	1092				682
1948	2141	1427				892
1949	2738	2464	273			1369
1950	3106	2795	310			1553
1951	3178	2797	381			1589
1952	3374	2834	539			1687
1953	2896	2292	555	486	625	1736
1954	4015	3364	1627	542	1302	2713
1955	4051	2778	1736	1041	1967	1968
1956	3811	2772	1732	693	2541	2987
1957	4113	2513	2171	571	2056	3006
1958	3935	3241	2199	578	1620	3093
1959	4458	3315	1829	571	1257	3108
1960	4847	3604	1988	621	1367	3407

Source: Barletta (1970, 14) (figures are estimates derived by distributing total expenditures on the basis of the percentage distribution of salaries paid for research in each crop).

promote a particular approach to a given problem through building new institutions, nurturing existing ones, providing additional funding to key centres so as to gain influence over its work, funding conferences, and setting up organisations for training purposes. Typically, it was hoped that the public purse would take up the funding of a given endeavour, ensuring that the state took on the role of maintaining a system which allowed the expansion of private enterprise and investment opportunities. It was in the area of institution building that RF staff became experts. Jennings and Cleaver agree that Rockefeller-funded science was directed with commercial ends in mind. In turn, both refer to attempts to foster a particular social order. But whereas for Cleaver, the end – the promotion of a capitalist system of free enterprise throughout the world – governed the programme's orientation, for Jennings, it is the means through which this was achieved that is most worthy of attention. In his analysis, scientific knowledge and its generation take centre stage. The key for him is the role that the Foundation played in directing the development of scientific disciplines.

Jennings' analysis, however, becomes distorted through a desire to equate motives with results, there being inadequate discussion of the relationship between ends and means. Surely even Jennings would agree that Rockefeller-funded work on the eradication of hookworm, on syphilis and on infantile paralysis contributed to the understanding of important problems. It may be quite legitimate to view the work of the philanthropists in terms of an expression of the interests of their sponsors, but their interventions were at least enlightened ones, designed to have appeal beyond the sponsors. Yet the impression given by Jennings, and indeed Cleaver, is one of an overtly commercial exercise in corporate greed. Neither author fully comes to terms with the fact that the Rockefeller programme in philanthropy offered tangible improvements to the lives of many with whom they came into contact. Indeed, one could argue that this was the reason why philanthropic work expanded without considerable controversy.

Even so, the RF's involvement in social engineering deserves mention. When the Foundation was reorganised in 1928, Warren Weaver became Head of the Natural Sciences Division. Weaver was beguiled by the potential of scientific understanding to influence society, writing of the 'possibility, based on knowledge of physical laws and understanding of natural processes, of bringing individuals into a more intelligent, a more accurately adjusted, and a happier relationship with our modern scientific civilisation' (cited in Jennings 1988, 36). A year later, he called for the Foundation to commence a programme in psychobiology, expressing his desire to breed a superior race of men (Jennings 1988, 36–40).

When Fosdick, who himself saw the RF's aims as 'promoting the rationalisation of life,' sought advice concerning the MAP, he turned to Weaver. Weaver was later quite candid as to the nature and goals of the programme. Commenting on the degree of co-operation with the Mexican Government, he suggested that talk of formal collaboration was 'very misleading ... they had little or no idea as to what we were talking about, or what we intended to do ... the programme was initially more or less stuffed down their throats' (Weaver 1950, 1–2). Most revealing of all, however, was his opinion that:

> I do not think that it is proper for the RF to think of this activity as 'Mexican' in the strict and limited sense that Mexico gets all the benefits of it and should therefore pay a large fraction of the bill. Actually, it is something of an accident, from our point of view, that this project happens to be located in Mexico. *The Rockefeller Foundation wanted to try an experiment. Mexico provided a favourable location. In the first instance this project is going to produce more and*

> *better corn and beans for Mexicans; but from a much larger and important view, it is a basic experiment in international technical cooperation.* (Weaver 1950, 3, my emphasis)

In short, the history of Rockefeller philanthropy makes it quite conceivable that the MAP was consistent with the RF's desires to modify society through applying science and technology. Yet such a rationale does not make attempts to improve people's lives completely unworthy. Questions remain, however, as to whether these improvements could have been accomplished through different approaches.

Genes in the bank

Chapter 5 highlighted the significance of the replacement by hybrid maize of a number of well adapted maize varieties in the United States. Such a process in Mexico would have been a quantitatively different phenomenon. A substantial body of opinion supports the view that maize was first domesticated in Mexico and, along with Guatemala, it is a centre of diversity for the crop itself and its wild relatives, of which teosinte is of the greatest agronomic significance.[5] One possible motivating factor behind the MAP, hitherto unexplored, relates to the diversity of maize residing there. The importance of diversity in corn, both for breeders and in the field, was underlined in Chapter 5 above (see also Jenkins 1978, 15). Brown (1975, 467) writes:

> When it became obvious that hybrid corn was rapidly replacing the open-pollinated varieties of the US Corn Belt, federal funds were made available to several Mid-western states for the purpose of collecting old varieties still existing in those states ... efforts were fragmentary, sporadic, and inadequate.

In support of the thesis I am tentatively proposing, three things are known:

1. Henry A. Wallace was in a position to appreciate the significance of genetic variability of maize in Mexico (see Chapter 5);
2. one of the early aims of the programme was to collect local varieties of maize which would constitute the genetic basis of the breeding programme which would follow; and
3. even if the germplasm being collected, or sought, was not for use in US hybrids, the growing number of hybrid seed companies would have had a close eye on new markets in Latin American where maize was the staple crop.

As early as 1923, a US researcher was working on hybrid maize in Argentina. In 1945, a Rockefeller affiliate invested in AGROCERES, a company set up by two breeders at the University of Viscosa to produce hybrid corn seed in Brazil (Jaffé and Rojas 1994, 7). Other transnational companies followed into Argentina in 1948 (Cargill), Mexico in 1955, and Chile in 1964 (these were direct investments and do not fully reflect company involvement) (Dalrymple and Srivavastava 1994, 194).

Brandolini (1970, 287) reports that the RF collection of maize germplasm in Mexico took place principally in the years 1945–52, and that Edgar Anderson was a key figure in early collection and classification of maize varieties. The RF then set up a programme in Colombia in 1950 along the lines of the MAP and a second Latin American maize genebank was established in Medellín. In recognition of the need for more systematic efforts, the National Research Council-National Academy of Sciences Committee on Preservation of the Indigenous Strains of Maize was set up in 1951 (Brown 1975, 474; Harrar 1954, 105). This was funded by the US State Department and administered through the Institute of Inter-American Affairs, whose head was Nelson Rockefeller. Germplasm was stored at Fort Collins, which would later be established as the heart of the US germplasm collection for all crops. Ten publications, the so-called 'race bulletins' (after 'races of maize' – see for example Wellhausen *et al.* (1952)) resulted from the work undertaken throughout Latin America and the Caribbean.

Note also that this work, leading to germplasm collections in the US, was potentially of enormous value, but its removal from the country of origin appears to have raised few if any eyebrows. A number of breeders have pondered the potential of exotic germplasm to contribute to maize improvement in the US (Hallauer 1978, 243–5; Stringfield 1964, 129; Mangelsdorf 1951, 138–9). One study (Cox *et al.*, 1988) suggests exotic germplasm has contributed little to maize improvement in the US, yet another takes the view that when it has done so, its effects may have been disproportionately large. Thus, RAFI (1994a) notes amongst contributions to US maize production, that of Mexican teosinte, worth an estimated $4.4 billion per annum, and that of West African germplasm, which provided the only resistance to Southern Corn Leaf Blight (a disease which caused $1 billion of damage in 1970). In addition, one third of all maize germplasm requests to the Centro Internacional de Mejoramiento de Maiz y Trigo (CIMMYT), which arose from the MAP, come from private companies, whose use of exotic germplasm is said to be increasing rapidly. If the choice of Mexico was, as Weaver mentioned, an accident, it was a happy one for maize breeders in the US, and a particularly happy one for Wallace.[6]

Competing approaches to agricultural development in Mexico

Introduction

Rockefeller philanthropy offered improvements in some people's lives, yet it is criticised both on the basis of its covert aims and the means for achieving them. The question is whether these improvements could have been made in other ways. If other paths were either ignored or marginalised, it is important to understand why. In addressing this question, one appreciates why the practice of scientific research is an appropriate subject for scholars of political economy.

Some of those engaging in the practice of science are either unaware of, or prepared to deny, the possibility that their work is being, or will be used to fight political battles. Yet to the extent that their work seeks to address a particular problem, the very fact that a number of solutions, or approaches to finding solutions, may exist lends to their work a potentially political dimension. In particular, it is important to acknowledge the difference between justifying the results arising from a given approach (which might be done empirically), and the need to justify the underlying assumptions that give rise to a particular approach to a problem (which takes us into the sociology and politics of science). In this context, the propagation and promotion of a particular view or approach, becomes as much a political and sociological programme as it is a scientific one.

Over time, a particular view has the capacity to become assimilated as orthodoxy, and dissenting or competing approaches and interpretations tend to be marginalised, forgotten, or dismissed. Any political dimension to the original battle for acceptance slips away with the passage of time. The significance of these discussions will become clear below. The process which is described represents, in the terminology of Chapter 3, the selection of one approach from a number of possibilities through progressive deformation of the fitness landscape. As will be seen, the choice of personnel plays an important role, as does the desire to present a particular approach to Mexican agricultural development on the part of the RF. The links with the US study of Chapter 5 are obvious through the fact that the same personnel are involved in the transformations under examination.

Moulding the MAP

When the idea of undertaking an agricultural programme in Mexico was considered by Raymond Fosdick, he consulted Henry Wallace and Warren Weaver. Weaver's position as Head of the Natural Sciences Division made him the key point of contact within the Foundation if a (non-medical) science-oriented programme was to be established in Mexico. As for Wallace, he was an acknowledged expert on corn, at least in the US context. Given Fosdick's choice of consultees, there was a clear possibility that any subsequent programme would be conceived in narrowly technical terms. Wallace was more than likely to extol the virtues of hybrid corn, whilst to the extent that a market for it could be found in Mexico, Wallace was well placed to see his own company, Pioneer Hi-Bred, benefit.

The area of enquiry began to narrow with the appointment by the Foundation of the committee to report on the situation in Mexico. According to Stakman *et al.* (1967, 22), one member of the three man committee, A. R. Mann, the best qualified of them in matters agricultural, had commented, 'Experience has shown that the greatest practical contributions to agriculture come through the fields of genetics and plant breeding, plant protection, soil science, livestock management, and general farm management.' It was therefore unsurprising that when the committee chose to select a three man Survey Commission to tour Mexico the disciplinary orientation of the men reflected Mann's comments. The members of the Survey Commission, writing of the 5 June 1941 meeting with the committee and others in New York, recall:

> The mandate was simple. 'Go to Mexico and find out whether you think the Foundation could make a substantial contribution to the improvement of agriculture, and if so, how?' There was no attempt to indoctrinate the Commission but every attempt to facilitate its work. (Stakman *et al.*, 1967, 24–5)

On the basis of their two-month tour, they made recommendations which were included in a Report that reached New York in October 1941. Though the Survey Commission's members spoke of an objective assessment of the situation in Mexico, this was coloured by their own experience:

> All three had seen the contributions of science, technology, and education to the phenomenal progress of agriculture in the United

> States during the quarter of a century before 1941 ... They had seen, and indeed one of them had helped, the increase of the acre yields of corn by upwards of 25 per cent through the development of hybrid lines. (Stakman *et al.*, 1967, 23)

Despite references to ill-fated attempts to 'modify' Mexican people, and to 'the need for specific knowledge derived from adequate regional and local experimentation,' their comments and recommendations betray an underlying desire to judge Mexico by the standards and norms of US agriculture. Mexican dietary standards were assessed through comparing Mexicans' consumption of meat and sugar with that of US citizens, whilst concern with agricultural education was expressed through comparing Mexican and US institutions (Stakman *et al.*, 1967, 30–1). Elsewhere, Bradfield (1957, 624) wrote:

> The underdeveloped countries, however, have not tested, modified nor applied much of the information which has been freely available in the United States for the last 25 to 50 years. This, of course, is the principal reason why the agriculture of these regions is underdeveloped.

Whilst the Commission advocated a programme of joint participation rather than preaching, their assessment of the prevailing situation would allow for little of the former:

> the schools can hardly be improved until the teachers are improved; extension work cannot be improved until extension men are improved; and investigational work cannot be made more productive until investigators acquire greater competence. (Stakman *et al.*, 1967, 37)

Reflecting on the desire of Ferrell to see an extension programme in place in Mexico, they opined:

> Extension alone, and other forms of education, can make great improvement only when there is a great reservoir of potentially useful but unused information, and there was no such reservoir in the Mexico of 1941. (1967, 34)

This conclusion was drawn despite their inability to communicate with 'back-country people', the assumption presumably being that these people could not contribute anything to the hypothesised information reservoir (Stakman *et al.*, 1967, 27).[7] Armed with these views, it would have been difficult to enter into a programme purely in the spirit of co-operation since the Commission's analysis recommended a one way transfer of knowledge. One begins to see why Weaver (above) was inclined to view the MAP as stuffed down the throats of the Mexicans.

The Survey Commission continues:

> the most acute and immediate problems, in approximate order of importance, seem to be the improvement of soil management and tillage practices; the introduction, selecting, or breeding of better-adapted, higher yielding and higher quality crop varieties; more rational and effective control of plant diseases and insect pests; and the introduction or development of better breeds of domestic animals and poultry, as well as better feeding methods and disease control. (in Stakman *et al.*, 1967, 32–3)

Little separates these views from those of A. R. Mann expressed earlier. Like Mann before them, and reflecting their own disciplinary expertise, they recommended that the Foundation initially send an agronomist and soils expert, a plant breeder, and an expert in crop protection, as well as an animal husbandman (which Mann also recommended), as a working commission to cooperage with the Mexican Ministry of Agriculture.

These recommendations were adopted in full by the Foundation's Board of Trustees in late 1941. Surely it is fair to say that the selection of particular people with their own subjective (disciplinary) perceptions of what needed to be done was a critical determinant of the MAP's early thrust. Is it really just coincidence that this commission, comprising a soils specialist, a geneticist and plant breeder, and a plant pathologist, recommended that the MAP initially be staffed by scientists with the same disciplinary orientations? I think not, and I believe, therefore, that decisions first of all to conceive the programme in narrowly technical terms, and then to proceed with men with specific disciplinary backgrounds, helped to elaborate the terms of reference of the MAP, as well as its initial staffing.

Dr George Harrar was chosen to lead the MAP and began in February of 1943. In the Autumn, the corn breeder Edwin Wellhausen joined Harrar. The following year, William Colwell, a soils expert, and Norman Borlaug, a plant pathologist, joined the programme. Stakman *et al.* (1967, 5) wrote, 'Subsequently many other men, from Mexico and the United States, contributed significantly to the program, but its general course was set during those first two eventful years.' I contend that in fact, its course was set even earlier.

Hybrids or open-pollinated varieties?

In the four years after Calles took power in Mexico (in 1925), previously favourable conditions began to change dramatically. A religious war lasting four years conspired with the saturation of world markets to weaken the Mexican economy. Having opined in 1928 that the period of agrarian reform was at an end, Calles was forced to eat his words as events conspired to force the question of agrarian reform back onto congress' agenda.

Martinez Saldaña (1991, 316–26) traces the changes instigated by the new Cardenas administration to a variety of causes, including the collapse of markets, lack of foreign exchange, dwindling resources of the exchequer and the growing rural discontent made manifest by armed movements in Veracruz and San Luis Potosi. When Mexicans were expelled from the USA, anti-US feeling in the country led to the expropriation of land held by US citizens. Cardenas became president in 1934 under Calles' tutelage. Two years later, Calles was expelled from Mexico as Cardenas sought to consolidate his position through mobilising more radical support. The years 1936 and 1937 saw the agrarianism of Cardenas reach its peak, and by the end of the decade, some twenty million hectares of land had been handed over to nearly a million campesinos.

Land reform in these years favoured the establishment of *ejidos*, a form of communal landholding system whose historical roots lay in prehispanic Mexico. In a concerted attempt to assist the *ejiditarios*, the Cardenas government promoted institutional change, transforming the credit system so that it supported small farmers and *ejidos*, and left large farmers isolated (Martinez Saldaña (1991, 316–8)). In addition, health services for *ejidos* and the rural technical schools were established:

> These institutions were the instruments of a transformation of agricultural and agrarian politics in Mexico. For the first time, the ejido was chosen as the path through which to develop agriculture. The hacienda system which had dominated since the colonial epoque began to be marginalised. (1991, 318, my translation)

Agricultural schools and research organisations were questioned concerning the beneficiaries of their work and their overall client orientation. In 1925, Calles had designed eight 'centrales agricolas' along US lines. Prior to the 1930s, research paid scant attention to food crops, being dependent on French technical assistance, and oriented to the study of the key export crops of pre-revolutionary Mexico (Hewitt de Alcantara 1976, 19). Yet the most important crop for Mexicans was maize, a crop whose cultural significance adds to its importance in dietary terms.[8] In the 1930s, a small Office of Experiment Stations (OES) was established in the Ministry of Agriculture, and began seeking higher yielding varieties of corn and wheat through selecting from varieties collected locally. Edmund Taboada, chief investigator in the Ministry of Agriculture at the time the MAP began, stated:

> Scientific investigation must consider the men who will put its findings into practice ... It is possible that a discovery can be made in a laboratory, in a hothouse, at an experiment station, but useful science, manageable, operable science, must grow out of the local laboratories of ... small farmers, ejiditarios, and indigenous communities. (cited in Hewitt de Alcantara 1976, 19)

To this end, in 1932, the centrales agricolas were transformed into regional rural schools, emphasising basic education for practical farmers. Those within the OES supported the view that increased food production was predicated on structural changes as much as technological ones (Hewitt de Alcantara 1976, 19).[9] The OES focused on open-pollinated varieties, and trusted farmers to carry out the most significant experiments in their fields (Hewitt de Alcantara 1976, 37–40).

The transfer of power from Cardenas to Camacho signalled an end to the era of unequivocal support for the *ejido*. It was this shift in emphasis, back toward the primacy of private enterprise, irrespective of size, and more significantly, ownership, that constituted the backdrop to the setting up of the Officina de Estudios Especiales (OEE, the name of the MAP's research organisation). Both Hewitt de Alcantara (1976) and Jennings (1988) note the importance of changes in the system of Mexican agricultural education and research following the establishment of the MAP. A system which sought to draw attention to the allegedly exploitative nature of the hacienda system and the injustice of land ownership by foreigners, could hardly co-exist with the policies followed by Camacho, which sought to assist those who, often by virtue of political connections, were in receipt of newly irrigated lands.

Having become more attentive to the condition of the majority of cultivators in the 1930s, the Mexican agricultural research system, significantly influenced by the MAP, shifted its emphasis during the 1940s and 1950s, a process more or less completed by the 1960s.

Since the Survey Commission had perceived that the Mexican system of agricultural education and research was inadequate, the MAP's initial staff were to be supplemented by Mexicans sent to US land grant colleges for training, funded by Rockefeller grants. Rockefeller money was also pumped into three key universities and used to influence the choice of staff, and through this choice, the curriculum taught at the universities. From his interviews with Harrar and Wellhausen, Jennings (1988, 104) believes that the rationale for this process was to ensure that agricultural research proceeded along scientific lines rather than political ones. MAP scientists sought to cleanse their scientific approach of the social and political issues which had affected Mexico's past and present, thereby lending to their own approach a political slant.

The first experiment station set up under the MAP was the El Horno station at Chapingo and it was here that research on maize was conducted. Edwin Wellhausen, with the assistance of Efrain Hernandez X., collected corn germplasm from all over Mexico. Wellhausen benefited from the earlier work of the OES and IIA in this area, receiving over two thousand experimental lines by 1947. Some of Wellhausen's early work focused on the dissemination of superior varieties and breeding synthetics which, like the open-pollinated varieties the OES and IIA worked with, would enable farmers to plant seed harvested from previous crops as they always had done. But, as Stakman *et al.* (1967, 58) point out:

> The long-term objective was to develop hybrid corn for Mexico that might revolutionize her production as hybrid corn had already revolutionized corn production in the United States ... The plan included four main steps so arranged that each step would not only increase corn production and provide more food but would also contribute to the attainment of the final goal.

Following the collection of germplasm, step one in the plan, step two, varietal testing, began in 1944, and it rapidly became clear that foreign germplasm was all but useless in Mexican conditions. In 1946, some varieties were distributed to farmers, so completing step two. The first synthetics were distributed in the same year, the third step in the plan. Inbreeding with the aim of producing hybrids, step four in the plan, also began in 1944. Early on in the programme, it was discovered that by growing maize in Progreso in Morelos, a winter crop could be grown, so speeding up the breeding process. By 1947, ten double-cross hybrids were ready for distribution.

In 1947, the OES was transformed into an official Instituto de Investgaciones Agricolas (IIA), thus apparently consolidating the position of OES scientists at just the time when the MAP was beginning to exert a most powerful influence over the agenda for agricultural research in Mexico. The IIA recognised that most corn was cultivated on rainfed land, and that access to inputs was limited for those growing the crop. The contrast in approaches of the IIA and the OEE is well captured by Hewitt de Alcantara (1976, 37):

> The IIA worked primarily on the development of improved corn seeds for the traditional small farming areas of Mexico, while the Office of Special Studies preferred to dedicate the major part of its resources to very high yielding varieties which were only likely to be used profitably by the best endowed farms of the nation.

This is not to imply that the more modest increases sought by the IIA would be any easier to achieve than more substantial increases in well-endowed areas. The ecological diversity of the Mexican countryside challenged researchers and farmers to find means of increasing production in non-irrigated areas.

Had the MAP acted so as to complement the IIA research programme, one could have better understood the emphasis on hybrids in the early days of the MAP. However, whilst in 1948, it was estimated that 80 per cent of all land planted to improved varieties consisted of open-pollinated IIA materials, by 1956, 96 per cent of all the Mexican Agriculture Ministry's research resources for corn were concentrating on hybrid production (Hewitt de Alcantara 1976, 40). The RF's influence was exerted principally through agricultural education:

> The entire staff at the School of Agriculture of the Institute of Monterrey are products of the Rockefeller Foundation program in Mexico, with the single exception of a veterinarian who teaches animal science. The agricultural curriculum of the College was developed in consultation with Dr Harrar and other members of the Office. Most of the faculty have received postgraduate training as Foundation scholars or fellows and doubtless others will be considered for similar experiences in the future. (Anon 1952, 22)

Quite apart from the influence that the Foundation was able to bring to bear within institutes of learning and research, increasingly one could find graduates of the OEE or Rockefeller scholars in positions of influence in the extension service, the Ministry of Agriculture, and private companies. The Foundation perceived competition for personnel between the latter two as problematic since it wanted its trainees to find positions of political importance.

When, in 1956, Foundation staff met to consider the future of the programme, it was to education that they turned their attention. It was stressed that those trained as researchers needed to be encouraged to teach if the MAP was ever to become self-sustaining. The Foundation's view was that agriculture was 'nothing more than the application of the principles of biology and other natural sciences to the art of growing food' (Stakman *et al.*, 1951, in Anderson *et al.*, 1991, p. 32). To ensure the adoption of such a view in the sphere of agricultural education in the wake of changes instigated in the Cardenas era required that education become the antiseptic that would cleanse the germ of politics from agricultural research. In Marglin's terms, *episteme* should dominate *techne*. The success of the Rockefeller programme can best be illustrated by the reorganisation of agricultural research in the wake of the official conclusion of the MAP in 1960. Superficially, the merging of the OEE and the IIA within the Instituto Nacional de Investigaciones Agricolas (INIA) in 1960 brought together two institutions that had followed different paths in seeking to develop Mexican agriculture. In reality, the merger marked the death of the IIA and its philosophy. Jennings (1988, 110–11) notes that the top six positions were held by individuals trained in the tradition of the MAP.

Notwithstanding the switch in effort away from OPVs to hybrids, Stakman *et al.* (1967, 70) noted that eighteen years after hybrids were made available in Mexico, they were being used on only 14 per cent of Mexico's corn acreage. They locate the problem in the refusal of the Mexican Government to liberalise the production and distribution of improved seed. The private sector would encourage the spread of seed to smaller farmers. Yet transnational seed corporations were already active in Mexico in the early 1960s. In 1961, subsidiaries of three companies, Northrup King, Asgrow and DeKalb began attempts to market hybrids in Mexico. Yet by the end of the 1960s, 25 companies were involved in the seed sector, half of them foreign (Barkin and Suarez 1986, 27). Stakman *et al.* (1967, 71) go on to note that:

> The corn-growing areas with dependable rainfall or irrigation probably offer the best prospect for further improvement. The knowledge necessary for doubling yields on these lands is available and needs only to be applied.

Scientists do what they know how to do, and the solution they best understood was that which they had been part of in the US. The desirability of sorghum as a substitute for maize in drought-prone areas is mentioned, but the authors make no reference to the fact that sorghum generally feeds livestock, whilst corn feeds people.[10] This prognosis implicitly requires Mexican agriculture to adapt to (a particular type of) research rather than the other way round.

Having visited IRRI in 1962, President Adolfo López Mateos suggested setting up an International Maize and Wheat Improvement Centre on the basis of the MAP in Mexico. CIMMYT (Centro de Mejoramiento de Maiz y Trigo) was established on 12 April 1966 as a 'nonprofit civil association under Mexican law, responsible to an international board of trustees' (CIMMYT 1992, 9). CIMMYT appears to have reversed the earlier emphasis on hybrids in favour of population improvement techniques mentioned in Chapter 5 above.[11] Only in the mid-1980s did CIMMYT start to orient some of its work in germplasm improvement to hybrid production, a decision which was controversial in view of CIMMYT's mandate to assist poorer farmers in the developing world.[12] However, even in 1990, population improvement was the main avenue for germplasm improvement:

> CIMMYT maintains a wide array of improved maize populations, some of which have been undergoing *full-sib* recurrent selection for well over a decade ... the currently available populations continue to be an extremely valuable source of improved germplasm for national programs and, even after many years of selection, appear to offer ample scope for further gains in yield and stress resistance. (CIMMYT 1991, 37)

By the early 1990s, 18 per cent of maize programme funds were devoted to hybrids compared with 30 per cent on population improvement and 13 per cent on development of open pollinated varieties (van Wijk 1994, 4). The increasing focus on hybrids *vis-à-vis* other methods is surely questionable both in terms of its function as an effective subsidy to private sector work, and the implicit favouritism thus accorded to cultivators in well-endowed environments.[13]

Carl Sauer in Mexico

Carl Ortwin Sauer was described, on his death in July 1975, as 'one of the most influential geographers of his generation' and 'the Dean of American geographers'. Prior to a period of field work in 1941, Sauer had made eleven field trips to Mexico, one of them, in 1935, funded by the Rockefeller Foundation (West 1979, 4). Though originally interested in geomorphology and archaeology:

> By the mid-1930s Sauer was beginning to observe rather closely aboriginal agriculture and related questions regarding the origin of domesticated plants in the American tropics ... At first, he was concerned chiefly with the aboriginal staples: maize, beans, and squash, and in subsequent trips to Mexico collected specimens of such crops mailing them in small packages to Berkeley, for examination by geneticists. (West 1979, 18)

Sauer was consulted by the RF prior to the visit of the Survey Commission (Jennings 1988, 50). Wallace had talked about a Mexican programme in health, nutrition and agriculture to Joseph Willits, Director of the Foundation's Social Science Division from 1939–54, whom Sauer knew from work they had done on a Committee of the Social Science Research Council (West 1979, 97).

Willits asked Sauer, on sabbatical in Mexico, his opinion of Wallace's views on a possible agricultural programme in Mexico:

> A grand job to be done or to be messed up beyond making good. The first step would be in economic geography. Identify the occurrence and usage of every domesticated plant form together with its utility in the kitchen and in agricultural practice (soil, climate, seasonal labor in planting and harvest, tolerance of extreme weather, of pests). Secondly, make sure that every genetically fixed form is preserved and grown in adequate quantity for experimental purposes. Thirdly, set up breeding centres for the development of better strains. Remember that the gene range of maize, beans, etc., is enormously beyond that available to the American plant breeder, that an individual plant like 'maize' has much more varied uses than in our commercial agriculture, that a large stock of native species is present which do not exist in the U.S., that a large number of old world Mediterranean plants are established. The possibilities of improvement by selection are enormous, but such a selection should proceed from local materials. A good aggressive bunch of American agronomists and plant breeders

> could ruin the native resources for good and all by pushing their American commercial stocks. The little agricultural work that has been done by experiment station people here has been making that very mistake, by introducing U.S. forms instead of working on the selection of ecologically adjusted native items. The possibilities of disastrous destruction of local genes are great unless the right people take hold of such work. And Mexican agriculture cannot be pointed toward standardization on a few commercial types without upsetting native economy and culture hopelessly. The example of Iowa is about the most dangerous of all for Mexico. Unless the Americans understand that, they'd better keep out of this country entirely. This thing must be approached from an appreciation of the native economies as being basically sound. (Sauer 1941a, 2)

In the same letter, Sauer stressed that the problems facing Mexico were economic and political rather than cultural, and that far from being lazy and ignorant, Mexicans were poor and carried a legacy of oppression. Regarding nutrition, and in contrast with the views of the Survey Commission, he advised:

> don't get the idea that they would eat better if they had nutrition experts to advise them. I've had a good deal of interest in the Mexican kitchen and if there is any other country in which sound nutrition is better practised as far as the pocketbook permits, I don't know it. (Sauer 1941a, 2)[14]

Six days after Willits received Sauer's letter, on Feb 18th 1941, a Foundation staff meeting initiated the process that led to the MAP. Willits had little time to circulate the letter from Sauer to others attending the meeting, though Jennings reports that Fosdick, Ferrell, Mann and Warren all received copies. Minutes of that meeting show that Willits did intervene to have Sauer's opinions considered by the Foundation. His intervention followed one of many contributions advocating a narrowly technological approach to the programme: 'JHW [Willits] makes one suggestion: hopes that before RF goes overboard too far it obtain formally invited criticism of Carl Sauer. Is in entire sympathy with importance of focusing on agriculture' (Anon 1941, 3).

The following month, though it is not clear that this was an officially sanctioned move on their part, Rockefeller staff visiting Mexico sought Sauer's opinion once more, and again Sauer wrote to

Willits. As well as targeting Wallace (see above), Sauer restated some of his fears:

> I fear the irreparable loss of native crop forms by a rapid Americanization. American tools may do a great deal of harm to soils by a too hasty introduction. (Witness what we have done to Puerto Rico!). Perhaps the native stubbornness is sufficient safeguard, but our agriculture is geared to a level of production that either involves a great using up of soil fertility or the inversion of sums in amelioration (fertilizer bills, engineering works) beyond – far beyond the capacity of most of these lands ... They must build on the preservation and rationalization of their own experience with slow and careful additions from outside. Only the exceptional American agricultural scientists will see that there are these hazards of cultural destruction. He will be under the pressure of quick results. He will hardly have the life-time of learning that native practices represent real solutions of local problems, partly because he comes as an alien to a strange situation. (Sauer 1941b,1–2)

Two questions arise; why were Sauer's views so different to those of the Survey Commission; and why were his views apparently dismissed? The answer to the first question relates to Sauer's training. His life's work was devoted to understanding landscape and culture. This forms part of the answer to the second question. Marglin suggests that Sauer's arguments were misconstrued as advocating maintenance of the status quo over any form of modernisation, and that viewed in this light, his arguments were contradictory:

> if the problems are essentially economic; if modernization equals economic development; and if tradition is opposed to modernization – then tradition itself becomes the problem. Development becomes the only way to alleviate the poverty that stands in the way of health and nutrition. (Marglin 1992, 23)

In the same breath, Marglin argues, Sauer could be considered to be advocating both maintenance of tradition, and its destruction.

Marglin's hypothesis suggests that the Foundation could find, in Sauer's views, support for their programme's rationale. After all, the Foundation was positively disposed to social transformation through modern science and technology, and if Sauer was critical of 'Americanization', Cleaver's (1975) history of the RF shows this to have been one of its goals. In the same way as US agricultural scientists believed that increasing production was a morally unambiguous duty, the philanthropists may have viewed modernization as an obligation. What led Sauer to urge caution could have led RF staff to accelerate the programme.

Even though it seems Sauer's criticisms were never really considered in such a way as they might change the course of the MAP in its early days, the RF still saw fit to seek his opinion. Indeed, over time, Sauer appears to have gained credibility within some parts of the Foundation, partly through the efforts of Willits, but as I have sought to show above, by this time, the course of the programme was already well established. Willits clearly admired Sauer, and when, in 1942, the Foundation arranged for a special grant of $50 000 to be used to help a small number of social scientists to carry out research in Latin America, Sauer was one of only four recipients of this funding (West 1979, 97–8; 1982, 1–3). Sauer was developing a reputation for Latin American scholarship and was an obvious candidate for an award, but what is more illuminating is the praise lavished by Willits on Sauer's letters to him during the Rockefeller-funded trip:

> I hope you understand how deep my personal appreciation has been and how much beyond the ordinary return the RF has received from this grant. Even though I have been almost silent as far as replies are concerned, the appreciation and gratitude to you has been very deep. (Willits 1942a)

Willits clearly tried to push Sauer's views into the minds of his colleagues. It may be that by doing this, he sought to improve his own status, or that of his Division, within the Foundation. On the other hand, Entrikin (1984, 407–8) writes that Sauer and Willits shared a 'common ideal of the nature of social science and a common dislike of the rationalization process in American society', so that Willits' efforts might have been borne more specifically out of admiration and a common purpose. In a letter to Chairman of the Board of Trustees, Walter Stewart (Willits 1942b), and in a memo to Foundation staff (Willits 1942c), he virtually pleaded the cause of Sauer. Regardless of his exact motives, Willits was obviously intrigued by Sauer's views, and respected them. But as far as the MAP's attitude towards applying US-style science to Mexican agriculture was concerned, Willits was 'the only skeptic in a hive of enthusiasts' (Marglin 1992, 19). Indeed, this, and the knowledge of what he could expect from Sauer, may have been the reason why Willits sought, then tried to project Sauer's views.

Once the programme was established, Sauer continued his work, accumulating more data on the practices of the indigenous cultures. With Willits help, the Foundation's opinion of Sauer soared, and the Foundation offered to underwrite any work Sauer wished to do in the 1944–45 academic year, so again Sauer returned to Mexico (West 1982, 4–5). There, Sauer met Wellhausen and Harrar and wrote another letter to Willits. In it, he wrote not only of the underestimated value of squash and their multiple uses in aboriginal diets, but also of the value of the *milpa* system of agriculture in Mexico. He wrote favourably of Wellhausen: 'He sees that they [the staff of the MAP] must work with the native corns; he is not a missionary for soybeans, and I suspect he sees the pitfalls in the wheat campaign' (Sauer 1945, 1), though he might have appreciated less Wellhausen's comments concerning the 'low economic and cultural level' of Mexican farmers (1950, 167–75).

Edgar Anderson, an expert in maize genetics, and a key figure in collecting maize germplasm in the late 1940s (Sauer sent him samples of corn from Mexico), made visits to Mexico in 1943 and 1944. Anderson made disparaging remarks to Warren Weaver regarding the MAP, leading Weaver to ask Paul Mangelsdorf, who had been a roommate of Anderson's whilst a graduate student, for his response to these criticisms:

> Without consulting him, I can say that Anderson, taking his cue from Carl Sauer, has been opposed to the Mexican agricultural programme from the beginning. His argument is somewhat as follows: If the program does not succeed, it will not only have represented a colossal waste of money, but will probably have done the Mexicans more harm than good. If it does 'succeed', it will mean the disappearance of many Mexican varieties of corn and other crops and perhaps the destruction of many picturesque folk ways, which are of great interest to the anthropologist. In other words, to both Anderson and Sauer, Mexico is a kind of glorified ant hill which they are in the process of studying. They resent any effort to 'improve' the ants. They much prefer to study them as they now are. (Mangelsdorf 1949, 1)

Mangelsdorf's view lends some support to my interpretation of why Sauer never managed to influence the course of the MAP. Modernisation was synonymous with improvement, and if Sauer could not accept this, he would never accept the programme's chief thrust, and the lives of

Mexicans could never be improved. Sauer was concerned with the nature of change, and its impact on culture:

> I am not interested in the Indians as museum pieces and I am also interested in non-Indian populations that have cultural values of their own as apart form the standardizing tendencies which are flowing out from the urban centers to strip the country of its goods and ablest men and pauperize it culturally as well as often economically. (Sauer 1945, 1)

Critical dimensions of change, for Sauer, were its pace, and the extent to which changes meshed with the prevailing way of doing things, a feature that would make such changes more appropriate for their needs. The MAP's view was that they knew what farmers wanted and needed. They had seen it in the United States.

Indigenous agricultural practices

Sauer alluded to the varied uses of maize in Mexican kitchens and elsewhere. Hernandez X. (1985) gives an excellent account of how farmers in Mexico make use of a variety of different maizes used for a plethora of different purposes. Furthermore, their decisions regarding the best combinations of seed and (micro-)environmental conditions reflect an intimate understanding of the ecological conditions in which they function. Both Nigh's (1976) thesis on Mayan farmers in Chiapas and Bellon's (1991) paper support the view of farmers as wise managers of their genetic resources, and they illustrate how important are diverse varieties, with at times, contrasting characteristics, to the farmers (see also Clawson 1985a, 57–9).

Yet, important as considerations of intra-specific diversity are, equally puzzling was the way the MAP ignored what centuries of farmers had been doing in Mexico, namely, intercropping. The MAP focused on maize and beans in isolation even though the two were commonly found growing side by side.[15] The MAP felt there was little to be learned from the way in which agriculture had been done in Mexico for centuries, such as the *chinampas*, whose productivity has been cited as enabling the Aztecs to develop from a small tribe to the most dominant Mexican group.[16]

Intercropping

Typically, maize has been intercropped with beans and squash in Mexico, with various permutations observable throughout Latin

America. Whether or not intercropping results in overyielding depends on the balance of what Vandemeer and Schultz (1990) call interference and facilitation. In by no means all cases does intercropping lead to overyielding. Yet, in Chapter 1, I cited Amador's thesis, and its findings that maize/beans/squash constitute an overyielding polyculture. Astoundingly little research is, or has been conducted with the aim of improving this system *as a system*. Jennings (1988, 68) contends that not only did the MAP ignore intercropping, but it promoted maize as cattle feed and industrial grain. Only in 1991 did CIMMYT report ongoing research in which, in combination with CIAT (Centro Internacional de Agricultura Tropical), it was seeking 'to ascertain whether maize and beans possess genetic variability for traits that would permit higher yielding combinations of these crops' (CIMMYT 1991, 32).

The most obvious agronomic interaction between these species is that beans are legumes and fix nitrogen in the soil. This helps fertilise the corn crop. Yet there are other interactions which appear to be important. Several factors come into play with regard to pest and disease control; physical barriers to aerial pathogens and their vectors; altered microclimate due to shading, changes in air movement and altered humidity and temperature; spatial effects within and between species; effects of pollen on spore germination; and altered host-pathogen interactions (Thurston 1992, 161–3; Perrin 1977).[17] In general, intercropping's impact on pests may be site, or even pathogen specific. However, the evidence with respect to maize and beans seems on balance to be favourable, though Thurston (1992, 164) cites one study in which one disease increased in severity in the intercrop (though in the same study, the severity of others was reduced), and Nordland *et al.* (1984) report that the corn, beans and tomato polyculture they researched performed less favourably than monocultures with regard to pest attack.

Furthermore, where squashes are used, two impacts on weeds have been noted. Leaf cover shades out many weed species, whilst the development of soil pathogens may be inhibited by allelopathic compounds washed from the leaves of squash plants (Amador and Gleissman 1990; Anaya *et al.*, 1992). The same leaves may reduce splash erosion and the transmission of spores through splashes. There is limited evidence on variability of harvest in these intercropped systems (some have been tempted to accept the controversial stability-diversity hypothesis (see Dover and Talbot 1987, 23–6) at face value), but to the extent that the facilitative effects are stronger than the interference effects, it is likely that risk is reduced in this system, particularly given the positive indications regarding disease, and the fact that a number of different crops are grown rather than just one.[18]

One of the most interesting points about the maize/bean combination is that the amino acid deficiencies in corn (lysine and methionine) are not the same as in beans (cystine) (Wilkes 1977a, 314). CIMMYT scientists spent much effort in the 1960s and 1970s developing quality protein maize (QPM) containing the *opaque-2* mutant gene, which doubled the content of lysine and tryptophan in maize. This gene is a recessive, so that it is quickly lost unless either the maize is grown in isolated plots, or grown as a hybrid with the understanding that the seed will be repurchased annually. Low adoption of QPM via the former recommendation led CIMMYT to develop QPM hybrids (CIMMYT 1992, 34–5). Far from improving the diets of poor people, most QPM has thus gone to feed pigs. This aims of this cutting-edge, though misguided science could have been met by encouraging intercropping.

Chinampas

The *chinampas* are a form of raised bed agriculture, and in pre-Colombian Mexico, were used to produce an abundance of corn, squash, tomatoes, beans, chilli and other vegetables. One commentator has described them as 'one of the most intensive systems of agriculture, past or present, in the Western Hemisphere, in terms of total production per unit of land' (Deneven 1970, 467; also Morales 1984). Mechanised commercial cultivation and the expansion of Mexico City have led to the demise of most of the best known of the *chinampas* in Xochimilco (Outerbridge 1987).

Chinampas are beds, usually rectangular in shape, raised above the level of surrounding channels of water. By dredging up nutrient-rich mud from the canal base, the canals are kept clear and the *chinampa* is fertilised. Aquatic weeds, animal manure, and sometimes night soil are also used as fertilisers (Altieri (1987, 87) reports that water hyacinths can supply 900kg/ha dry matter daily). Maize is planted directly into the *chinamapa*, but other crops are usually grown in small rectangular chapines, in which one seed is planted. These are cut from the almacigo, a layer of specially collected (often by canoe) mud spread over vegetation, and covered with a straw, or dried plant mulch. Selected plants are subsequently transplanted directly into the *chinampas* as soon as the previous crop has been harvested, giving them strong roots. Permanent moisture and favourable climate allows cropping all year round (Outerbridge 1987, 80; Thurston 1992, 112–3).

Maize yields as high as 6–7 t/ha have been reported in *chinampa* systems in Tabasco (Gómez-Pompa 1978). This compares favourably with those achieved in the US in 1991, and is more than three times the national average for Mexico (based on CIMMYT 1992, 47, 53). It is thought that both the health and the diversity of plants contribute to pest and disease management (and so, yield) in the system. Furthermore, Lumsden *et al.* (1990) give a highly favourable account of the system's impact on soil pathogens, a consequence of the high levels of organic matter stimulating biological activity in the soil. Similarly, Zuckerman *et al.* (1989) report reduced nematodal activity in the soil owing partly to the level of organic matter in the soil, and to the fact that this stimulated activity of nine organisms with antinematodal properties.

These examples illustrate that far from being backward, there were Mexican farmers who had a sophisticated understanding, based on experience, of their agricultural systems. Some of these systems were highly productive.

Conclusion

The RF MAP originated in the United States and it carried with it the hallmarks of United States agriculture and agricultural science. The United States *agronomos*, for all the expertise they possessed in US agriculture, were not experts in Mexican agriculture. Yet for the RF Survey Commission, 'Agriculture is nothing more than the application of the principles of biology and other natural sciences to the art of growing food' (Stakman *et al.*, 1951). Thus, those who possessed the scientific tools would be the best practitioners of agriculture wherever they went. The ecological diversity of conditions in Mexico, however, made the MAP's work irrelevant. *Episteme* would need to complement the *agronomos' techne* in any successful research programme for less well-endowed farmers.

There were alternative paths that could have been pursued. Neither Sauer nor Willits had the benefits of hindsight, yet both could appreciate the possible dangers of proceeding as though Mexico was, or could be, part of the US Corn Belt. Mexican researchers had already decided that hybrids were a dead end for most Mexican cultivators, and that incremental yield increases on large areas of land were likely to achieve as much, if not more than large increases in yield on small areas. Yet, the institutional changes that the RF promoted led to a concentration on research on hybrids in Mexican research organisations, though ironically, CIMMYT would stand apart from hybrids until 1985.

Official census figures showed that by 1960, only 9.5 per cent of the land planted to maize in Mexico was irrigated. In such conditions, hybrids were unlikely to diffuse widely. Newly irrigated land in Sonora and Sinaloa became the birthplace of miracle wheats. In large part, the institutional changes, more correctly viewed as a transfer, that occurred in the early days of the MAP, revealed a limited breadth of vision. How could maize cultivators appear other than backward when that adjective was applied to anything that did not resemble a farm in Iowa? The arrogance of researchers, fostered by their scientific training and the presumed superiority of *episteme* over *techne*, manifested itself in their refusal to consider the possibilities that, under the conditions facing them, farmers' decisions might not have been so backward. This prevented them from entertaining routes other than that of hybrid development and monocultures, a path that was pursued with limited success in the case of Mexico owing to the limited relevance of the technology.

Notes

1. Reports suggest that wheat yields have plateaued, or are actually falling in many parts of the world, including north-west Mexico, where the variety, Sonalika, released in 1967, has not been replaced due to lack of satisfactory replacements despite its vulnerability to a new race of leaf rust (Wilkes 1992, 47–8).
2. 'In the 1930–38 period, the US government expelled a million Mexicans who, on arrival in their country, provoked a wave of anti-Northamerican feeling, as well as demanding seizure of lands held by US citizens' (Martinez Saldaña 1991, 315–6, my translation; see also Galeano 1973, 134–40, 175–6).
3. A Standard Oil affiliate, American Cordage Trust, managed henequen plantations in the Yucatan. Galeano (1973, 136) describes these as, 'concentration camps where men, women and children were bought and sold like mules'. International Harvester, the agricultural machinery company, was also owned by the Rockefellers.
4. See also Marglin (1990).
5. It is still debated whether teosinte is merely a wild relative, or the true ancestor of domesticated corn (see Wilkes 1977b, 216–9; 1977c; Niebur 1993). Other centres of diversity exist, all in Latin America, but Central America is the only region where teosinte is also found (see also Wilkes 1997; Sánchez González and Ruiz Corral 1997).
6. Pioneer Hi-Bred, Wallace's company, grows out, and keeps copies of, CIMMYT germplasm (GRAIN 1989).
7. The Survey Commission's command of Spanish was such that Mangelsdorf asked a student to accompany them.
8. In the Mayan 'Bible', the Popul Vuh, man is descended from a maize ear. One can see depicitions of this in some of the murals of Diego Rivera and others.
9. The analysis above lends tentative support to this view.

10. Corn in the US, on the other hand, was already principally used for feed. Hence, support for US corn farmers in the 1930s came through a corn-hog programme, the two markets being intimately related.
11. I am indebted to Professor Stephen Biggs for pointing out that the reasons why CIMMYT switched its focus in this way would be worth exploring. Without having done so in detail, a number of possible explanatons offer themselves: (i) CIMMYT was an international centre with explicit objectives linked to poverty alleviation axcross continents; (ii) within Mexico, CIMMYT might have just duplicated the work of the IIA; and (iii) the growth of private sector interest in seed supply globally, being concentrated on hybrids, could be left to those organisations. CIMMYT could concentrate on improvement of populations. Certainly, the maize seed market in developing countries tends to be relatively clearly split between hybrid varieties (private sector) and the production of synthetics/open-pollinated varieties (public sector).
12. Whether a farmer purchases hybrid seed depends on the expected increase in production over and above the extra cost of seed. At low yields, the percentage yield increase over and above that of local varieties has to be higher to justify the investment (CIMMYT 1986, 14–19). Thus as CIMMYT (1986, 16) itself points out, 'For those lower yielding and more risky areas in Mexico, it appears that improved [open-pollinated] varieties and lower cost hybrids are more appropriate'.
13. The phenomenon of apomixis, which enables plants which usually reproduce sexually to reproduce asexually, would make it possible to breed hybrids which breed true. CIMMYT is believed to be close to doing this (Jefferson 1994; also, CIAT 1994). In purely economic terms, this brings hybrids closer to the poor, but only insofar as they are agronomically suitable for cultivators lacking access to irrigation. The danger would be that this would introduce another element of uniformity and vulnerability (the apomixis gene) into the agricultural system.
14. Compare this view of Mexican nutrition with that of the Survey Commission, based on meat and sugar consumption. West's account of Sauer's work shows that he often went out of his way to eat in the kitchens of poorer people as much for nutritional reasons as for those related to scholarship (West 1979, 11–14).
15. Thurston (1992, 155) points out that reports of maize/bean intercropping in Ecuador date from 1573.
16. It is thought this system originated with the great Mayan civilisations of Tikal and Palenque (Adams *et al.*, 1981). Other systems meriting attention include those of the Lacandon Maya (Nations 1987), and the Huastec Indians' (Alcorn 1984).
17. Altieri and Trujillo (1987, 198) note that planting designs also vary with the variety of bean used.
18. For a general discussion, see Vandermeer and Schultz (1990).

7
Biotechniques and the Neglect of Alternative Agriculture

Introduction

This chapter moves us from the past to the present, with an eye to the future. It makes the claim that, despite the emergence of alternatives discussed in Chapter 1, the trend towards genetic uniformity is likely to be perpetuated owing to the emphasis now placed upon new biotechniques.[1] Biotechniques provide more powerful tools to achieve genetically uniform ideal plants, but they could also be deployed as part of strategies aimed at promoting deployment of *in situ* diversity. The argument presented here is that biotechniques are closing the door which alternative movements managed to prise open, if only briefly, in the 1980s. This could be made in various ways, but here, the argument is developed through analysis of the intellectual property rights (IPR) issue, since this has radically affected the issue of genetic resource control in recent years.

Support for biotechniques takes place against the backdrop of a global economy in which concern to maintain 'technological leadership' and 'competitiveness' has prevented serious consideration of important questions concerning technological options, and the need to consider the matching of ends with technological means. The more deep-seated this concern, the more likely it becomes that institutional change will underpin it, further legitimising the same concern. This is the domain of Arthur's self-reinforcing mechanisms. For this reason I suggest that we are becoming increasingly locked in to genetic uniformity only moments after it seemed that the door was possibly opening for (again, possibly) genetically more diverse alternatives.

Genetic resource control in the 1970s and 1980s

The rescue of US corn – the benefits of free exchange

When southern corn leaf blight hit the US maize crop in 1970 (see Chapter 1), the solution was found not through re-introducing genetic diversity, but in genes from Mayorbala maize found in a West African field (RAFI 1994a).[2] West African germplasm was used at virtually no cost to save US farmers from a disease that had cost them $1 billion in 1970. That US farmers should have been assisted so freely owed itself to the system of free exchange of germplasm operating at the time, based on the principle that biodiversity was 'the common heritage' of humankind.

The UN's Food and Agriculture Organisation (FAO) had discussed genetic resources at its founding conference in Quebec in 1946, and with the International Biological Programme (IBP) hosted a conference on plant genetics in 1961. Little was being done to address the issue of genetic erosion. Most collecting of germplasm was done by academics in universities, and virtually all was done at a sub-national level, an exception being that conducted on maize (see Chapter 6). The issue of crop germplasm conservation was highlighted by the work of Erna Bennett, who in 1967 organised the second IBP/FAO Conference on genetic resources and subsequently set up the Crop Ecology Unit at the FAO in Rome (Fowler and Mooney 1990, 149–50).

Only a year after southern leaf blight destroyed much of the US crop, a cold winter followed by a dry spring drastically affected the wheat crop of the Soviet Union. Much of the land was planted to the Besostaja variety which was neither cold tolerant nor drought resistant. Wheat prices shot up, and when famine hit the Sahel region in Africa, Malthusianism became fashionable once more (see for example Paddock and Paddock 1967; Ehrlich and Ehrlich 1970; and Meadows *et al.*, 1974). Against this backdrop, Sir Otto Frankel, Bennett's co-editor for a path-breaking book on the subject (Frankel and Bennett 1970), persuaded the 1972 Stockholm Conference on the Human Environment to adopt a resolution calling for concerted action on genetic resources. The need to conserve genetic resources was at last being taken seriously as the advancing BCM mode hastened their disappearance.

Hambridge and Bressman (1936, 131) recognised long ago the tensions between free exchange in germplasm, and unfree exchange of the product derived from it:

> From its rivals a nation may get the wheat germ plasm that enables it to supply its own needs or overwhelm those rivals in international

> trade ... Will nations have the wisdom to deal with this situation, or will it lead to more bitter rivalries and more deadly conflicts, as the beneficent science of chemistry has enormously increased the deadliness of war? In his use of modern science, man has proved again and again that he is a bright child playing with fire.

The Wardian case, invented in 1829, facilitated an exodus of germplasm to one of a growing number of botanical gardens. The 'botanical chess game' (Mooney 1983, 85; Brockway 1988) has played an important role in shaping the international division of labour, a fact recognised by Marx:

> You perhaps believe, gentlemen, that the production of coffee and sugar is the natural destiny of the West Indies. Two centuries ago, nature, which does not trouble herself about commerce, had planted neither sugarcane nor coffee trees there. (Marx u.d., 207)[3]

Metropolitan powers appreciated that control over commodity trade depended on restrictions on the movement of germplasm. History is therefore replete with examples of the heroic efforts of plant explorers in overcoming embargoes on the movement of seeds, the breaking of which was in many cases punishable by death (Juma 1989, Chapter 2; Raeburn 1995, 64–70).

Only in the post-colonial world did 'free exchange' reach its truly international apogée. Some interpret this as allowing gene-poor northern countries in the developed world to maintain access to germplasm residing in the gene-rich, financially poor, countries of the developing south (Kloppenburg and Kleinman 1988).[4] Yet if, as Galeano (1973, 77) writes, the international division of labour was organised 'not by the Holy Ghost but by men', so it was with the system of germplasm exchange. As such, it could be changed by men.

Intellectual property rights and the seed industry[5]

Unless they were crossed inbreds (see Chapter 5), farmers could save seed for planting the following year without appreciable yield loss. The first claims for plant patents were made in 1885 (Kloppenburg 1988, 132). In 1922, lawyers met in London to discuss patent protection for plant varieties but no action followed (Fowler *et al.*, 1988, 252). It was the nursery industry which was primarily responsible for the passage of the Plant Patent Act of 1930 in the United States. For nurserymen, the obstacle to proprietary ownership of varieties lay in competition from other nurserymen, not farmers, since trees could easily be propagated in competing nurseries.

The Plant Patent Act of 1930 (see Fowler 1994, 74–90) made it possible for asexually reproduced plants to be patented, with the exceptions of potatoes and Jerusalem artichokes. The rhetoric used in support of the act, that breeding had made such significant advances over the past decades, was actually completely irrelevant as far as asexually reproduced plants were concerned (Fowler 1994, 74). Most of the work done by nurserymen lay in multiplying varieties that had been discovered by chance, and that were the product of insect or wind pollination, raising issues as to whether they should have been eligible for patents. Fowler concludes that: 'The PPA did not recognise the individual inventor or the creative act as much as it recognised and rewarded the system that produced the new variety, whether by luck or by design' (Fowler 1994, 89).

In France, ever since the turn of the century, rose breeders had been seeking the same recognition as inventors of machines. Early attempts were rejected by lawyers on grounds that even full disclosure would not make it possible for breeders to reproduce a variety. By 1928, however, there existed in the Ministry of Agriculture *de facto* protection of breeders' rights through an 'identity and purity service' (Berlan and Lewontin 1986, 785). The Italian High Court declared plant varieties patentable in 1948, but confusion led to calls for a special plant patent law. By 1957, with the view that plants and animals should not be patented in the ascendancy, the International Association of Plant Breeders for the Protection of Plant Varieties (ASSINSEL) accepted an invitation to host a conference in Paris on plant breeders' rights (PBR), leading to the establishment of the Union for the Protection of New Varieties of Plants (UPOV) in 1961 (Fowler *et al.*, 1988, 239–40), whose International Convention was revised in 1972, 1978 and 1991.

For the most part, PBR legislation has been true to the UPOV Convention, requiring that plants pass the DUS test (see Chapter 4). The 1960s and 1970s saw several countries either joining UPOV or implementing a system of PBR of their own.[6] Key to the passage of these acts was the belief that sexually reproduced plants could breed true, which, for sexually reproducing plants, is only the case for pure-line varieties (those having undergone four to nine generations of selfing). Also critical was the definition (thought by many to be impossible) of the term 'variety'. The UPOV resorted to 'nothing other than a description of the steps of the method of breeding' (Berlan and Lewontin 1986, 787), or more accurately, pure-line (Mendelian) breeding. Hence, the extension of IPR to plants through PBR was an institutional innovation shaped by, and made possible by, changes in breeding techniques and technology respectively. However, these were not institutional changes waiting to be implemented as soon as these techniques emerged. Just as there were technical options open to breeders, so the institutional changes made represented a choice from myriad possibilities.

PBR facilitated an increasingly international outlook on the part of the seed industry. Modern varieties were spreading across the globe, and notwithstanding some efforts to improve disease resistance in new varieties, new seeds made increased use of other inputs more likely. In the 1980s, policy-related, or structural adjustment lending undertaken by the World Bank advised privatisation of input supply industries, and an expanded role for private sector seed research and distribution, especially in the development of hybrids. The emphasis began to shift, as it had done in the developed world, away from the public system in favour of reduced public sector involvement.

As late as the 1960s, there were few multinational companies in the seed industry. A wave of acquisitions occurred in the 1970s as seed companies were bought up by transnational corporations, mainly food trading and petrochemical companies.[7] Food traders, seeking to open up new export markets in the era of US 'food power', sought to extend their activities upstream. The development of high-input seeds by international agricultural research centres had also led (agro-) chemical companies to seek new markets in the developing world, so these companies sought to market seeds through the same channels. With PBR legislation in place in many developed countries, seeds were no longer a weak link in the input supply industry. UPOV, by creating a degree of harmonisation in PBR legislation, fostered the emergence of a global seed industry, whilst the horizontal integration of agricultural input supply has deepened the inter-relatedness of inputs over time.

The challenge to the BCM paradigm

The BCM mode has come under, and continues to operate in the face of, considerable pressures for change. These are due to its:

1. ecological impact;
2. impact on food quality and health (in farming, and in consumption);
3. impact on rural communities; and
4. being supported by state policies, and high levels of farm support, which effectively exacerbate the problems mentioned above.

It is beyond the scope of this book to address each of these in detail. Suffice to say that, in the words of Almås and Nygård (1994, 168), the BCM mode has produced 'some of its own executioners'.

Aims to reform agricultural and farm support policies seek, increasingly, to re-direct support towards environmentally sound practices. These are generally believed to imply reduced use of agrochemicals (Harvey 1991), and also seeds.[8] To the extent that it continues to be allowed (see below), seed-saving becomes more economically attractive in times of low output prices (van Wijk 1993, 14).[9] In the spirit of challenges to the BCM mode, it is to the issue of seeds and germplasm that we now turn.

Sowing the seeds of discontent

As the Rockefeller and Ford Foundations expanded their efforts in international agricultural research, they began to seek public funds to support their work. Following a meeting in Belaggio in April 1969 organised by the Rockefeller Foundation, 15 governments attended the first meeting of the Consultative Group for International Agricultural Research (CGIAR) in January 1971 (Ravnborg 1992, 2). The World Bank would provide a secretariat and administer finances. International agricultural research, the year after Norman Borlaug received a Nobel Prize for his work on dwarf wheats in Mexico, had come of age.

In 1972, the CGIAR's Technical Assistance Committee (TAC) convened a meeting in Beltsville, USA, to formulate an international strategy for genetic resources conservation. After much debate, the International Board for Plant Genetic Resources (IBPGR) was set up as part of the CGIAR network. Under the IBPGR's system, the majority of the world's genebank accessions, mostly from developing countries given to believe that genetic resources were the common heritage of humankind, were stored in genebanks in developed nations increasingly predisposed to the notion of IPR over germplasm. In the late 1970s, developing countries were increasingly concerned by IBPGR statistics showing that of more than 1.9 million samples stored, 55 per cent were in developed countries and another 14 per cent were held in the Northern (donor) dominated CGIAR system (GRAIN 1993a, 6). Collections were clearly biased towards crops of interest to the developed countries, the top cereal crops representing more than 75 per cent of accessions in the pre-1980 period. At a 1981 IBPGR/FAO/UNEP conference, Latin American countries pressurised the IBPGR, successfully, to increase collections of crops less prominent in international trade (Fowler and Mooney 1990, 157–60).

By the end of 1981, despite opposition from US and UK representatives, a Resolution tabled by the Mexican delegation had been passed at the FAO calling for the FAO Director General to draft elements of an international convention on plant genetic resources, and investigate the feasibility of establishing a new international gene bank. Two years later, at the FAO's biennial conference, Jose Ramon Lopez Portillo, son of the former Mexican President, forced another vote which led to an International Undertaking on Plant Genetic Resources (IUPGR) and the creation of an International Commission on Plant Genetic Resources (ICPGR). The aim of these moves was transfer of control over genetic resources from the developed countries, and IBPGR in particular, to the United Nations.

Under the IUPGR, the notion of free exchange was to be respected, and it was not just landraces that were to be 'available without restriction', but also 'special genetic stocks (including elite and current breeders' lines and mutants)' (FAO 1983, 5). This angered the American Seed Trade Association (ASTA), who were represented in the US FAO delegation. They charged that the IUPGR struck:

> at the heart of free enterprise and intellectual property rights ... The definition includes unimproved and obsolete varieties, land races, wild and weedy species, all of which the seed industry believes appropriate to be preserved and freely exchanged. However, it also includes improved elite varieties and breeding lines within the definition of plant genetic resources ... This puts the Undertaking in direct conflict with the rights of holders of private property ... The anti-private business bias of the Undertaking is clear. (ASTA 1984, 1–2, in Fowler 1994, 190)[10]

The IUPGR also proposed establishing a network of base collections under the jurisdiction of the FAO (FAO 1983, 7). Yet the Undertaking was a mild and voluntary agreement rather than a legally binding convention.

The emergence of agricultural biotechnology and patents on life

The birth of biotechnology[11]

In the second half of the twentieth century, enormous strides have been made in the life sciences, particularly in the discipline which has come to be known as molecular biology.[12] As a result of this growing body of knowledge, new commercial opportunities appeared on the horizon based on the use of tools developed through new discoveries in this field. The idea that plants could be made resistant to pesticides was no new idea. Wiebe and Hayes (1960) discussed it decades ago with regard to the reaction of barley varieties to the application of DDT. Yet work in biotechnology brought such a strategy closer to hand, raising the possibility of breeding plants designed to tolerate applications of proprietary chemicals.

The use of plasmids, in 1973 by Cohen and Boyer, to mediate gene transfer made possible a new alchemy (Kloppenburg 1988, 195; Kenney 1986, 23). In the immediate aftermath, biotechnologists in the US showed awareness of public unease regarding this new technology by proposing a moratorium on certain types of research. Since 1977, guidelines laid down by the National Institutes of Health (NIH) have been progressively relaxed.[13] The desire to regulate the industry has dwindled as authorities were persuaded of the commercial significance of the new technologies (Krimsky 1982; Yoxen 1983, 63; Kenney 1986, 27). Reduced regulation of the biotechnology industry began to be perceived, and not just in the United States, as a means through which a country could maintain or improve its position in emerging bio-industries. Field (1988, 20) reported that 'industrial competitiveness appears to represent the central and overriding concern of national strategies'. From a different perspective, the United Nations Centre on Transnational Corporations opined that comparative advantage and the international division of labour were increasingly being shaped by technological prowess (UNCTC 1988, 1–2). Increasingly, regulation shied away from determining which technologies should be allowed for use, and the imperative of allowing new technologies to develop began to shape which regulations were considered acceptable.

The emergence of biotechniques for technology generation makes it possible to circumvent the constraints imposed upon genetic recombination by species incompatability. I argue below that formal agricultural research is undergoing a transition from the BCM mode to a biotechniques-mechanisation-legislative (BML) mode. This is not to imply that traditional plant breeding and chemical inputs are about to disappear from view, either now or in the near future. The role of breeders, where they do not disappear altogether, is likely to undergo a change such that their work complements that of the biotechnologists,[14] whilst the fact that to date, herbicide tolerance in crops is the most widely tested trait to date testifies to the likelihood of continued use of chemicals into the future.[15] Nevertheless, the genetic determinants of interactions between the plant and various chemical and biological inputs are likely to become the focus of innovation in crop (and livestock) agriculture. The pivotal institutional innovation in enabling such a strategy to become privately profitable is IPR legislation.

Patenting genetic materials

In 1976, the first of the new biotechnology companies, Genentech, was formed by Herbert Boyer and venture capitalist Robert Swanson. In 1980, Genentech placed a share offering on the New York Stock Exchange, the prices of which shot up from $35 to $89 per share in twenty minutes, a record rate of increase (Kloppenburg 1988, 195–6). This was due to the fact that three months earlier, General Electric had successfully challenged an earlier decision by the US Patent and Trademark Office (PTO) which had ruled that an oil-degrading micro-organism developed by their scientist, Ananda Chakrabarty, was not patentable subject matter. The new ruling held that whether or not an invention was alive or dead was irrelevant to patent law (Fowler 1994, 149).[16]

In the PTO's ruling on Chakrabarty, the legal principle of 'pre-emption' disqualified materials protectable under the PPA or the PVPA from patent protection. But this ruling was also overturned in the 1985 *Ex parte Hibberd* case, in which Hibberd was granted patents on the tissue culture, seed, and whole plant of a corn line selected from tissue culture. Breeders could now choose the form of protection most suitable to them, including utility patents. The gene was being commodified.

TRIPS of the General Agreement on Tariffs and Trade (GATT)

Anxious to preserve its lead in the biotechnological race, the United States has moved fastest in bringing institutions into line with industry's desires. Employing both bilateral and multilateral channels, it has sought to harmonise standards across nations in line with its own structures, thus opening the way for global marketing of proprietary products of biotechniques. In November of 1982, at a ministerial level meeting held at the GATT at US insistence, the US proposed that the new round debate issues never before considered in earlier GATT rounds (Raghavan 1990, 70–73). One such issue was trade in counterfeit goods, such as 'fake' Rolex watches, but the scope of this particular area was widened at the behest of the US and others to include the issue of IPR. Raghavan (1990, 72) notes that this was 'thanks mainly to the negligence of the disorganised Third World countries, most of whom thought that it did not affect them'. This would not have been so critical had it not been for the fact that what many countries saw as a preparatory discussion was subsequently proposed by the US and others as the agenda for a new round. The inclusion of many new issues was given justification through addition of the prefix 'trade-related'.

Many countries hoped that by stifling their objections to the inclusion of new issues such as 'services' and IPR, they would be rewarded with concessions on 'old' issues, such as tariffs on tropical products (and escalating tariffs on processed products thereof), textiles, and a continuation of benefits under the Generalised System of Prefences (GSP). A compromise text was agreed at Punta del Este in Uruguay at the end of 1986 which included IPR. Even then, developing countries refused to negotiate on the subject before the mid-term review in Montreal in December 1988, where agreement was reached concerning the negotiating agenda (Siebeck 1990, 1; WTO 1996, 23).

The GATT agreement was finally signed at Marrakesh in April 1994 despite the fact that market access negotiations had not been verified, and with many developing country negotiators complaining that they had seen the texts only weeks before (ICDA 1994). Ratification in many countries was rushed through with little debate (ICDA 1995), and on 1 January 1995, the new World Trade Organisation (WTO) came into existence. This would co-exist with the GATT until the end of 1995. The final text of the TRIPs (Trade Related Intellectual Property Rights)[17] agreement establishes new multilateral rules on IPR based on uniform minimum standards for their protection and enforcement, including their availability, use and scope.

As regards plant materials, the treaty does allow for exemptions on grounds of perceived environmental or public order impacts, yet at the same time, the treaty states that plant varieties shall be protected by patents 'or by an effective *sui generis* system or by any combination thereof' (Article 27.3 (b)). Although developing countries and least developed countries are allowed, respectively, five and ten years to implement the agreement, the '*sui generis*' clause is to be reviewed four years after the date of entry into force of the WTO agreement (Article 27.3b). The TRIPs agreement offers little encouragement to communities that might seek to protect innovations which are the property of, as it were, the collective. The agreement recognises IPR as private rights, and also requires products to be 'capable of industrial application'. There is no mention of communities and their rights (WTO 1996, 23).

Before the GATT negotiations even began, the US had made its intentions in respect of IPR abundantly clear through applying pressure bilaterally. Mexico was targeted as early as the 1970s, but little progress was made until, in the mid-1980s, the US began to link the issue of IPR reform to expansion of GSP concessions. By the end of 1986, Mexico had adopted a revised Patent and Trademark Law, though as a result of the efforts of domestic lobbying, this was deemed inadequate by the US administration (van Wijk 1991). In 1984, the US Trade and Tariff Act had been revised, S.301 of Title III of which invested the prevailing Administration with coercive powers aimed at righting 'unfair' trading practices.[18] In 1985, cases against Brazil and South Korea were initiated, the former concerning, *inter alia*, copyright issues regarding software, the latter concerning failure to protect intellectual property (Raghavan 1990, 84). The same issue led to talks with China (van Wijk 1991), whilst India has also come under pressure to reform its IPR legislation in the past (Economic and Political Weekly 1988a,b).

In January 1987, Mexico was informed of President Reagan's intention to withdraw $200 million of GSP benefits unless the perceived inadequacies of its new legislation were corrected. In 1988, the Omnibus Trade and Competitiveness Act was passed in the US. This included sections which came to be known as Super 301 and Special 301 respectively, retrospectively strengthening the coercive powers vested in the administration by the Trade and Tariff Act's 1974 revision. Under Special 301, a procedure was set up whereby the US Trade Representative could identify and initiate proceedings against countries considered to be offering inadequate IPR protection. Within this list of countries, a Priority Watch List of countries was to be specified annually, and Mexico was on that first list in May 1989.[19] When President Salinas de Gortari began his programme of liberalisation, and plans for a North American Free Trade Agreement were materialising, Mexico's stance on the IPR issue altered quite radically, and in 1990, when Mexico introduced a proposal for a TRIPs agreement at the GATT, it slipped off the 301 lists. Mexico's patent law was revised in June 1991 to explicitly allow for the patenting of plant varieties. It specifically addresses innovations likely to arise from the deployment of biotechnology.

As Fowler (1994) has made clear, GATT and the growing concerns of the US over intellectual property issues generally, enabled transnational corporations involved in agricultural biotechnology to have their concerns *vis-à-vis* the patenting of life addressed in new fora. In the case of the GATT, it became possible for the issue to be bundled up not only with concerns over patents and trademarks as they related to mechanical innovations, but also, since this was a take-all-of-it-or-leave-all-of-it package, with fourteen other areas with which the Uruguay Round was concerned. Significantly, Watkins (1992, 91–2, my emphasis) notes,

'The major actors in [the TRIPs] exercise have been the US-based Intellectual Property Coalition – a grouping of 13 major companies, including IBM, DuPont and General Motors – *and European agro-chemical giants such as Unilever, Hoechst and Ciba Geigy.*' [20] Thus, IPR was being simultaneously harmonised and extended across the globe.

UPOV

Paralleling the moves to enhance intellectual property protection under the auspices of the GATT were moves on the part of UPOV to bring the Convention into line with developments elsewhere, and particularly with respect to biotechnology. PBRs' research exemption made them inadequate for protecting biotechnically engineered plants since they offered protection at the level of the whole plant when what was required was protection at the level of the gene (Lesser 1990, 67; Van Wijk *et al.,* 1993, 7). But by 1987, it was clear that UPOV would be strengthened. According to Fowler *et al.,* (1988, 241), UPOV's members had been divided between small seed houses and the integrated genetics supply industry, the former fearing gene patenting, the latter favouring new initiatives in this respect. UPOV was revised in March 1991.

There were some critical changes made to the Convention which are outlined in Table 7.1. Note that the right of farmers to save seed from one harvest for planting in the next, what the American Seed Trade Association had referred to as the 'farmers' right' in hearings on the PVPA, had become known as the 'farmers' privilege' and was no longer secure (RAFI 1993b; Fowler 1994, 153).[21] Section 15.2 of the new Convention allows, as an optional exception, seed saving 'subject to the safeguarding of the legitimate interests of the breeder', implying that royalties should be paid to breeders where seed is saved (UPOV 1991, 14). On the other hand, there is a compulsory exemption for breeding other varieties (Article 15 (1) (iii)). However, the interests of the breeder are, in general, strengthened since protection applies to 'Essentially derived and certain other varieties' as defined under Article 14 (5) (b) and (c). Lesser (1991, 129) expresses concern that, since the definition is unclear, this will lead to quasi-IPR being granted to a breeder over thousands of attributes of a variety which he/she did nothing to create. The 'essentially derived' clause would appear to apply to genetic insertion, giving the owner of PBR the right to demand royalties from innovations based on insertion of one or two genes into a plant over which the right is held.

Increasingly, developed countries are bringing their PBR legislation into line with the 1991 UPOV Convention.[22] Furthermore, although

Table 7.1 Comparison of main provisions of PBR under the UPOV Convention and patent law

Provisions	*UPOV 1978*	*UPOV 1991*	*Patent law*
Protection coverage	Plant varieties of nationally defined species	Plant varieties of all genera and species	Inventions
Requirements	Distinctness; uniformity; stability	Novelty; distinctness; uniformity; stability	Novelty; inventiveness; nonobviousness; industrial application; and usefulness
Protection term	Min. 15 years	Min. 20 years	17–20 years (OECD)
Protection scope	Commercial use of reproductive material of the variety	Commercial use of all material of the variety	Commercial use of protected matter
Research exemption	Yes	Not for essentially derived varieties	No
Farmers' privilege	In practice: yes	Up to national laws	No
Other	Species eligible for PBR cannot be patented		

Source: Adapted from van Wijk *et al.* (1993, 8).

the TRIPs '*sui generis*' clause appears to offer room for manoeuvre in designing IPR for plant varieties, most believe that this translates into UPOV type standards and nothing less. Since accession to UPOV 1978 was closed at the end of 1995, this now implies the standards of UPOV 1991.[23] Increasingly today, companies can choose which combination of protection they prefer, although European patent law currently forbids patenting of plant varieties.

Biotechniques: locking in to uniformity

Introduction

Reflecting the above-mentioned events, the Crucible Group (1994, 2) reported: 'Those who reviewed patent law a few decades ago may not recognize it today'. Probably no formal agricultural research organisation in the world has not at least cogitated upon the changes considered above, not all of whose are implications are, as yet, clear.[24] It is important to understand at least some of these, and to contemplate the relevance of the changes to the existing BCM mode of agriculture.

Biotechniques and environmental critiques of agriculture are reported to be bringing about a reshaping of the technological development of agriculture. This is jumping the gun slightly. There are still unresolved questions concerning, in particular, consumer acceptance of genetically engineered products, which are already having an impact on the industry's development. However, Sharp (1990, 109) may be right to to talk of the laying of a new set of 'ground rules', making it 'inconceivable that those developing new drugs, new herbicides or pesticides, or new plant species, should not, somewhere *en route*, make use of gene cloning and sequencing techniques.' The implications would be that research which did not require such techniques might fall by the wayside.

Biotechniques – old wine in new bottles or paradigmatic shift?

A number of authors have commented on the paradigm-like shift that biotechnologies could achieve (Orsenigo 1989, Otero 1995, Roobeek 1995, Freeman 1995). Much of this discussion considers the issue at the macroeconomic level, and takes the view that it will not be biotechnology alone that leads to a new mode of accumulation, but biotechnology, the development of new materials, and information technologies working synergistically to form a new techno-economic paradigm along lines discussed in Chapter 3.

There is no doubting that there could well be some revolutionary changes about to occur in the way in which the agro-food system functions. Most interesting of all are potential developments in the food processing industry, where some authors have speculated as to the possible emergence of a 'generic biomass inputs sector' as a result of technologies which allow biological materials to be fractionated into component parts for the final manufacture of food products (Goodman and Wilkinson 1990, 135, 134–9; also, Goodman *et al.*, 1987, 123–44). The implications for commodity markets as they are currently understood could be far-reaching (see Ruivenkamp 1992; Hobbelink 1991, ch. 6). Other potentially revolutionary techniques relate to so-called novel products, which will affect the ways in which agriculture interacts with other sectors of the economy.

Yet, whilst certain techniques used to create new products are certainly emerging, there appears to be substantial continuity with the past with respect to:

- Increasing horizontal integration across agricultural inputs – breeding for responsiveness to inputs and to facilitate harvesting will give way to herbicide tolerant varieties.
- Deepening of vertical integration – breeding has facilitated mechanised harvesting and handling of the final product. Biotechniques are increasingly geared towards downstream aspects of food production, representing higher value-added, and greater opportunities for profit, in upstream sectors of the chain of value in food (Lacy and Busch 1991, 160). Lamola (1995; also RAFI 1997a) speaks of end-use tailored, or identity-preserved varieties.
- The actors involved are, in many cases, one and the same as those who prospered through the BCM paradigm (erstwhile agrochemical and seed companies).[25]
- Emerging products take their cue from their supposed ability either to replace, or alter the functioning of, elements of the BCM paradigm which have been heavily criticised in the past (Buttel 1995, 32–5).
- A continuing lack of emphasis, in private sector breeding, on pest resistance – although biotechniques provide tools for reducing pesticide use, current trends seem likely to increase, rather than reduce their use. Where resistance breeding is undertaken, it is of the gene-for-gene, vertical resistance type.

In many respects, therefore, the goals remain rather similar to those in the BCM paradigm. In particular, the attractions of the new techniques are seen principally in terms of the increased control that can be exerted over the transformation of organisms through recombinant DNA techniques. Indeed, Richards (1994) speaks of biotechniques as heralding a 'second designer phase' for agriculture. Whereas the Green Revolution focused on ideotypes for monocropping in controlled *physical* environments, the second phase seeks to shape 'econotypes' to meet the need of future *economic* environments. As with the BCM mode, an emphasis on control within the laboratory[26] has tended to obscure and marginalise the significance ecological issues concerning the functioning of biotechnically engineered products in the field.

The most obvious break with the past is the ability of biotechniques to extend the genepool available to breeders and biotechnicians beyond the primary and secondary genepoools into the hitherto unexplored (because of species incompatibility) tertiary genepool (Plucknett 1994, 357). Less immediately obvious, are the changes which have already been wrought by biotechniques on our perception of the nature of life itself, and the significance of these for our perception of the nature of food and agriculture in the longer-term.

Paradoxically, therefore, we are witnessing changes which are simultaneously profound, and incremental. The mode of agricultural research is changing through use of powerful new techniques, but its roots remain in the BCM mode. This is to be expected if one accepts that agriculture had become locked in to the BCM way of doing things. As Teece (1988, 264) points out, one aspect of the locking in process is that firms tend to do best what they have done in the past. If the emergence of biotechnology constituted a radically new paradigm, learning advantages accumulated over time by established FAROs would have lost much of their significance. However, the fact that biotechnology is very much a process technology has meant that much of the significance of learning, particularly in the downstream operations of private multinational corporations, has been retained. Furthermore, as I have suggested above, the techniques are deployed in pursuit of a familiar goal, that of the genetically uniform ideal plant.

Changing modes

The transition that is occurring can be understood through appealing to the framework developed by Freeman and Perez discussed in Chapter 3. As noted above, the BCM mode has come under fire for a variety of reasons, principally those associated with food quality and the environment. If the limits to the expansion of this mode had not yet been reached, such expansion was clearly under threat. The world market for agrochemicals saw three years of decline in the years 1991–93 before recovering somewhat in 1994 and 1995 (British Agrochemicals Association 1996, 24). Farm support schemes, in the European Union and elsewhere, have begun to shift away from price support, which led to elevated levels of use (relative to that which would prevail with prices at world market levels) of agrochemical inputs, and towards conservation, often rewarding farmers for using fewer inputs.

For Freeman and Perez (1988, 59), the transition from one techno-economic paradigm to another, brought on by the onset of recessionary trends, is characterised by:

> the increasing degree of mismatch between the techno-economic sub-system and the old socio-institutional framework. It shows the need for a full-scale reaccommodation of social behaviour and

institutions to suit the requirements and the potential of a shift which has already taken place to a considerable extent in some areas of the techno-economic sphere. This reaccommodation occurs as a result of a process of political search, experimentation and adaptation.

Once the socio-institutional framework matches the techno-economic sub-system, investment moves forward and growth is restored. For the BML mode to flourish, its techno-economic sub-system requires an appropriately matching institutional framework, including:

- a sympathetic regulatory framework for undertaking relevant research, including risk assessment methodologies as applied to the release of the products of gene technologies, and food safety legislation regarding genetically engineered food;
- political, social and environmental acceptance of the technologies and their end-products, reflecting confidence in the regulatory framework; and
- appropriate IPR protection, the importance of which is confirmed by, amongst others, Thelwall and Clucas (1992, 42), Caulder (1991, 73), Lamola (1995, 88), and Duffey (1987, 30).

A lack of institutional change will delay any upswing, and indeed, resistance from consumers concerning issues of health and environmental risk has been strong. Consequently, products have been slow to reach the market. Yet, for reasons elaborated below, it is the issue of IPR legislation which has greatest bearing on the issue of biological diversity in use in agriculture, and thus, the environmental risks posed by new biotechnologies in terms of vulnerability and the continued use of pesticides.

Those in the vanguard of the BML mode have sought to project it as environmentally friendly. In doing this, they have stressed the biological, *ergo* natural, characteristics of the work they are undertaking. The semantics involved have been illuminating, at one and the same time suggesting radical new possibilities (the economic and environmental attractions) and on the other, in an attempt to downplay the risks associated with the products of biotechniques (and the need for regulation), suggesting continuity with the past. Critics of biotechniques turn the matter around completely. Whilst not disputing the fact that there is money to be made, they argue that the new possibilities should be reflected in the need for new forms of regulation, whilst continuity is likely to be reflected in the continuation of environmental problems. Their criticisms relate mainly to:

- The uncertainties in *ex ante* risk assessment associated with release of genetically engineered organisms into the environment, not least the difficulties in extrapolating from small-scale trials to large-scale field use, and problems associated with trade in the products concerned (Mellon 1991a; Nilsson 1992; Munson 1993; Fogel and Meister 1994; de Kathen 1996).
- The nature of individual products and their possible environmental and health consequences (Doyle 1995; Hindmarsh 1991, 1992; Rissler and Mellon 1993; Green Alliance 1994; Fogel and Meister 1994; GRAIN 1994a)
- The environmental consequences of possibly increased uniformity (see below) (Kloppenburg 1988; Fowler and Mooney 1990; Hobbelink 1991; The Crucible Group 1994).
- The political economy of the research being undertaken (who is it done by, and for?) (Kloppenburg 1988; Clunies-Ross and Hildyard 1992, ch. 7; Kenney 1986, 230–7; GRAIN 1994b).
- The impact in terms of research not undertaken (a point well made by Rachel Carson in 'Silent Spring' (1965); see also McGuire 1996).

These issues are not unrelated. The nature of the organisation funding research will determine the degree to which a notional social welfare function is reflected in their activities. Private organisations need not be concerned with social welfare, or only insofar as it affects profits.

The key to deepening private sector involvement in agrofood biotechnical research has been the extension of IPRs to living organisms.[27] Governments have welcomed private sector participation in research, and have tended either to move the focus of their research away from near-market, and towards more basic, research, and/or to seek to take advantage of the patent system themselves to make financial gains from ongoing research.[28]

A future for diversity?

As mentioned above, one of the criticisms levelled at those who believe that biotechniques herald a new 'sustainable agriculture' is that existing problems of uniformity will be exacerbated. Can the transition from BCM to BML mode re-introduce diversity into a system based on uniformity? From a purely technical view, biotechniques' capacity to draw upon genetic material from the tertiary genepool would suggest that additional genetic variation might be introduced. Thus, Bassett (1995, 42) argues that new varieties 'will simply coexist with the old varieties: diversity will have been increased, not decreased'.

But this approach has two major shortcomings. Firstly, the basic research and the application are inextricably linked (Field 1988, 17; Balmer and Sharp 1993, 465), so much so that possible applications are driving the direction of basic research.[29] Thus, a growing proportion of public sector research is supporting commercially oriented research. Secondly, what actually happens is a subset of what could happen, as the earlier case-studies have argued. Kloppenburg (1993) bluntly states: 'the baby of biotechnology is not so easily separated from the corporate bath water,' which is exactly why, as Lesser (1990, 63) points out, trends towards uniformity predate the existence of PBR and patents.

Strengthening IPR as applied to living material has encouraged private industry to engage in biotechnical research at levels above those that would have existed in their absence (Scott-Ram 1995). Furthermore, because Governments now see biotechnical prowess as important for maintaining competitiveness, public research is beginning to resemble privately sponsored research, either through its increasingly subservient position to private industry (only highest priority 'basic' research receives funding – McGuire 1996), or through more overt aims to generate revenue from patentable research outputs (Duvick 1996, 16). Erstwhile President of Harvard University, Derek Bok, has expressed concern that Universities will increasingly 'differ from corporations only because there are no shareholders and no dividends' (in Lacy and Busch 1991, 159).

An element of historical contingency is at work here, since many governments are cutting state spending on education and research where it is perceived to be of low market value (see Anon 1993). To the extent that environmental issues are intimately related to issues of social welfare, and because many environmental costs are not captured in market transactions, one assumes that private organisations are, notwithstanding their own public relations, less likely to integrate environmental issues into their research programmes. Indeed, one survey, aimed at eliciting the ranking of breeding companies' priorities, placed the environment at the bottom of seven criteria (Thelwall and Clucas 1992, 85). However, the same is increasingly true of publicly funded research. Institutional changes are making the public / private distinction irrelevant in predicting the social welfare goals which will be pursued by one or other form. This is one reason why, notwithstanding some of the claims made for the efficacy of biotechnologies (most of which, incidentally, are made with the BCM technologies as the implied baseline), IPR will if not increase, then maintain, the vulnerability of agriculture in the field. Other reasons include the following:

Distinctness of varieties

Already, anecdotal evidence suggests that seed companies rely on a few elite cultivars in their research programmes and new varieties are developed through minor modifications to these (Vellvé 1992, 42). Very little hard evidence is available, but it is clear that the number of varieties available (i) does not reflect genetic diversity, and (ii) masks the concentration in varietal use out in the field.

The future strategy for breeders will be structured by the IPR legislation in place. The combination of UPOV 1991 and patents ensures that the work of the breeder, and the value of an identified gene, are both recognised. The 'essentially derived' clause was introduced to deal with the problem of genetic distancing (Correa 1992; Thiele-Wittig 1992). Biotechniques make it possible to reduce the genetic distancing required to discriminate between varieties, in which case, the genetic variation existing in the field would become even further divorced from consideration of the number of varieties grown (see Plant Varieties and Seeds 1989). For example, since distinctness could now be measured at the level of the gene, a superfluous (in agronomic terms) gene could be spliced into a variety thus making it, potentially, distinct. As Smith (1992, 193) points out, the practice of reverse engineering of varieties is increasingly common and would make the aforementioned practice more likely, rendering the granting of PBR meaningless. The 'essentially derived' clause, though of theoretical value, raises important questions of definition. For both Espenhain (1992, 172) and Smith (1992, 195), who, in his excellent account, notes that numerous controversies in this regard are in no one's interest, case law will provide the answers.[30]

Because UPOV 1991 extends the breeder's right to the commercialisation of essentially derived varieties (the principle of dependence), companies using genetic transformation techniques will either work with their own PBR-protected varieties, or license genes of interest to other companies for incorporation into their varieties. Strategic alliances between those specialising in biotechniques and those with greater specialism in traditional breeding seem likely. Hence, Pioneer markets both soyabeans containing Monsanto's Roundup Ready gene (tolerant of Monsanto's glyphosate herbicide) and soyabeans containing the DuPont-owned sulfonylurea tolerant gene (Pioneer 1997). DEKALB also has cross-licensing agreements with both Monsanto and DuPont (DEKALB 1997). Those licensing technologies will seek to gain from the technology premium which biotechnically developed products aim to attract (Sehgal 1996, 19).

To a significant degree, varietal make-up will remain as before, but with genes spliced into a particular variety's background, and with plants themselves being made more uniform. However, since industrial structure will be affected by the evolving IPR framework, as well as the techniques themselves, and developments in individual sectors, there may be implications for the diversity of what is offered to farmers. [31] This is considered in what follows.

Research concentration

Patent enforcement prevents companies from carrying out research on a patented genetic sequence or process without paying royalties to the patent holder, whilst essentially derived varieties are subject to PBR under UPOV 1991.[32] The strategic importance of patents for any company depends to a considerable degree on the extent of exclusion implied by the text of the patent. In this context, the current trend towards granting broad patents to companies is worrying indeed.[33] For example, Agracetus, a subsidiary of W. R. Grace (and recently taken over by Monsanto), was granted patent rights in the US over any genetically manipulated variety of cotton, and by the European Patent Office over all genetically transformed soybeans (with those for rice, groundnut and maize pending) (GRAIN 1993b; Bijman 1994; RAFI 1994a, 7).[34] Agrigenetics' patent on high oleic acid sunflowers effectively stopped all such work in this area outside the company. The prevalence of 'driftnet patenting' is at odds with the view held by many that patents encourage innovation. It raises the possibility that the seed industry for any one crop may ultimately become dominated, or at least hostage to, one commercial enterprise.

Patent enforcement is an important tool in building corporate empires and eliminating competition. Monsanto, which has staked much on its quest to become the 'Microsoft of engineered foods' (Morse 1996), has acquired companies, and stakes in others, reflecting its belief that patents will be a key source of competitive advantage in coming years.[35] Pioneer's moves to patent its in-bred lines on grounds that this would prevent other companies carrying out research on them illustrates that even the 'biological patent' which hybrid corn varieties are endowed with is being superseded by strengthened IPR legislation (Freiberg 1996a, 8).

As patents proliferate, it will become increasingly difficult for any enterprise to conduct research in the full knowledge that it is legitimate, especially since burden-of-proof legislation makes it incumbent upon those accused of patent violation to prove that they are innocent. The possibility arises in which a company carrying out research happens, by accident, to be working with a variety in which a patented gene sequence exists. Such a company would be unwittingly breaking the law, and would be expected to provide proof of its innocence.

There is a suggestion that IPR-related concerns are driving strategic alliances in the industry. Since, increasingly, more than one form of IPR will be involved in developing a given variety, such alliances reduce the likelihood of any one IPR-holder blocking development of the product concerned.[36] IPR-based restructuring can be expected to produce 'many casualties, some survivors, and a few successes' (Sehgal 1996, 21). The growing significance of IPR, appears to be leading to greater concentration in breeding effort, which is unlikely to promote diversity in agriculture.

Loss of farmers' privilege

UPOV 1991 appears to deny that farmers might also be breeders (Butler and Pistorius 1996, 8). Industry estimates suggest farmer-saved seed accounts for between one and two thirds of all seed planted in the world (Hobbelink 1991, 113; Thelwall and Clucas 1992, 39), though the proportion for developing countries is believed to be of the order 85 per cent (Cromwell 1993, 3). Indeed, in developing countries, the exchange of seed from farmer to farmer is probably the main avenue for diffusion of new seed varieties. Erstwhile Director General of GATT, Peter Sutherland, has suggested that such informal practices are 'generally not of interest to the owners of protected varieties' (Sutherland 1994, 4), yet already in the US, court actions have been taken against farmers involved in such activity, and it will not have been lost on IPR owners that interfarm sales constitute 62 per cent of all seed purchases in India where Sutherland made his speech (Pandey and Chaturvedi 1993, 11).

In the US, companies such as Monsanto require growers of their Roundup Ready Soybeans to be licensed to grow the material (Seed and Crops Industry 1996; Leach 1996), though other companies, including AgrEvo and DuPont, do not require licensing since, in the words of one commentator, 'they seem to feel that the additional chemical usage that's tied in with [DuPont's] STS beans is enough' (Freiberg 1996b, 3). Monsanto states that 'if necessary, the terms of the [Roundup Ready] contract will be enforced under the P.V.P.A., US Patent Law and general contract law' (Seed and Crops Industry 1996, 5; RAFI 1997a). One study estimates that within a few years, 40 per cent of all US farmers will be contract growers, or renters of germplasm from the same companies to whom they sell their product (RAFI 1994a, 10; also see Lamola 1995, 91).

Farmers will also need to be alive to the possibility, especially when growing outbreeding crops (those which cross-pollinate) of falling foul of patent legislation. It is not clear, given reverse burden of proof, how the law would interpret a situation where a farmer grew a variety which through cross-pollination, contained a patented sequence. Potentially, the onus will be placed upon farmers to ensure no such cross-pollination takes place.[37] Furthermore, the possibilities for farmers to experiment with varieties covered by UPOV 1991 seem limited. As with the case of 'essentially derived' varieties, it seems likely that case law will determine what is and is not allowed, but in the meantime, farmer experimentation with new varieties, which can create new races of outbreeding crops (see Chapter 1), may be a risky enterprise. This may have implications for diversity. More recently, the patent awarded to Delta and Pine Land Company on so-called 'terminator technology' (which prevents seeds from germinating in the next generation) provides a technological means through which to prevent farmer seed-saving (Lehmann 1998).

Wide use of agronomic genes

It has been stated that the introduction of novel genes will increase genetic diversity in agriculture. However, if the same gene is licensed to several companies for use in a large proportion of varieties in use, and in different crops, the gene becomes a component of uniformity. The possibility arises of the occurrence of a southern corn leaf blight on a more global scale. In China, 15 million hectares are planted to hybrid rice, each plant possessing, as with US corn varieties in 1970, a common gene for cytoplasmic male sterility (FAO Secretariat 1996, 23). Genes from the bacterium, *Bacillus thuringiensis,* which produce a protein that is lethal to some insects upon ingestion, are of great interest to corporations involved in biopesticides. Yet there are already concerns for insects' resistance to a number of strains of the bacterium (Hobbelink 1991, 62–4; Commandeur and Komen 1992, 7).

Herbicide tolerant genes could quite conceivably be transferred into vast areas of crop land, potentially increasing the vulnerability of crops globally, and leading to heightened problems with herbicide resistant weeds. Indeed, the relay race mentality of the BCM mode is accepted as a matter of course: 'in 50 years, biotechnologists will almost certainly still be developing new batteries of pest- and disease-resistant genes' (de Greef 1996, 72). The aim to engineer plants with tolerance to herbicides is a goal of companies that have integrated crop protection and seed production. In this way, purchasers of seed would be locked-in to the purchase of proprietary chemicals. This strategy reflects the fact that the costs of developing new agrochemicals is increasing owing to costly approvals processes. The costs of engineering the seed to lengthen the effective life of a given chemical compound are less, whilst it is also possible to extend the patent life of chemicals coming 'off patent' by specifying use of the proprietary form of a generic compound. The industry claims to have shifted emphasis to compounds of lower toxicity, although much attention has focused on glyphosate, which was listed by the US National Academy of Sciences as a potential carcinogen in 1987 (Doyle 1995, 233).

Ecological interactions

One of the major fears of environmentalists is that a gene which has been transferred to a variety to enhance its competitive performance may be transferred sexually into wild relatives, especially where maize, potatoes, rice, chickpeas and common beans are concerned. In these crops a wild-weed-crop complex is observed in which there is continual gene flow between wild and cultivated forms. More speculatively (the processes are poorly understood), horizontal gene flow mediated by micro-organisms may occur. In either case, the transfer of one gene may be sufficient for a plant to become invasive (de Kathen 1996, 12–13).

Lack of research on diversity

Quite apart from the tendencies remarked upon above which might exacerbate trends towards uniformity, the simple fact remains that little if any research associated with the use of biotechniques is being undertaken to encourage the use of diversity in the field. Indeed, biotechniques make it possible to clone plants and seeds so eliminating what residual variability there may have been in a crop bred using traditional methods.

Interaction between IPR and existing seed marketing legislation

Perhaps most importantly, and what may in time become the most concrete expression of the way in which biotechniques affect the offerings of the seed industry to farmers in the fact that over time, varieties which are not the product of genetic manipulation will slip off existing National Lists of seed varieties. Since those on the list are the only ones which can enter into commerce, slowly it will become impossible to purchase seeds which are not genetically engineered. This will force farmers and consumers alike, irrespective of concerns regarding genetic engineering, to purchase genetically engineered seeds. To the extent that genetic engineering concentrates on the integration of specific sequences within existing elite cultivars, irrespective of the number of cultivars made available to farmers, the genetic diversity within farming may decline, and certainly seems unlikely to increase.

Possibly, IPR will not increase uniformity. One scenario would see patented genes integrated into the background of existing varieties, and no change other than those created by the introduced genes. But this scenario assumes an unlikely scenario in which changes wrought by IPR and biotechniques will leave the seed industry unchanged in other respects, a scenario which current trends suggest is unlikely.

Our inability to measure diversity, and the fact that we do not know where we stand today, makes it impossible to assess change on the basis of any reliable baseline. However, it seems reasonable to suggest that biotechniques are taking agriculture in different directions to those which would be implied by the alternative approaches discussed in Chapter 1. If the BCM mode was environmentally damaging, and if there remain unanswered questions regarding the impact of the BML mode, why does this new mode appear to be gaining the support of most FAROs, public and private, especially when alternative paths exist?

Alternative agriculture, alternatives to IPR

In Mellon's (1991, 67–8) words 'the hype surrounding biotechnology diverts our attention from those [pesticide free] solutions by focusing attention on technologically dazzling new products ... By setting the proper goal, we will avoid the danger of spending millions trying to genetically engineer ten "better" pesticides, when for far less we could have taken our agriculture systems off the pesticide treadmill forever.' In particular, the alternatives described in Chapter 1 that (re-)emerged with some force in the 1980s on the back of environmental concern seem likely to remain alternatives in terms of the resources devoted to them (Merrigan 1995, 61). The transition from BCM to BML mode must be understood in this context. This was not a transition that was inevitable. The BCM was under fire, and alternative approaches beyond biotechniques were available. If IPR was essential to the BML mode, they may be decidedly unhelpful for those seeking to do research outside that mode. Sederoff and Meagher (1995, 71) opine that IPR 'are having a dramatic negative effect on the progress of non-profit research'.[38] The following sections consider debates concerning genetic resources which have taken place outside GATT and UPOV.

Continuing debates at the FAO: recognition for farmers' rights

The aim to resist IPR strengthening has been closely related to attempts to encourage the use of diversity in agriculture. Whilst the agricultural biotechnology companies had their sights on institutional changes allowing for the granting of IPR over life through the GATT, the FAO was debating the issue of the rights of farmers over germplasm, especially from 1987 onward. Farmers' Rights would be the counterbalance to the spread of PBR. In 1987, an International Fund for Plant Genetic Resources was set up, and was legally established in 1989. The Fund was designed for genetic conservation and utilisation work, and administered by the ICPGR. Mexico argued that donations to such a Fund should be mandatory in the same way as are royalties to a patent holder, but no agreement was forthcoming.

By 1989, developing country governments had let it be known through negotiations at the FAO that failure on the part of developed country governments to acknowledge the concept of Farmers' Rights would result in those countries being denied access to developing country genetic resources. Similarly, developed country patents would not be honoured. In what was effectively an exercise in horse-trading, at the 1989 meeting, developed countries insisted on an additional Resolution (4.89) modifying the International Undertaking such that it recognised PBR (McDougall and Hall 1994, 11). 'Free access' to germplasm explicitly would not mean 'free of charge' (as had previously been stated – see Solleiro 1995, 110). In return, a Resolution (5.89) on Farmers' Rights was passed recognising the rights of farmers in respect of their work conserving, improving, and making available genetic resources.

In the midst of an increasingly polarised debate, a notable event was the Keystone Dialogue held at the Keystone Center in Colorado in 1988, at Madras in 1990, and Oslo in 1991. Major transnational corporations, NGOs, IBPGR, national genetic resource programmes, the academic community, and the Rockefeller Foundation reached the following consensus conclusions after the second meeting:

- IPR were credited with encouraging development of new varieties, but also with encouraging genetic uniformity and erosion;
- attempts to include IPR for plants under the auspices of the GATT negotiations were criticised; and
- recognition of Farmers' Rights, and commendation of the idea of a Fund such as that extant at the FAO as a means of providing a form of concrete recognition thereof.

Corporate delegates refused to sanction a compensation mechanism, merely a fund recognising Farmers' Rights, a position which some participants would have found unacceptable were it not for the fact that it was agreed the fund should be mandatory. Furthermore, rather than the figure of $150 million proposed by NGOs at the FAO, the consensus figure arrived at was $500 million.

In November 1991, another Resolution (3.91) concerning genetic resources was passed at the FAO. This amendment to the Undertaking upheld 'that nations have sovereign rights over their plant genetic resources and that breeders' lines and farmers' breeding material should only be available at the discretion of their developers during the period of development' (in Solleiro 1995, 11). Although Farmers' Rights were recognised, they were given no substance. The International Fund, legally established in 1989, failed to materialise. Only in India have attempts been made to give substance to the concept through taxation of seed industry income (Crucible Group 1994, 34–5). Farmers' Rights still amount to little more than a polite thank you to farmers who have conserved genetic diversity *in situ*. This was most clearly illustrated in discussions at the June 1996 Leipzig Conference where the issue of Farmers' Rights showed that developed country donors were reluctant to support *in situ* conservation (Nemoga-Soto 1996; Jaura 1996), partly, one suspects, because of issues related to sovereignty in respect of genetic resources (see next section).

The Convention on Biological Diversity (CBD)

National sovereignty over genetic resources was a feature of the Convention on Biological Diversity (CBD), which entered into force as a legally binding international treaty at the end of 1993. [39] The CBD is a framework convention whose objectives, stated in Article 1, are:

> the conservation of biological diversity, the sustainable use of its components and *the fair and equitable sharing of the benefits arising out of the utilization of genetic resources, including by appropriate access*

> *to genetic resources and by appropriate transfer of relevant technologies*, taking into account all rights over those resources and to technologies, and by appropriate funding. (My emphasis)

Article 3 of the CBD lays down the principle that:

> States have, in accordance with the Charter of the United Nations and the principle of international law, the *sovereign right to exploit their own resources* pursuant to their own environmental policies, and the responsibility to ensure that activities within their own jurisdiction or control do not cause damage to the environment of other states or of areas beyond the limits of national jurisdiction. (my emphasis)

This effectively confirms that the free exchange principle is something of the past.

For corporations, sovereignty appeared to cede too much control to governments over genetic resources which were of increasing value to biotechnology companies. For non-government organisations, this debate seemed to miss the point that it was not states who were really responsible for maintaining biological diversity within their borders, but local communities (Colchester 1994; Lohmann 1991; Brush 1994). The CBD offers little for local communities, and does not explicitly recognise Farmers' Rights (Mugabe and Ouko 1994, 6), though the role of indigenous communities in conserving biodiversity is recognised in the preamble.

Articles 12, 17, 18 and 19 each refer to aspects of the biotechnology debate, but Article 16 of the CBD, dealing with transfer of technology, was the most heavily negotiated (WTO 1996, 6–10). Essentially, the debate centred around the fact that developing countries would most likely be providing raw materials for a biotechnology industry seeking to patent innovations. Article 16.2, suggests that:

> In the case of technology subject to patents and other intellectual property rights, such access and transfer shall be provided on terms which recognise and are consistent with the adequate and effective protection of intellectual property rights. The application of this paragraph shall be consistent with paragraphs 3, 4 and 5 below.

This Article was essentially a compromise meant to defuse the situation as regards the way in which technology transfer should account for

IPR. These latter paragraphs make provision for the transfer of technology to developing countries on 'mutually agreed terms'. Article 16.5 suggests that the Contracting parties:

> recognizing that patents and other intellectual property rights may have an influence on then implementation of this Convention, shall cooperate in this regard subject to national legislation and international law in order to ensure that such rights are supportive of and do not run counter to its objectives.

The CBD's equivocation on IPR issues was perhaps the main reason why George Bush's US delegation felt unable to support the CBD in Rio. Ever since the late 1980s, when methods of screening increased in sophistication and fell in cost, plant research has acquired new significance for pharmaceutical companies (Axt *et al.*, 1993, 6–7; Lewis 1991). Thus, head of the US delegation, William Reilly, stated: 'We have negotiated in the Uruguay Round of GATT to try to protect intellectual property rights. We're not about to trade away here in an environmental treaty what we worked so hard to protect there' (Pistorius 1992, 8–9).

The CBD does not apply retrospectively. Thus, the legal status of *ex-situ* collections of resources donated by developing countries but housed in the developed world was left unclear. In 1994, the World Bank appears to have sought to prevent the CGIAR's *ex situ* collections from falling under intergovernmental control by taking control of these itself in exchange for new funds for the CGIAR. However, 112 governments unanimously, and ultimately successfully, called for establishment of intergovernmental control over the CGIAR *ex situ* collections. The status of other collections held in developed country genebanks was discussed at the FAO's Leipzig Conference, but the outcome gave little encouragement to those countries which donated germplasm in the first place, as exemplified by the attempts of the company PHYTERA to gain access to genetic resources collected in developed country botanical gardens (Shiva 1996, 8).

Indigenous peoples and IPR

As so often in the past, indigenous peoples have been forgotten in the bulk of negotiations which affect a resource maintained largely by them. Their concerns span both the prime focus of the CBD and the issue of Farmers' Rights, as well as the TRIPs negotiations. Increasingly, the contributions made by indigenous peoples in terms not just of germplasm, but also knowledge regarding its use, is recognised. Yet this recognition has led to little concrete action to protect their interests.

In response principally to the heated debate generated by issues related to the patenting of indigenous people's cell lines, the Indigenous Peoples Biodiversity Network was formed with the objective of safeguarding their interests with respect to biodiversity and their knowledge thereof (RAFI 1994a, 24; 1993b).[40] Out of the debate concerning Farmers' Rights and IPR as they affect indigenous peoples has come an awareness that current formal systems do not adequately recognise indigenous knowledge systems, in which knowledge is often held at the level of the community. More and more companies are screening plant materials for useful products, yet it is estimated that the chances of finding useful products are at least doubled if indigenous, or folk knowledge is utilised (Axt *et al.*, 1993, 11).

An authority on the issue of IPR and indigenous peoples, Darrell Posey (1995, 7–8), notes that IPR pose seven problems for indigenous peoples:

1. they do not grant rights to collective entities;
2. they protect unique acts of discovery rather than transgenerational knowledge (from, for example, spirits, vision quests, or oral transmission) which tends to be public;
3. they do not recognise non-western systems of ownership, access and tenure;
4. they aim to promote commercialisation whereas the aims of indigenous peoples may be to prevent such activity;
5. they recognise market values only and not spiritual, aesthetic, or cultural value;[41]
6. they are, as is clear from the above, intimately bound up with power relations;[42] and
7. they are expensive to obtain and difficult to defend.

Posey goes on to cite a number of examples where indigenous peoples have displayed almost uniform hostility to what they perceive as an insidious trend in IPR legislation which seeks to deepen the exploitation of their resources and their knowledge. They perceive simple recognition of their contributions, as exemplified by the Farmers' Rights issue, as typically patronising in the face of a continued absence of legal mechanisms adapted to meet their needs and concerns.

RAFI (1994a) explores a number of ways in which indigenous communities could find space within existing legislation to protect their innovations. They argue that:

> There is a strong case to be made that the uncompensated appropriation of farmers' varieties and medicinal plants constitutes real theft and that the parties responsible should be pursued under criminal law at the expense of national law enforcement agencies in the country where the theft occurs (the patenting country). (RAFI 1994a, 31)

In the face of the forces mentioned above, it seems unlikely that much room for manoeuvre exists for those who would seek to place Farmers' Rights on the same level as IPR. Some countries have, however, been exploring the potential for exploiting the '*sui generis*' clause mentioned above in designing alternative IPR regimes which allow, for example, for communities, and not just individuals, to make IPR claims (RAFI 1994; Posey 1995; Butler and Pistorius 1996; Seiler 1998). As pointed out by Allen (1983), much innovation is the product of collective rather than individual efforts. In India, the concept of Collective Intellectual Property Rights (Shiva and Holla-Bhar 1993; Shiva 1994) is gaining credibility, whilst the Andean Pact is committed to developing a regime on collective rights of indigenous peoples.

Conclusion

This chapter has sought to show, through examining evolving IPR regimes, that a new mode of development of agricultural technologies is emerging. Some, indeed most of its most vocal and powerful supporters are drawn from the leaders of the BCM mode. The commercial possibilities presented by biotechniques saw pressure to extend IPR schemes. Following the 1980s, a decade which saw recognition gained in an international forum for the concept of Farmers' Rights, the rights of farmers to save seed first underwent conversion to a 'privilege', and were then consigned to history.

The old BCM mode still prevails despite the attacks of environmentalists and the growing awareness of available alternatives. If the possibilities for alternatives to thrive alongside the BCM mode seem limited, they are likely to be more so as the BCM mode is superseded by BML techniques. Although the BML mode's supporters have often appeared as keen as environmentalists to see the back of the old BCM mode, their agenda is not an alternative based on bringing diversity back into the picture, but the ushering in of new techniques aimed at increasing the potential for achieving a genetically uniform ideal plant. Major suppliers of agrochemicals condemn the technologies for which they themselves have been responsible as manifestly unsustainable. In answer to his own question whether such companies are 'Planetary patriots or sophisticated scoundrels?', Kloppenburg (1993) writes: 'Having been recognised as wolves, the industrial semioticians (and you thought they were only manipulating genes!) are now redefining themselves as sheep, and green sheep at that.'

Notes

1. Sharp (1990, 98; see also OTA 1988, 78) points out that biotechnology is essentially a process technology, for which reason, I prefer the term biotechnique to biotechnology.
2. Attempts have been made, through *in vitro* mutant induction, to find T-cytoplasm plants which are resistant to *H. maydis* toxins, but all toxin resistant mutants are male fertile (van Marrewijk 1995, 118).
3. Marx (u.d. 207) went on to note that this 'alleged natural destiny' of the West Indies was being undermined by means of cheaper production elsewhere. The potential for biotechnological methods to enable import-substitution of tropical products is a very real threat to many countries with narrow, agriculturally-based export portfolios (see especially Nana-Sinkam *et al.*, 1992; Junne 1992).
4. There exist relatively gene-rich countries in the developed world and gene-poor countries in the developing world.
5. This section draws fairly extensively on material in Fowler *et al.* (1988, 239–56), Juma (1989, ch. 5), Hobbelink (1991, ch. 7) and Brush (1993) (see also Nott 1992 and Price 1992). On the history of patents, see Machlup and Penrose (1950), Penrose (1951) and Primo Braga (1990).
6. See Fowler (1994, 106–18, 135–46) and Kloppenburg (1988, 134–51) on the US's Plant Variety Protection ACT (PVPA). Both are of the opinion that the PVPA was very much a marketing tool for the seed industry wishing to conform to UPOV standards and offer reciprocal protection for other countries in an increasingly global industry. On the UK, see Bould and Kelly (1992). Clunies-Ross (1995, 29–33) expresses the view that the UK's Plant Varieties and Seeds Act, replacing the 1920 Seeds Act, shifted the focus of its protection away from the consumer (the grower) and towards the breeder, the intention being, as a 1957 Committee reviewing legislation reported, to keep 'unsuitable varieties and strains off the market' (31).
7. See Table 6.3. in Kloppenburg (1988, 148) and Goodman *et al.*, 1987 (109–15). Fowler and Mooney's (1990, 123) research indicated that since 1970, more than 1000 independent seed companies had been acquired by, or were under the control of transnational corporations. However,

Kloppenburg (1988, 136; also Barton 1982, 1072) notes that industry concentration was already underway in the US by 1959 as seed companies increasingly began to make outlays for research, but also for marketing as a means of differentiating their products from those of public research organisations (though in some cases, they were probably the result of public research).

8. There is not universal agreement here. Environmentalists revile the Common Agricultural Policy, but are unsure of how reduced support for agriculture would affect the environment. In general, the question is less whether, more what, to support. Trade negotiations, however, are likely to limit what can be supported.
9. More obviously affecting seed purchases are the set-aside schemes and crop reduction programmes operating in the EU and US.
10. The ASTA's recommendations were adopted wholesale as the position of the US Government (Fowler 1994, 190).
11. For the purposes of this thesis, biotechnology is defined as knowledge of tissue culture, and gene transfer techniques, though the emphasis is on the latter. These form, respectively, the first and second phases of biotechniques (OTA 1995, 209).
12. The term molecular biology was first used by Warren Weaver of the Rockefeller Foundation (see Chapter 6), who funded several centres of excellence in pre-World War II years (Kenney 1986, 17–22; Yoxen 1983, 34–46).
13. There is an interesting debate surrounding both the exact motivations of these scientists, and the impact that their concerns had on the subsequent development of biotechnology as an industry (see for example Yoxen 1983; Kenney 1986; Kloppenburg 1988; Campbell 1991; Levin 1991; Wildavsky 1991).
14. 'One of the most dramatic negative effects of biotechnology over the past decade has been that the lure and glamour of high technology has detracted attention and funding from traditional breeding ... no application of biotechnology can reach the farmer without the intermediate step of an intermediate breeding system' (de Greef 1996, 74; also, Lacy and Busch 1991, 156).
15. A study by the OECD of field-testing of biotechnically engineered crops showed that of 1257 traits tested between 1986 and 1992, 740 were for herbicide resistance or tolerance (GRAIN 1994a). As of September 1996, 4 of 7 approved products, and 6 of 9 products up for approval in the European Union were engineered for herbicide tolerance (CEC 1996). James (1998) estimates that 54 per cent of the area planted to transgenic crops in 1997 was for crops expressing the trait of herbicide tolerance.
16. There is continued exasperation at the application of patent law to living matter. Patents are granted on the basis that what is patented is new, involves an inventive step (the non-obviousness criterion), and is useful for something. Hobbelink (1991, 106) cites one plant breeder's response: 'Who will have the guts to declare a gene novel and non-obvious? Would anyone know enough of genetics and nature to claim such arrogance?'
17. The final agreement refers to Trade-Related Aspects of Intellectual Property Rights, Including Trade in Counterfeit Goods.
18. For an illuminating discussion, see Finger (1991).

19. Eight countries were on the priority watch list and seventeen others featured on the wider list. GSP benefits were routinely used as leverage in negotiations with these countries (van Wijk 1990).
20. In 1883, Geigy (later Ciba-Geigy, now Novartis) denounced patents as 'a paradise for parasites', and 'a playground for plundering patent agents and lawyers' (cited in Hobbelink 1991, 99).
21. See Knudson and Hansen (1990, 4) and RAFI (1994b) on the *Asgrow* v *Kunkle Seed Co.* and *Asgrow* v *Winterboer* cases respectively.
22. On the US, see RAFI (1993). On the UK situation, see MAFF (1994a,b). The UK approach is based on that of the Council of the European Union (1994). See van Wijk (1993) for a perspective on the impact on seed saving in the European Union.
23. In the US's Statements of Understanding of the relationship between GATT and the Biodiversity Convention, it is stated that 'we [the US] will stress the benefits of our trading partners of adopting standards consistent with the 1991 revision' (in Wirth 1994). Before the GATT was concluded, CONASUR countries were considering PBR along lines of UPOV 1978, whilst some Andean countries were following UPOV 1991 (Correa 1993; Silva 1995).
24. For a slightly dated discussion of the technical difficulties with respect to the application of patent law, see Crespi (1985). For the approach taken by the CGIAR system (which is not uniform across its centres), see Persley (1990, ch. 8), Hobbelink (1991, ch. 8), GRAIN (1991), Komen (1992); CGIAR Secretariat (1993) and Barton and Siebeck (1994). For developing countries, see van Wijk *et al.* (1993), Commandeur (1993).
25. Takeovers have been facilitated by what some see as the industry's premature commercialisation on the basis of hype and speculation (Buttel 1995, 38–9; see also Maitland 1996). Even established boutiques, such as Genentech, are being taken over (Junne 1990). Eight years after the flotation of Genentech in 1988, no biotechnology company in the US was reporting a profit (OTA 1988, 81).
26. This degree of control is, in any case, illusory (see, e.g., Mellon 1991a).
27. Persley (1990, 49) estimated that in 1985, of an estimated $900 million devoted to agriculture-related biotechnology, $550 million was from the private sector. She adds: 'The major reason for the greatly increased role of the private sector in modern biotechnology is that, for many of the new technologies, the process and/or the product is protectable.'
28. In the US, the Stevenson–Wydler Technology Innovation, and Bayh–Dole Acts were passed in 1980, enabling 'government contractors, small businesses and nonprofit organizations to retain certain patent rights in government sponsored research and [permitting] the funded entity to transfer the technology to third parties' (Tribble 1995, 97). For the US case, see Kenney (1985, 27–125, 230–7); for the UK, see Field (1988, 30) and Anon (1993), and for the CGIAR , see Hodgson (1995, 352–3). Field (1988, 20) comments that UK research follows industry's priorities, and that 'industrial competitiveness appears to represent the central and over-riding concern of national strategies.'
29. This is nothing new. As Mowery and Rosenberg (1989, 14) point out, the distinction between basic and applied research had always been nonsensical in matters of health, medicine and agriculture.

30. WTO (1996, 13) reports attempts by UPOV to harmonise the distinctness criterion.
31. This question is behind the differing views of the Ministry for Trade and Industry and the Ministry of Agriculture Fisheries and Food in Japan, the latter fearing that stregthening IPRs will have a detrimental impact on breeding, due to its concentration in the hands of large corporations (Commandeur 1995, 11–12).
32. In fact, EU patent law allows for use of patented intellectual property for true research purposes, but US utility patent legislation allows no such exemption, although judicial rulings have made allowance in the past for 'non-commercial' research (Bijman 1994, 9).
33. US patent examiners must grant the claims of the inventor unless specific reasons for denying them can be established (Lesser 1991, 125). This would have to be prior knowledge or inventions, which is rare in newer fields of endeavour.
34. The patents were subsequently cancelled after a legal challenge. The Patent and Trademark Office of the US noted that the patents concerned 'would have allowed Agracetus to demand royalties on virtually any type of cotton with gene modifications' (Aharonian 1994).
35. RAFI (1994a, 4) envisions a Futures Market in intellectual property rights stocks. See also Sehgal 1996. After the initial draft of this work, Monsanto merged with American Home Products (Lehmann 1998).
36. An innovation may involve patents related to selectable marker genes, the gene for a given trait, the promoter, and various elements required to ensure expression, as well as PBR on the background variety (Sehgal 1996, 19; Evans 1996, 72–4). In the future, more than one trait is likely to be involved in a given plant, for which PBR may already have been granted.
37. There was a ruling by the European Patent Office in February 1995 which effectively upheld the farmers' privelege in the context a patent granted to Plant Genetic Systems and Biogen (inc.) for plants resistant to glutamine synthetase inhibitors (herbicide resistant plants). However, the EU Directive for the Legal Protection of Biotechnological Inventions would reverse this situation if passed (Greenpeace 1995).
38. In stark contrast to those who trust the patent system to call forth new ideas, Sederoff and Meagher (1995, 77) point to research which will not get done as a result of the higher costs of conducting it (due to the existence of IPRs and the need to pay royalties on patented processes); 'With the decline in public spending and the increasing focus of the private sector upon short-term results, we are in danger of failing to build the foundation for discoveries in the long term' (see also Day (1995, 85)).
39. For a brief history of the drafting of the CBD, see WTO (1996, 6–10).
40. The US Secretary of Commerce Ron Brown applied for a patent on a Panamian woman's cell line. The implications, following this, of the Human Genome Diversity Project (see Cavalli-Sforza 1990; Bowcock and Cavalli-Sforza 1991), have aroused great suspicion amongst indigenous peoples (see Roberts 1992; RAFI 1993a). Although the concept of piracy was used as a rationale to strengthen IPRs under the GATT, RAFI's (1994a) study estimates that the developed world benefits more from reverse piracy than it loses from piracy.

41. In fact, some patent laws do, as currently constituted, exclude innovations on moral grounds. This should constitute a clear obstacle to harmonization of patent law across nations since few would claim to be able to discern a transcendent morality where 'national religions' are closely matched, let alone where they are quite distinct.
42. This is less a problem with IPR facing indigenous peoples, more the reason why they face the others.

8
Conclusion

'one cannot step twice into the same river'
Heraclitus

Results

Chapter 1 asked the question why agriculture seemed to be standardised across such a large area despite the fact that there are problems associated with genetic uniformity, which sits at the heart of this standardisation. Of these problems, those associated with pesticide use are rather better understood than that concerning crop vulnerability, though in fact any attempt to estimate the social and environmental costs of either are subject to huge uncertainties. Though citing selectively, I have attempted to show that even on the production criterion alone, alternatives do not suffer tremendously by comparison with modern agriculture. One must not lose sight of the impact of the relative distribution of research funding over time. This has been massively skewed in favour of high external input techniques, and given this, it is somewhat surprising that alternatives can still even hold a candle to high input approaches, let alone lay claim to being viable competing approaches (as, I suggest, they can).

After briefly exploring the theories of the classical economists, I have also examined two orthodox theories of technical change in agriculture, that of Ester Boserup, and the induced innovation theory of Hayami and Ruttan (and Binswanger). Both were shown to suffer from shortcomings. Whilst Boserup's thesis is very much in the 'stages' mould, Hayami and Ruttan's theory is structured almost entirely around markets, and a strictly marginalist interpretation of the response of technical and institutional change to movements in relative factor prices. Both, however, appear to postulate industrial agriculture as an endpoint, a sort of agricultural equivalent of the 'end of history' debate. There is inadequate consideration, in Boserup's theory, of how agriculture might be practised outside the linear progression she hypothesises. Hayami and Ruttan, on the other hand, postulated a 'greening' of modern agriculture through inducement mechanisms, but a quarter of a century after their writing, any changes which have occurred are at best reformist in nature and have occurred principally at the margins. Thus, whilst Boserup's linear stages model would *implicitly suggest* that lock-in to a trajectory characterises the whole history of agriculture, the failure of the inducement mechanisms hypothesised by Hayami and Ruttan to create the institutional changes mentioned could lead one *to deduce* that lock-in has occurred. Neither approach appreciates the significance of competing technologies and approaches, and how these either 'win out' or 'lose' over time.

Drawing on literature from evolutionary economics and from science and technology studies, I have proposed a framework for understanding technical change using the concepts of technological trajectories, path-dependence, lock-in, and the fitness landscape. The framework suggests that ways of looking at the world, and the formal and informal institutions which embody these and are structured by them, as well as economic interests, act so as to deform an imaginary landscape from which, gradually, a particular technology or technique comes to be selected (in the form of a peak on the landscape). The sharpness of the peak increases as more interests coalesce around this way of doing things.

In agriculture, breeders have used the concept of a plant ideotype to guide their work. For breeders, an ideal plant has been shaped by the interests and institutions which have co-evolved with the efforts of the breeders. These interests have become increasingly powerful and have evolved such that the farmer is often no longer the primary client of the breeder (if indeed, the farmer ever was the primary client, for it is possible to argue that whilst the farmer has necessarily been the proximate target for researchers, the ultimate aim has been to assist other interests, such as bankers, merchants, retailers and food processors, in the achievement of their goals).

Chapter 4 reviewed the many factors which can contribute to the systemic process of lock-in. In the first half of the chapter, the argument was developed through looking at how research organisations emerge, and how they are staffed. These decisions effectively determine what an organisation will choose to do, and what methodological approach will be developed. Typically a methodological approach has a bearing on who the researcher or research organisation sees as its clients. The second half of the chapter looked at the ecological factors that reinforce the lock-in process, these being traceable, in part, to the design of institutions governing the certification and marketing of seed. These factors combine to 'select' from the fitness landscape a particular way of doing things.

The first case study, that of hybrid maize, showed that a number of historical factors, not least of which was the desire of professional plant breeders to take breeding out of the hands of farmers, led to the crossing of inbreds being favoured over selection-based techniques in the wake of the rediscovery of Mendel's laws. The weight of support for this technique increased as it became clear that the hybrid technique conferred *de facto* proprietary rights over the product of the breeders' work. Changes in personnel in key positions facilitated this process, and as a result, a major industry grew up on the basis of this new technique, the foundations of which had been laid in public agricultural research organisations. Thus breeding was not only taken out of the hands of farmers, but over time, it became concentrated in private sector companies, with the public sector supporting their work through 'basic' research. Hybrids, uniform in the field, replaced the numerous varieties grown in the United States in the 1930s, and made possible the mechanisation of harvesting through synchronisation of tasks. The limitations of this approach were revealed much later in the 1970 southern corn leaf blight episode.

The early days of the Green Revolution were characterised by the Rockefeller Foundation's (RF's) approach to a problem (which, I have sought to show, may not have been a problem at the time) based upon the use of science to increase production. This narrowly scientific approach was reflected in the make up of Committees and the Survey Commission which prevaricated over the design of the Mexican Agricultural Programme (MAP). Reflecting their training, and the work they had done in the United States, they made recommendations for Mexico which suggested that Mexico should, in essence, become more like the United States. Competing interpretations of the problem were on hand. These came from the rural regional schools in Mexico, through the geographer, Carl Sauer, and from the fields of Mexico itself. However, the RF sought to set the agenda for research through influencing the curricula taught in organisations for agricultural education in Mexico, Carl Sauer's views were effectively ignored, and Mexican agriculture was deemed backward and ill-informed. Mexican agriculture was not based on the scientific principles that were the tools of the trade of the agronomists who journeyed to Mexico in the wake of the triumphant march of hybrid corn in the US.

The MAP became 'stuck' in what was to be the pre-hybrid phase, that of breeding synthetics. The impact of the MAP on Mexican diversity may not, therefore, have been great. What is significant is the impact the RF has had on the agenda of research organisations in Mexico. Increasingly, these have tended to follow the hybrid route rather than developing varieties appropriate for farmers cultivating in diverse ecological conditions (although CIMMYT pushed the hybrid route to one side in its early years, only to return to it, controversially, at a later date).

The final case study looks to the future. However, it is guided by the experience of the past. IPR for plants have been strengthened due to the desire of companies to move the products of agricultural research into the world of high-technology. Governments, in turn, are increasingly seeking to foster a conducive framework for this industry based on the view that leadership in high-technology industries is a key determinant of country competitiveness. The chapter suggests that, on the basis of the past trajectory, uniformity, if it is not about to increase (which, in any case, is difficult to assess), will not decrease either. These developments took, and are taking place against the backdrop of the re-emergence of alternatives, as well as a fight to maintain a free exchange of germplasm, and the right of farmers to save seed, in the face of the encroachment of IPR.

Conclusion

The historical continuity displayed through the case-studies itself illustrates the path-dependent nature of technical change in agriculture, which is intimately connected with FAROs' approaches to agricultural research. In the case of hybrid maize, the fitness landscape was initially distorted through a combination of the desires of scientists to become experts in plant breeding, and the rediscovery of Mendel's laws. The discovery of heterosis by Schull, East and Jones further distorted the landscape through the promise of control over breeding that it offered, and the possibilities for profit that could arise. This discovery acted, for all intents and purposes, like a blinkering mechanism for agricultural scientists. The necessary staffing and institutional changes required to galvanise research in the required direction, and eliminate the near-market work of the public sector, effectively led to the selection of one method – the crossing of inbreds – for the production of seed, notwithstanding the advances made in recurrent selection techniques. Once this route was taken up, breeding for fertiliser responsiveness, for machine harvesting, and *ergo* uniformity began to lock farmers in to a technique in which they were required to purchase seed year on year (a change to which, by all accounts, farmers took time to adapt). Mechanisation proceeded rapidly, and pesticides became used more widely. Many varieties (we do not know how many) were lost as hybrids spread across the US.

It was the choice of staff which one can reasonably state led to the shaping of the fitness landscape in the case of the MAP. This, allied to the fact that the RF had something of a track record in, as well as a desire to carry out, experiments in social engineering through the application of science led to a conceptualisation of Mexican problems in narrowly technical terms. The choice of personnel, their training, and their country of origin increased the likelihood of their focusing on hybrid corn as the ultimate goal of the programme. Hence, even though Carl Sauer recommended a cautious approach to Mexican agriculture, which he knew was well adapted (in dietary as well as 'agronomic' terms) to the prevailing constraints, the MAP proceeded on the basis that it knew what Mexico needed and how best to achieve it. Because of the MAP's lack of success, its impact on genetic diversity may not have been that significant. *Ex-situ* collections were established through efforts in Mexico and Latin America, providing a reservoir of useful germplasm for breeders. I have argued that this may have been one of the rationales for launching the programme in Mexico.

Recombinant-DNA techniques ushered in the era of the commercialisation of the gene. However, a major institutional change, long fought for by breeders, was necessary to encourage new investments in these techniques. Through the courts and through the GATT, despite growing opposition to such trends expressed at the FAO, corporations and developed country governments succeeded in changing IPR institutions in the interests of those who would invest in a new way of doing plant breeding. On reflection, it is fair to say that the changes in IPR legislation as applied to plants were piggy-backing on qualitatively different IPR issues related to, for example, matters of copyright regarding the production of fake Rolex watches and 'boot-legged' audio cassettes.

However new the techniques, and however novel the genes they introduce, there does appear to be substantial continuity with the past, and the revolutionary claims made for this technology seem misplaced. The breeders' quest for a genetically uniform ideal plant is, I have argued, simply being extended through the use of biotechniques, in which patented genes encoding for specific traits are being spliced into varieties over which breeders have acquired PBR. One can say that technological changes in the life sciences have prompted (admittedly at the behest of substantial investors) institutional changes which have deformed the fitness landscape so as to promote a new technique which is a more powerful means to achieve a well-established goal. These new techniques, as with the ones which they seek to supercede, select against informal ways of doing things. Biotechniques and the institutions supporting them are co-evolving in such a way as to effectively lock-out (competing) alternative approaches to agriculture.

In the context of the argument presented in this book, these case-studies are best viewed as part of a larger study in the transformation of agriculture. I have sought to show that agriculture has become locked in to a particular way of doing things, at the heart of which is genetic uniformity. Chapters 5 and 6 show how the concept of an ideal and uniform plant, the hybrid composed of crossed inbreds, was developed, and then transferred to Mexico. With it were transported the supporting research infrastructure, the ideas behind the plant, and the focus on control of the production process. Over time, the nature of the ideal plant has co-evolved with the growth in industries which the uniformity of the plant, and the quest for control implied by the ideal form, have in turn made possible. Chapter 7 showed how today biotechniques are providing more powerful tools to fashion ideal plants with genes expressing for specific traits transferred into background varieties.

These have altered the constraints which limit the form of the ideal plant, and have encouraged a transition towards the BCL mode, which is occurring despite pressures for change in alternative, more environmentally benign directions. The transition has been made easier by the fact that, since it is still broadly the same uniform ideal plant that is being sought (albeit with a different box of tools), the same players are involved in its making. This would not have been the case had the techniques been used to develop alternatives, a scenario which is made all the less likely given our knowledge of how research and technological development has evolved as a path-dependent process.

We have reached a pessimistic conclusion. Throughout the writing of this book, I have been considering whether, and if so how, the alternatives which I discussed in Chapter 1 can create greater space for themselves in a world where the over-riding tendency is towards homogenisation of agriculture and an apparently conscious attempt to select against techniques which encourage the use of diversity in the field. In other words, how could alternatives become something other than alternatives? It is clear that some parts of the world will remain untouched by any system which has uniformity at its core. As Cromwell and Wiggins (1993, 114–5) state:

> If environmental conditions are relatively stable but complex, the need is for a large number of distinct varieties, to maximise productivity by slotting each into a given micro-zone. However, if growing conditions are highly variable (in terms of rainfall, etc.) rather than highly complex, a large number of intra-varietal variation is more useful than a large number of distinct varieties, to increase the chances of part of the crop producing a harvest.

They also note (117) that breeding for uniformity and stability may simply waste time and resources since these features are the ones farmers in marginal environments may be least concerned about.

At one level, therefore, there would appear to be a real possibility of agricultural research splitting into two systems, one whose outputs are highly uniform, the other, less so. Yet in some resource rich areas too, farmers are bucking the trend. A number of authors writing in a 'post-Fordist' tradition emphasise the re-assertion of individual farming styles in the face of homogenising tendencies (for example van der Ploeg 1992; Buttel 1994; Pretty 1998). The fact that alternatives are observable is testimony to their resilience and, arguably, to the misguided nature of so much research by FAROs. Groups of farmers are carving out ways of doing agriculture that stand apart from the dominating tendency of high external input techniques. Even at the level of the nation state, in some European countries (such as Sweden) there are real efforts being made to promote, on a nation-wide scale, organic agricultural techniques, although no doubt the purists will comment that such 'alternatives', to the extent that they become co-opted by the mainstream, lose their transformative potential.

I have not sought to answer how these alternatives might prosper in the future, though on the positive side, one can point to the emergence of new social movements, the rise in prominence of non-state actors in global politics, and in the growth in schemes falling under the heading, 'community supported agriculture'. One can also speculate as to the potentially enormous consequences that the imposition of a punitive tax on transport/fuel might have for agriculture and its patterns of production (and the consequences for regional crop genetic diversity). But for all these signals, alternative agriculture looks set to remain largely that, an 'alternative' in the sense of falling outside the mainstream approaches. It is entirely possible that, with seed production increasingly concentrated in the hands of corporations which are vertically integrated with agrochemical input supply, the choice of seeds which farmers have available to them will become increasingly restricted over time as seeds which are not genetically engineered slide off the National Lists, and catalogues which determine what can and cannot enter into commerce become dominated by genetically engineered varieties. These possibilities are potentially powerful testimony to the ability of the mainstream, for whom progress and biotechnology appear to have become synonyms, to lock-out, and slam the door upon, the very alternatives which may hold the key to making agriculture truly sustainable.

Bibliography

ABN (Agricultural Biotechnology Network) Bulletin (1992) 'Diverse Control: Cereal Variety Mixtures and the Options They Offer for Disease Control', *ABN Bulletin*, issue 2, January, pp. 5–6.

Achilladelis, B., A. Schwarzkopf and M. Cines (1987) 'A Study of Innovation in the Pesticide Industry: Analysis of the Innovation Record of an Industrial Sector', *Research Policy*, vol. 16, pp. 175–212.

Adams, R. and J. Adams (eds) (1992) *Conservation of Plant Genes: DNA Banking and In Vitro Biotechnology*, London: Academic Press.

Adams, R. E. W., W. E. Brown Jr. and T. P. Culbert (1981) 'Radar Mapping, Archaeology and Ancient Maya Land Use', *Science*, vol. 213, pp. 1457–63.

Aggarwal, P. C. (1990) 'Third World Network Features: Natural Farming Succeeds in Indian Village', in Third World Network, pp. 461–4.

Agliardi, E. (1991) 'Essays on the Dynamics of Allocation Under Increasing Returns to Adoption and Path Dependence', unpublished PhD Thesis, University of Cambridge.

Aharonian, G. (1994) 'PATNEWS: Patent Office Revokes Third Patent', E-mail publication, 8 December .

Ahmad, S. (1966) 'On the Theory of Induced Invention', *Economic Journal*, vol. LXXVI, pp. 344–57.

Ahmed, I. (ed) (1992) *Biotechnology: A Hope or a Threat*, London: Macmillan.

Ahmed, I. and V. W. Ruttan (eds) (1988) *Generation and Diffusion of Agricultural Innovations: The Role of Institutional Factors*, Aldershot: Gower.

Alcorn, J. B. (1984) 'Development Policy, Forests and Peasant Farms: Reflections on Huastec-managed Forests' Contribution to Commercial Production and Resource Conservation', *Economic Botany*, vol. 38, no. 4, pp. 389–406.

Allaby, M. (1973) 'Miracle Rice and Miracle Locusts', *The Ecologist*, vol. 3, no. 5, pp. 180–85.

Allen, P. M. (1988) 'Evolution, Innovation and Economics', in Dosi *et al.* (eds) (1988), pp. 95–119.

Allen, P. M. and J. McGlade (1987) 'Modelling Complex Human Systems: A Fisheries Example', *European Journal of Operational Research*, vol. 30, pp. 147–67.

Allen, R. C. (1983) 'Collective Invention', *Journal of Economic Behaviour and Organization*, vol. 4, pp. 1–24.

Almås, R. and B. Nygård (1994) 'European Values and the New Biotechnologies: Post-materialism or a New Arena of Rural-urban Conflict?', in D. Symes and A. Jansen (eds) (1994), pp. 167–79.

Altieri, M. (with contributions by R. Norgaard, S. Hecht, J. Farrell and M. Liebman) (1987a) *Agroecology: The Scientific Basis of Alternative Agriculture*, London: Intermediate Technology.

— (1987b) 'The Significance of Diversity in the Maintenance of the Sustainability of Traditional Agroecosystems', *ILEIA Newsletter* vol. 3, no. 2, reprinted in Third World Network (1990), pp. 502–6.

— (1991) 'Traditional Farming in Latin America', *The Ecologist*, vol. 21, no. 2, pp. 93–6.

— (1992) 'Avances de CLADES en la Investigacion Agroecologica', *Agroecologica y Desarrollo*, vol. 2/3, Julio, pp. 61–3.

Altieri, M. A. and L. C. Merrick (1988) 'Agroecology and In Situ Conservation of Native Crop Diversity in the Third World', in K. Wilson (ed.) (1988), pp. 361–9.

Altieri, M. A. and J. Trujillo (1987) 'The Agroecology of Corn Production in Tlaxcala, Mexico', *Human Ecology*, vol. 15, no. 2, pp. 189–220.

Amador, M. F. (1980) *Comportamiento de Tres Especies (Maiz, Frijol, Calabaza) en Policultivos en la Chontalpa, Tabasco, Mexico*, Tesis Profesional, Colegio Superior de Agricultura Tropical, Tabasco, Mexico.

Amador, M. F. and S. R. Gleissman (1990) 'An Ecological Approach to Reducing External Inputs Through the Use of Intercropping', in S. R. Gleissman (ed.) (1990), pp. 146–59.

Anaya, A. L., R. C. Ortega and V. Nava Rodriguez (1992) 'Impact of Allelopathy in the Traditional Agroecosystems in Mexico', in Rizvi and Rizvi (eds) (1992), pp. 271–301.

Anderson, E. (1967) *Plants, Man and Life*, Berkeley: University of California Press.

Anderson, J. R. (ed.) (1994) *Agricultural Technology: Policy Issues for the International Community*, Wallingford, Oxon: CAB International in Association with the World Bank.

— (1994) 'Agricultural Technology: Contemporary Issues and Opportunities for the World Bank', in Anderson (ed.) (1994), pp. 3–14.

Anderson, M. (1994) 'Economics of Organic and Low-Input Farming in the United States of America', in Lampkin and Padel (eds) (1994), pp. 161–84.

Anderson, P. W., K. J. Arrow and D. Pines (eds) (1988) *The Economy As An Evolving Complex System*, Redwood City, California: Addison-Wesley.

Anderson, R. S., P. R. Brass, E. Levy and B. M. Morrison (eds) (1982) *Science, Politics, and the Agricultural Revolution in Asia*, Boulder: Westview, for the American Association for the Advancement of Science.

Anderson, R. S., E. Levy and B. M. Morrison (eds) (1991) *Rice Science and Development Politics: Research Strategies and IRRI's Technologies Confront Asian Diversity (1950–1980)*, Oxford: Clarendon Press.

Anon (1941) *Staff Meeting Minutes*, February 18, Rockefeller Foundation Archives, 1.2/323/10/63.

Anon (1952) *Mexico: Agricultural Education*, June 20 1952, Rockefeller Foundation Archives, RG 1.2/323/4/25.

Anon (1993) *Realising Our Potential: A Strategy for Science, Engineering and Technology*, Presented to Parliament by the Chancellor of the Duchy of Lancaster by Command of Her Majesty, May, London: HMSO.

Antle, J. M. and P. Pingali (1994) 'Health and Productivity Effects of Pesticide Use in Philippine Rice Production', *Resources*, no. 114, Winter, pp. 16–9.

Arce, A. (1993) *Negotiating Agricultural Development: Entanglements of Bureaucrats and Rural Producers in Western Mexico*, Wageningen Studies in Sociology, no. 34, Wageningen: Wageningen Agricultural University.

Arce, A. and Norman Long (1994) 'Re-Positioning Knowledge in the Study of Rural Development', in Symes and Jansen (eds) (1994), pp. 75–86.

Arnold, M. H. *et al.* (1986) 'Plant Gene Conservation', Nature, vol. 319, February, p. 615.

Arntzen, C. (1991) 'Plant Agriculture', in Davis (ed.) (1991), pp. 108–16.

Arrow, Kenneth J. (1962a) 'Economic Welfare and the Allocation of Resources for Inventions', in NBER, pp. 609–26.

— (1962b) 'The Economic Implications of Learning by Doing', *Review of Economic Studies*, vol. 29, pp. 155–73.

Arthur, W. (1988a) 'Self-reinforcing Mechanisms in Economics', in Anderson *et al.* (eds), pp. 9–31.

— (1988b) 'Competing Technologies: An Overview', in Dosi *et al.* (eds), pp. 590–607.

— (1989) 'Competing Technologies, Increasing Returns, and Lock-in by Historical Events', *Economic Journal*, vol. 99, pp. 116–31.

— (1990). 'Positive Feedbacks in the Economy', *Scientific American*, February, pp. 92–9.

Arthur, W. B., Y. M. Ermoliev and Y. M. Kaniovski (1987) 'Path-dependent Processes and the Emergence of Macro-structure', *European Journal of Operational Research*, vol. 30, pp. 294–303.

Ashford, N. A. (1994) 'An Innovation-Based Strategy for the Environment', in Finkel and Golding (eds) (1994), pp. 275–314.

Ashford, T. and S. D. Biggs (1992) 'The Dynamics of Rural and Agricultural Mechanization: The Role of Different Actors in Technical and Institutional Change', *Journal of International Development*, vol. 4, no. 4, pp. 349–74.

Axt, J. R., M. L. Corn, M. Lee and D. M. Ackerman (1993) *Biotechnology, Indigenous Peoples, and Intellectual Property Rights,* Congressional Research Service Report for Congress, April 16, Washington, DC: Congressional Research Service.

Balmer, B. and M. Sharp (1993) 'The Battle For Biotechnology: Scientific and Technological Paradigms and the Management of Biotechnology in Britain in the 1980s', *Research Policy*, vol. 22, pp. 463–78.

Barber, L. (1996) 'Brussels Stalls on Genetic Maize', *Financial Times*, Wednesday November 13.

Bardhan, P. (1989) 'New Institutional Economics and Development Theory', *World Development*, vol. 17, no. 9, pp. 1389–95.

Barkin, D. and B. Suárez (1986) 'The Transnational Role in Mexico's Seed Industry', *Ceres*, vol. 19, no. 6, pp. 27–31.

Barletta, N. A. (1970) *Cost and Social Returns of Agricultural Research in Mexico,* Ph.D. Thesis, Department of Economics, University of Chicago.

Barton, J. H. (1982) 'The International Breeders' Rights System and Crop Plant Innovation', *Science*, vol. 216, 4 June, pp. 1071–75.

Barton, J. H. and W. E. Siebeck (1994) *Intellectual Property Issues for the International Agricultural Research Centers,* Issues in Agriculture no. 4, Washington, DC: CGIAR.

Bassett, R. (1995) 'Biodiversity and Intellectual Property: Some Questions and Answers', in BIA (ed.), pp. 41–4.

Beachell, H. M. and P. R. Jennings (1965) 'Need for Modification of Plant Type', in IRRI, pp. 29–35.

Bean, L. H. (1935) 'Crop Adjustment Needed to Prevent Return to General Overproduction', in USDA, pp. 163–4.

Beaumont, P. (1993) *Pesticides, Policies, and People: A Guide to the Issues*, London: The Pesticides Trust.

Beck, U. (1992) *Risk Society: Towards a New Modernity* (Originally Published in German in 1986), London: Sage.

Becker, M. and N. Ashford (1995) 'Exploiting Opportunities for Pollution Prevention in EPA Enforcement Agreements', *Environmental Science and Technology*, vol. 29, no. 5, p. 220–6.

Becker, S. L. (1976) *Donald F. Jones and Hybrid Corn,* Conneticut Agricultural Experiment Station Bulletin no. 763, April 19.

Behmel, F. H and I. Neumann (1981) 'An Example of Agroforestry in Tropical Mountain Areas', in MacDonald (ed.) (1981), pp. 92–8.

Bell, M. (1979) 'The Exploration of Indigenous Knowledge: Whose Use of What for What?', *IDS Bulletin*, vol. 10, no. 2, pp. 44–50.

Bellon, M. R. (1991) 'The Ethnoecology of Maize Variety Management: A Case Study from Mexico', *Human Ecology*, vol. 19, no. 3, pp. 389–418.

Bellon, M. R. and S. B. Brush (1994) 'Keepers of Maize in Chiapas, Mexico', *Economic Botany*, vol. 48, no. 2, pp. 196–209.

Belshaw, D. (1979) 'Taking Indigenous Knowledge Seriously: The Case of Intercropping Techniques in East Africa', *IDS Bulletin*, vol. 10, no. 2, pp. 24–7.

Benedict, M. (1953) *Farm Policies of the United States, 1790–1950* New York: The Twentieth Century Fund.

Berardi, G. M. (1978) 'Organic and Conventional Wheat Production: Examination of Energy and Economics', *Agro-Ecosystems*, vol. 4, pp. 367–76.

Berland, J-P. and R. Lewontin (1986) 'Breeders' Rights and Patenting Life Forms', *Nature*, vol. 322, pp. 785–88.

Bernstein, H., B. Crow and M. Mackintosh (eds) (1990) *The Food Question: Profits Versus People?* London: Earthscan.

Berry, W. (1986) *The Unsettling of America: Culture and Agriculture,* San Francisco: Sierra Club Books.

BIA (BioIndustry Association) (1995) *Innovation from Nature: The Protection of Inventions in Biology*, London: BioIndustry Association.

Bierck, H. A. (1968) *The United States and Latin America, 1933–1968,* London: MacMillan.

Biggs, S. (1982) 'Institutions and Decision-Making in Agricultural Research', in Stewart and James (eds) (1982), pp. 209–22.

— (1990) 'A Multiple Source of Innovation Model of Agricultural Research and Technology Promotion', *World Development*, vol. 18, no. 11, pp. 1481–99.

Biggs, S. and E. Clay (1981) 'Sources of Innovation in Agricultural Technology', *World Development*, vol. 9, no. 4, pp. 321–36.

— (1988) 'Generation and Diffusion of Agricultural Technology: Theories and Experiences', in Ahmed and Ruttan (eds) (1988), pp. 19–67.

Bijker, W. E. (1995) *Of Bicycles, Bakelites and Bulbs: Toward a Theory of Sociotechnical Change,* Cambridge, MA and London: MIT Press.

Bijker, W. E., T. P. Hughes and T. J. Pinch (eds) (1987) *The Social Construction of Technological Systems: New Directions in the Sociology and History of Technology,* London: MIT Press.

Bijman, J. (1994) Agracetus: Patenting All Transgenic Cotton, *Biotechnology and Development Monitor*, no. 21, December, pp. 8–9.

Binswanger, H. P. (1978a) 'Induced Technical Change: Evolution of Thought', in Binswanger *et al.* (eds) (1978), pp. 13–43.

— (1978b) 'The Microeconomics of Induced Technical Change', in Binswanger *et al.* (eds) (1978), pp. 91–127.

Binswanger, H. P., V. W. Ruttan and others (eds) (1978) *Induced Innovation: Technology, Institutions, and Development,* London and Baltimore: Johns Hopkins University Press.

Blackhouse, R. (ed.) (1994) *New Directions in Economic Methodology,* London: Routledge.

Bockenhoff, E., *et al.* (1986) 'Analyse der Betriebs und Produktionsstruturen Sowie der Naturalertrage im Alternativen Landbau', *Berichte Uuber Landwirtschaft,* 64 (1), 1–39.

Bonnieux, F. and P. Rainelli (1997) 'Agricultural Use of Sewage Sludge and Municipal Waste and the Environment', in E. Romstad, J. Simonsen and A. Vatn (eds) (1997), *Controlling Mineral Emissions in European Agriculture,* Wallingford, Oxon: CAB International, pp. 157–73.

Boserup, E. (1965) *The Conditions of Agricultural Growth,* London: Allen and Unwin.

— (1981) *Population and Technology,* Oxford: Basil Blackwell.

Bould, A. and A. F. Kelly (1992) 'Plant Breeders' Rights in the UK', *Plant Varieties and Seeds,* vol. 5, no. 2, pp. 143–150.

Bowcock, A. and L. Cavalli-Sforza (1991) 'The Study of Variation in the Human Genome', *Genomics,* vol. 11, pp. 491–98.

Bowers, J. K. and P. Cheshire (1983) *Agriculture, the Countryside and Land Use,* London: Methuen.

Boyer, R. (1988) 'Technical Change and the Theory of "Régulation"', in Dosi *et al.* (eds) (1988), pp. 67–94.

Bradfield, R. (1957) 'The Agronomists' Accomplishments and Opportunities for Future Contributions in the International Field', *Agronomy Journal,* vol. 49, no. 12, pp. 621–5.

Brandolini, A. (1970) 'Maize', in Frankel and Bennett (eds) (1970), pp. 273–309.

Braverman, H. (1974) *Labor and Monopoly Capital,* Now York: Monthly Review Press.

Bray, J. O. and P. Watkins (1964) 'Technical Change in Corn Production in the United States, 1870–1960', *Journal of Farm Economics,* vol. 46, pp. 751–65.

Bressers, J. Th. A. and D. Huitema (1996) *Politics as Usual: The Effect of Policy-making on the Design of Economic Policy Instruments,* CSTM Studies en Rapporte, October.

British Agrochemicals Association (1996) *Annual Review and Handbook 1996,* Peterborough: British Agrochemicals Association.

Brockway, L. (1988) 'Plant Science and Colonial Expansion: The Botanical Chess Game', in Kloppenburg (ed.) (1988), pp. 49–66.

Bromley, D. W. (1989) *Economic Interests and Institutions: The Conceptual Foundations of Public Policy,* Oxford: Basil Blackwell.

Browder, J. O. (1989) *Fragile Lands of Latin America: Strategies for Sustainable Development.* Boulder: Westview.

Brown, A. D. H., O. H. Frankel, D. R. Marshall and J. T. Williams (eds) (1989) *The Use of Plant Genetic Resources,* Cambridge: Cambridge University Press.

Brown, W. L. (1975) 'Maize Germplasm Banks in the Western Hemisphere', in Frankel and Hawkes (eds) (1975), pp. 467–72.

Browning, J. A. and K. J. Frey (1969) 'Multiline Cultivars as a Means of Disease Control', *Annual Review of Phytopathology,* vol. 7, pp. 355–82.

Brush, S. B. (1991) 'A Farmer Based Approach to Conserving Crop Germplasm', *Economic Botany*, vol. 45, no. 2, pp. 153–65.

— (1992) 'Reconsidering the Green Revolution: Diversity and Stability in the Cradle Areas of Crop Domestication', *Human Ecology*, vol. 20, no. 2, pp. 145–67.

— (1994) *Providing Farmers' Rights Through In Situ Conservation of Crop Genetic Resources,* A Report to the Commission of Plant Genetic Resources, August.

Brush, S. B., J. E. Taylor and M. R. Bellon (1992) 'Biological Diversity and Technology Adoption in Andean Potato Agriculture', *Journal of Development Economics*, vol. 39, pp. 365–87.

Bull, D. (1982) *A Growing Problem: Pesticides and the Third World Poor,* Oxford: OXFAM.

Bunders, J. F. G. and J. E. W. Broerse (1991) *Appropriate Biotechnology in Small-Scale Agriculture: How to Reorient Research and Development,* Wallingford, Oxon: CAB International.

Burdon, J. J. and A. M. Jarosz (1989) 'Wild Relatives as a Source of Disease Resistance', in Brown *et al.* (eds) (1989), pp. 280–96.

Burmeister, L. L. (1987) 'The South Korean Green Revolution: Induced or Directed Innovation?', *Economic Development and Cultural Change*, vol. 35, pp. 766–90.

— (1988) *Research, Realpolitik and Development in Korea: The State and the Green Revolution.* Boulder: Westview.

— (1995) 'Induced Innovation and Agricultural Reserach in South Korea', in Koppel (ed.) (1995), pp. 39–55.

Burrow, J. W. (1968) 'Editor's Introduction', in Darwin (1968), pp. 11–48.

Busch, L. and W. Lacy (1983) *Science, Agriculture and the Politics of Research,* Epping: Bowker.

Busch, L., W. B. Lacy, J. Burkhardt and L. R. Lacy (1991) *Plants, Power, and Profit: Social, Economic, and Ethical Consequences of the New Biotechnologies,* Oxford: Basil Blackwell.

Butler, B. and R. Pistorius (1996) 'How Farmers' Rights Can be Used to Adapt Plant Breeders' Rights', *Biotechnology and Development Monitor,* no. 28, September, pp. 7–11.

Buttel, F. H. (1990) 'Biotechnology and Agricultural Development in the Third World', in Bernstein *et al.* (eds) (1990), pp. 163–80.

— (1995) 'Biotechnology: An Epoch-Making Technology', in Fransman *et al.* (eds) (1995), pp. 25–45.

Campbell, A. (1991) 'Microbes: The Laboratory and the Field', in Davis (ed.) (1991), pp. 28–45.

Cantley, M. (1995) 'Popular Attitudes, Information, Trust and the Public Interest', in Fransman *et al.* (eds) (1995), pp. 311–26.

Capra, F. (1983) *The Turning Point: Science, Society and the Rising Culture* (first published in 1982), London: Fontana.

Carson, R. (1965) *Silent Spring* (first published in 1962, 1982 printing), Harmondsworth: Penguin in Association with Hamilton.

Carter, J. (1995) 'Alley Farming: Have Resource-Poor Farmers Benefited?', *ODI Natural Resource Perspectives*, no. 3, June.

Caseley, J. C., G. W. Cussan and R. K. Atkin (1991) *Herbicide Resistance in Weeds and Crops,* Oxford: Butterworth-Heinemann.

Castillo, G. (1983) *How Participatory is Participatory Development? A Review of the Philippine Experience,* Manila: Philippine Institute of Development Studies.

Caulder, J. (1991) 'Biotechnology at the Forefront of Agriculture', in MacDonald (ed.) (1991), pp. 71–5.

Cavalli-Sforza, L. L. (1990) 'Opinion: How Can One Study Individual Variation For 3 Billion Nucleotides of the Human Genome?', *American Journal of Human Genetics*, vol. 46, pp. 649–51.

Cavigelli, M. and J. Kois (1988) *Sustainable Agriculture in Kansas: Case Studies of Five Organic Farms,* Whiting, Kansas: Kansas Rural Center.

CEC (Commission of the European Communities) (1996) *Report of the Review of the Directive 90/220/EEC in the Context of the Commission's Comunication on Biotechnology and the White Paper,* COM(96) 630 Final, Brussels, 10.12.96.

CGIAR (Consultative Group on International Agricultural Research) (1980) *Consultative Group on International Agricultural Research,* Washington, DC: CGIAR.

— (1993) *Intellectual Property Issues and Implications for IARCs,* Paper Prepared for the International Centers Week, Washington DC, October 25–29) Washington, DC: CGIAR.

Chacon, J. C. and S. R. Gleissman (1982) 'Use of the "Non-weed" Concept in Traditional Agroecosystems of Southern Mexico', *Agro-Ecosystems,* vol. 8, pp. 1–11.

Chambers, R. (1983) *Rural Development: Putting the First Last,* Harlow: Longman.

— (1993) *Challenging the Professions: Frontiers for Rural Development,* London: Intermediate Technology.

Chambers, R. and B. P. Ghildyal (1985) 'Agricultural Research for Resource-Poor Farmers: The Farmer First and Last Model', *Agricultural Administration,* vol. 20, no. 1, pp. 1–30.

Chambers, R. and J. Jiggins (1986) 'Agricultural Research for Resource-Poor Farmers: A Parsimonious Paradigm', *IDS Discussion Paper,* no. 220, Institute of Development Studies, Brighton, Sussex.

Chambers, R., A. Pacey and L. A. Thrupp (eds) (1989) *Farmer First: Farmer Innovation and Agricultural Research,* London: Intermediate Technology.

Chandler, R. F. Jr. (1968) 'Dwarf Rice: A Giant in Tropical Asia', in USDA 1968, pp. 252–5.

— (1972) 'The Impact on the Improved Tropical Plant Type on Rice Yields in South and Southeast Asia', in IRRI (1972), pp. 77–84.

Chang, T. T. (1989) 'The Case for Large Collections', in Brown *et al.* (eds) (1989), pp. 123–35.

CIAT (Centro Internacional de Agricultura Tropical) (1994) 'Apomixis Gene May Allow Farmers to Plant Their Own Hybrid Seed', *CIAT Press Release*, May.

CIMMYT (Centro Internacional de Mejoramiento de Maíz y Trigo, or, International Centre for Maize and Wheat Improvement) (1990) *1989/90 CIMMYT World Maize Facts and Trends: Realizing the Potential of Maize in Sub-Saharan Africa,* Mexico, D.F.: CIMMYT.

— (1991) *CIMMYT (1990 Annual Report. Sustaining Agricultural Resources in Developing Countries: Contributions of CIMMYT Research,* Mexico, D.F.: CIMMYT.

— (1992a) *World Maize Facts and Trends: Maize Research Investments and Impacts in Developing Countries,* Mexico, D.F.: CIMMYT.

— (1992b) *Enduring Designs for Change: An Account of CIMMYT's Research, its Impact, and its Future Directions,* Mexico, D.F.: CIMMYT.

— (1992c) *CIMMYT (1991 Annual Report. Improving the Productivity of Maize and Wheat in Developing Countries,* Mexico, DF: CIMMYT.

Clapp, J. (1994) *Dumping on the Poor,* Global Security Programme Occasional Paper no. 5, August, Cambridge: GSP.

Clark, N. (1987) 'Similarities and Differences Between Scientific and Technological Paradigms', *Futures,* vol. 19, no. 1, pp. 26–42.

— (1990) 'Evolution, Complex Systems and Technological Change', *Review of Political Economy,* vol. 2, no. 1, pp. 26–42.

Clark, N. and C. Juma (1987) *Long-run Economics: An Evolutionary Approach to Economic Growth,* London: Pinter.

— (1988) 'Evolutionary Theories in Economic Thought', In Dosi *et al.* (eds) (1988), pp. 197–218.

Clawson, D. L. (1985a) 'Harvest Security and Intraspecific Diversity in Traditional Tropical Agriculture', *Economic Botany,* vol. 39, no. 1, pp. 56–67.

— (1985b) 'Small Scale Polyculture: An Alternative Development Model', *Philippine Geographic Journal,* vol. 29, pp. 3–4.

Clay, E. J., H. D. Catling and P. R. Hobbs (1978) 'Yields of Deepwater Rice in Bangladesh', *IRRI Newsletter,* vol. 3, no. 5, pp. 11–2.

Cleaver, H. M. Jr. (1975) *The Origins of the Green Revolution,* Ph.D. Thesis, Stanford University, Ann Arbor, Michigan: Xerox University Microfilms.

Clunies-Ross, T. (1995) *Seeds, Crops and Vulnerability: A Re-examination of Diversity in British Agriculture,* Discussion Paper, July, Sturminster Newton: The Ecologist.

Clunies-Ross, T. and N. Hildyard (1992) *The Politics of Industrial Agriculture,* London: Earthscan.

Cochrane, W. (1957) *Farm Prices: Myth or Reality,* Minnesota: Minnesota University Press.

— (1979) *The Development of American Agriculture: A Historical Analysis,* Minnesota: Minesota University Press.

Cohen, J. I. (1994) 'Biotechnology Priorities, Planning, and Policies: A Framework For Decision Making', *ISNAR Research Report* no. 6, The Hague: ISNAR.

Cohen, J., J. B. Alcorn and C. S. Potter (1991a) 'Utilization and Conservation of Genetic Resources: International Projects for Sustainable Agriculture', *Economic Botany,* vol. 45, no. 2, pp. 190–9.

Cohen, J., J. T. Williams, D. L. Plucknett and H. Shands (1991b) 'Ex Situ Conservation of Plant Genetic Resources: Global Development and Environmental Concerns', *Science,* vol. 253, pp. 866–72.

Colchester, M. (1994) *Salvaging Nature: Indigenous Peoples, Protected Areas and Biodiversity Conservation,* January 1994 Draft of UNRISD Discussion Paper, Presented at a Global Security Programme seminar at Emmanuel College, Cambridge University, 'Parks or People: Indigenous Peoples in Global Triage', organised by the author.

Collier, G. W. (1928) 'Corn-Borer Control Adds to Farm Costs But is Worth While', in USDA, pp. 199–202.

Commandeur, P. (1993) Latin America Commences to 'Biotechnologize' its Industry, *Biotechnology and Development Monitor,* no. 14, March, pp. 3–5.

— (1995) 'Public Acceptance and Regulation of Biotechnology in Japan', *Biotechnology and Development Monitor,* no. 22, March, pp. 9–12.

Commandeur, P. and J. Komen (1992) 'Options for Biological Pest Control Increase', *Biotechnology and Development Monitor,* no. 13, Dec., pp. 6–7.

Connor, S. (1988) 'Genes on the Loose', *New Scientist*, vol. 118, no. 1614, pp. 65–8.

Conway, G. R. and E. B. Barbier (1991) *After the Green Revolution: Sustainable Agriculture For Development*, London: Earthscan.

Conway, G. R. and J. N. Pretty (1991) *Unwelcome Harvest: Agriculture and Pollution*, London: Earthscan.

Coombs, R., P. Saviotti and V. Walsh (1987) *Economics and Technical Change*, London: Macmillan Education.

Cooper, D., R. Vellvé and H. Hobbelink (eds) (1992) *Growing Diversity: Genetic Resources and Local Food Security*, London: Intermediate Technology.

Correa, C. (1992) 'Biological Resources and Intellectual Property Rights', *European Intellectual Property Review*, vol. 14, no. 5, pp. 154–7.

— (1993) 'Breeders' Rights Protection: New Proposals and Current Trends', *European Intellectual Property Review*, vol. 15, no. 2, pp. D-35–36.

Council of the European Union (1994) 'Council Regulation (EC) no. 2100/94 of 27 July 1994 on Community Plant Variety Rights', *Official Journal of the European Communities* no. L 227/1, September 1.

Cowan, R. (1987) *Backing the Wrong Horse: Sequential Technology Choice Under Increasing Returns*, D.Phil. Dissertation, Stanford University, July, Ann Arbor, Michigan: Xerox University Microfilms.

— (1991) 'Tortoises and Hares: Choice Among Technologies of Unknown Merit', *Economic Journal*, vol. 101, no. 4, pp. 801–14.

Cox, G. and M. D. Atkins (1979) *Agricultural Ecology: An Analysis of Food Production Systems*, San Francisco: W. H. Freeman.

Cox, T. S., J. Murphy and M. M. Goodman (1988) 'The Contribution of Exotic Germplasm to American Agriculture', in Kloppenburg (ed.) (1988), pp. 114–44.

Crabb, A. R. (1947) *The Hybrid-Corn Makers: Prophets of Plenty*, New Brunswick, NJ: Rutgers University Press.

Crespi, R. S. (1985) 'Patenting Plants: A Problem Area', *Annual Proceedings of the Phytochemical Society of Europe*, no. 26, pp. 197–203.

Cromwell, E. (1993) 'Patenting Plants: The Implications for Developing Countries', *ODI Briefing Paper*, November, London: ODI.

Cromwell, E. and S. Wiggins, with S. Wentzel (1993) *Sowing Beyond the State: NGOs and Seed Supply in Developing Countries*, London: ODI.

Crossley, G. (1984) 'Farmers Weekly Adviser Survey', *Farmers Weekly*, 6 January.

Crucible G. (1994) *People, Plants, and Patents: The Impact of Intellectual Property on Trade, Plant Biodiversity, and Rural Society*, Ottawa: International Development Research Centre.

Dalrymple, D. G. and J. P. Srivastava (1994) 'Transfer of Plant Cultivars: Seeds, Sectors and Society', in Anderson (ed.) (1994), pp. 180–207.

Danbom, D. (1979) *The Resisted Revolution: Urban America and the Industrialization of Agriculture 1900–1930*, Ames, IA: Iowa State University Press.

Darwin, C. (1968) *The Origin of Species* (1985 Reprint of the First Edition, First Published in 1859), London: Penguin.

Dasgupta, P. and P. Stoneman (eds) (1987) *Economic Policy and Technological Performance*, Cambridge: Cambridge University Press.

Da Silva, E., C. Ratledge and A. Sasson (eds) (1992) *Biotechnology: Economic and Social Aspects. Issues for Developing Countries*, Cambridge: Cambridge University Press.

Datta, S. C. and A. K. Banerjee (1978) 'Useful Weeds of West Bengal Rice Fields', *Economic Botany*, vol. 32, no. 3, pp. 297–310.

David, P. A. (1975) *Technical Choice, Innovation and Economic Growth,* Cambridge: Cambridge University Press.

— (1985) 'Clio and the Economics of QWERTY', *Economic History,* vol. 75, no. 2, pp. 332–37.

— (1987) 'Some New Standards for the Economics of Standardization in the Information Age', in Dasgupta and Stoneman (eds) (1987), pp. 206–39.

Davis, B. (ed.) (1991) *The Genetic Revolution: Scientific Prospects and Public Perceptions,* London: Johns Hopkins University Press.

Davis, B. (1991) 'The Issues: Prospects Versus Perceptions', in Davis (ed.) (1991), pp. 1–27.

Day, P. (1995) 'The Impact of Patents on Plant Breeding Using Biotechnology', in MacDonald (ed.) (1995), pp. 79–85.

Dazhong, W. and D. Pimentel (1990) 'Energy Flow in Agroecosystems of Northeast China', in Gleissman (ed.) (1990), pp. 322–36.

Deiaco, E., E. Hörnell and G. Vickery (1990) *Technology and Investment: Crucial Issues for the 1990s,* London: Pinter.

de Candolle, A. L. P. P. (1886) *Origin of Cultivated Plants,* New York: Hafner.

de Greef, W. (1996) 'A Perspective on Crop Biotechnology for the First Half of the 21st Century', in OECD, pp. 65–74.

de Janvry, A. (1978) 'Social Structures and Biased Technical Change in Argentine Agriculture', in Binswanger *et al.* (eds) (1978), pp. 297–323.

de Janvry, A. and E. P. LeVeen (1983) 'Aspects of the Political Economy of Technical Change in Developed Economies', in Piñeiro and Trigo (eds) (1983), pp. 25–44.

de Janvry, A., M. I. Fafchamps and E. Sadoulet (1995) 'Transaction Costs, Public Choice, and Induced Technological Innovations', in Koppel (ed.) (1995), pp. 151–65.

DEKALB (1997) From DEKALB website, page http://www.dekalb.com/highlights/seed/html#patent.

de Kathen, A. (1996) 'The Impact of Transgenic Crop Releases on Biodiversity in Developing Countries', *Biotechnology and Development Monitor,* no. 28, September, pp. 10–14.

Dempsey, G. (1992) 'Diversity and Crop Defence: The Roles of Farmer and Breeder', *Biotechnology and Development Monitor,* no. 13, December, pp. 13–15.

Dennis, J. V. (1987) *Farmer Management of Rice Variety Diversity in Northern Thailand,* Ph.D. Thesis, Cornell University, Ann Arbor, Michigan: Xerox University Microfilms.

Denslow, J. S. and C. Padoch (eds) (1988) *People of the Tropical Rainforest,* Berkeley: University of California Press.

Di Castri, F., F. W. G. Baker and M. Hadley eds. (1984) *Ecology in Practice, Part II: The Social Response,* Dublin: Tycooly.

Dietrich, M. (1997) 'Strategic Lock-in as a Human Issue: The Role of Professional Orientation', in Magnusson and Ottosson (eds) (1997), pp. 79–97.

Dinham, B. (1998). 'The Costs of Pesticides', *Pesticides News,* no. 39, March, p. 4.

Dogra, B. (1983) 'Traditional Agriculture in India: High Yields and No Waste', *The Ecologist* vol. 13, no. 2/3, pp. 84–7.

Donald, C. M. (1968) 'The Breeding of Crop Ideotypes', *Euphytica,* vol. 17, pp. 385–403.

Dosi, G. (1982) 'Technological Paradigms and Technological Trajectories: A Suggested Interpretation of the Determinants and Directions of Technical Change', *Research Policy*, vol. 11, pp. 147–62.

— (1988) 'The Nature of the Innovative Process', in Dosi *et al.* (eds) (1988), pp. 221–38.

Dosi, G. and L. Orsenigo (1988) 'Coordination and Transformation: An Overview of Structures, Behaviours and Change in Evolutionary Environments', in Dosi *et al.* (eds) (1988), pp. 13–37.

Dosi, G., C. Freeman, R. Nelson, G. Silverberg and L. Soete (eds) (1988) *Technical Change and Economic Theory*, London: Pinter.

Dover, M. and L. Talbot (1987) *To Feed the Earth*, Washington DC: World Resources Institute.

Doyle, J. (1985) *Altered Harvest*, New York: Penguin Books.

Dragun, A. and K. Jacobsson (eds) (1997) *New Horizons of Environmental Policy*, Cheltenham: Edward Elgar.

Dudley, N. (1987) *This Poisoned Earth: The Truth About Pesticides*, London: Piatkus.

Duffey, W. (1987) 'The Marvelous Gifts of Biotech: Will They Be Nourished or Stifled By Our International Patent Laws?', in WIPO (ed.) (1987), pp. 28–35.

Duvick, D. N. (1996) 'Utilization of Biotechnology in US Plant Breeding', *Biotechnology and Development Monitor*, no. 27, June, pp. 15–17.

Easlea, B. (1981) *Science and Sexual Oppression: Patriarchy's Confrontation with Woman and Nature*, London: Weidenfeld and Nicholson.

East, E. and D. Jones (1919) *Inbreeding and Outbreeding: Their Genetic and Sociological Significance*, Philadelphia: J. B. Lippincott.

Echeverría, R. (1990) *Inversiones Públicas y Privadas en la Investigación Sobre Maíz en México y Guatemala*, Documento de Trabajo 90/03 del Programa de Economía del CIMMYT, Mexico, DF: CIMMYT.

Economic and Political Weekly (1988a) 'The Patent Question', *Economic and Political Weekly*, October 29, pp. 2244–5.

— (1988b) 'Tailoring Patent Law to Suit MNCs', *Economic and Political Weekly*, October 29, pp. 2257–8.

Edwards, C. A., R. Lal, P. Madden, R. H. Miller and G. House (1990) *Sustainable Agricultural Systems*, Ankeny, Iowa: Soil and Water Conservation Society.

Ehrlich, P. (1968) *The Population Bomb*, New York: Ballantine.

Ehrlich, P. R. and A. H. Ehrlich (1970) *Population, Resources, Environment: Issues in Human Ecology*, San Francisco: W. H. Freeman.

Ekins, P. and M. Max-Neef (eds) (1992) *Real-life Economics: Understanding Wealth Creation*, London: Routledge.

Eliasson, G. (1988) 'Schumpeterian Innovation: Market Structure and the Stability of Industrial Development', in Hanusch (ed.) (1988), pp. 151–99.

Emerson, R. A. and E. M. East (1913) *The Inheritance of Quantitative Characters in Maize*. Nebraska Agricultural Experiment Station Research Bulletin no. 2.

ENDS (1996) 'Unilever and Nestlé Give Ground on Genetically Modified Soya', *ENDS Report*, no. 261, October, pp. 27–8.

Engel, P. G. H. (1993) 'Basic Configurations in Agricultural Innovation: Towards an Understanding of Leadership and Coordination in Complex Innovation Theaters', Draft of Paper Presented to the Working Group on the Social Construction of Agrarian Knowledge, XVth European Congress of Rural Sociology, Wageningen, 2–6 August 1993.

Entrikin, J. N. (1984) 'Carl O. Sauer, Philosopher in Spite of Himself', *The Geographical Review*, vol. 74, no. 4, pp. 387–408.

ERM (Environmental Resources Management) (1997) *Cadmium in Fertilizers: Volume I*, Final Report to DGIII, European Commission, December.

Erwin, W. (1993) 'Risk Assessment: A Farmer's Perspective', in NABC 1993, pp. 65–72.

Espenhain, F. (1992) 'Development of Plant Breeders' Rights in Denmark', *Plant Varieties and Seeds*, vol. 5, no. 2, pp. 167–73.

Europe Environment (1996a) 'GMOs: Alarm Over Genetically-Modified Soya Beans', *Europe Environment*, no. 485, October 8, pp. 5–6.

— (1996b) 'Foodstuffs: Consumers Have Right to Choose Genetically-Modified Soya', *Europe Environment*, no. 486, October 22, pp. 6.

— (1996c) 'Biotechnology: Commission Postpones Transgenic Maize Commission', *Europe Environment*, no. 489, December 3, pp. 17.

Evans, D. A. (1996) 'Genetic Engineering of Flavor and Shelf Life in Fruits and Vegetables', in NABC (1996), pp. 69–76.

FAO (Food and Agriculture Organisation of the United Nations) (1983) *International Undertaking on Plant Genetic Resources*, Document C 83/II REP/ 4 and 5, 22 November, Rome: FAO.

FAO Secretariat (1996) 'The State of the World's Plant Genetic Resources: Diversity and Erosion', *Third World Resurgence*, no. 72/73, Aug/Sep, pp. 20–3.

Farnsworth, N. (1988) 'Screening Plants for New Medicines', in Wilson (ed.) (1988), pp. 83–97.

Farrell, J. (1990) 'The Influence of Trees in Selected Agroecosystems in Mexico', in Gleissman (ed.) (1990), pp. 169–83.

Farrington, J. (1997) 'Farmers' Participation in Agricutural Research and Extension: Lessons from the Last Decade', *Biotechnology and Development Monitor*, no. 30, March, pp. 12–5.

Faucheux, S. (1997) 'Technological Change, Ecological Sustainability and Industrial Competitiveness', in Dragun and Jacobsson (eds) (1997) pp. 131–49.

Feenberg, A. (1991) *Critical Theory of Technology*, Oxford: Oxford University Press.

Felker, P. and R. S. Bandurski (1979) 'Uses and Potential Uses of Leguminous Trees for Minimal Energy Input Agriculture', *Economic Botany*, vol. 33, no. 2, pp. 172–84.

Fellner, W. (1961) 'Two Propositions in the Theory of Induced Innovation', *Economic Journal*, vol. 71, pp. 305–8.

Ferrell, J. A. (1941) *Memorandum of Conference with Vice President Henry Wallace*, February 10, 1941, Tarrytown: Rockefeller Foundation Archives: 1.2/323/1/2/.

Feyerabend, P. (1978) *Science in a Free Society*, London: Verso.

Field, N. (1988) *Biotechnology R&D: National Policy Issues and the Changing Role of Government*, Paris: OECD

Finger, J. M. (1991) 'The Meaning of "Unfair" in U.S. Import Policy', *World Bank Policy, Research, and External Affairs Working Papers*, WPS 745, August.

Finkel, A. M. and D. Golding (eds) (1994) *Worst Things First: The Debate Over Risk-based National Environmental Priorities*, Washington, DC: Resources for the Future.

Fischer, F. (1990) *Technocracy and the Politics of Expertise*, London: Sage.

Fischer, K. and J. Schott (eds) (1993) *Environmental Strategies for Industry: International Perspectives on Research Needs and Policy Implications*, Washington, DC: Island Press.

Fitzgerald, D. (1990) *The Business of Breeding: Hybrid Corn in Illinois, 1890–1940*, Ithaca and London: Cornell University Press.

Fogel, C. and I. Meister (1994) *Biotechnology and the Convention on Biological Diversity*, E-mail Communication to Newsgroup *igc:biodiversity*, October 18.

Fowler, C. (1994) *Unnatural Selection: Technology, Plants and Plant Evolution*, Yverdon, Switzerland: Gordon and Breach.

Fowler, C. and P. Mooney (1990) *The Threatened Gene: Food, Politics, and the Loss of Genetic Diversity*, Cambridge: The Lutterworth Press.

Fowler, C., E. Lachkovics, P. Mooney and Ho. Shand (1988) 'The Laws of Life', *Development Dialogue*, no. 1–2.

Francis, C. A. (ed.) (1986) *Multiple Cropping Systems*, London: Collier Macmillan.

Frankel, O. H. (ed.) (1973) *Survey of Crop Genetic Resources in their Centres of Diversity*, Rome: FAO International Biological Programme.

Frankel, O. H. and E. Bennett (1970) *Genetic Resources in Plants: Their Exploration and Conservation*, International Biological Programme, Handbook no. 11, Oxford: Blackwell Scientific Publications.

Frankel, O. H. and J. G. Hawkes (eds) (1975) *Crop Genetic Resources for Today and Tomorrow*, International Biological Programme no. 2, Cambridge: Cambridge University Press.

Frankel, R. (ed.) (1983) *Heterosis: Reappraisal of Theory and Practice*, New York: Springer-Verlag.

Fransman, M., G. Junne and A. Roobeek (eds) (1995) *The Biotechnology Revolution?* Oxford: Blackwell.

Freedman, S. M. (1980) 'Modifications of Traditional Rice Production Practices in the Developing World: An Energy Efficiency Analysis', *Agro-Ecosystems*, vol. 6, pp. 129–46.

Freeman, Christopher (1988) 'Introduction', in Dosi *et al.* (eds) (1988) pp. 1–8.

— (1992) *The Economics of Hope: Essays on Technical Change, Economic Growth and the Environment*, London: Pinter.

— (1995) 'Technological Revolutions: Historical Analogies', in Fransman *et al.* (eds) (1995) pp. 7–24.

Freeman, C. and C. Perez (1988) 'Structural Crises of Adjustment, Business Cycles and Investment Behaviour', in Dosi *et al.* (eds) (1988) pp. 38–66.

Freeman, C., J. Clark and L. Soete (1982) *Unemployment and Technical Innovation: A Study of Long Waves and Economic Development*, London: Frances Pinter.

Freeman, Peter and Tomas Fricke (1984) 'The Success of Javanese Multi-Storied Gardens', *The Ecologist*, vol. 14, no. 4, pp. 150–52.

Freiberg, B. (1996a) 'Welcome to the Era of Patented Hybrids', *Seed and Crops Industry*, vol. 47, no. 2, pp. 8 and 30.

— (1996b) 'Capitalism to the Rescue', *Seed and Crops Industry*, vol. 47, no. 3, p. 3.

Fukuoka, M. (1978) *The One-Straw Revolution*, Emmaus: Rodale Press.

— (1985) *The Natural Way of Farming*, Tokyo and New York: Japan Publications.

Galeano, Eduardo (1973) *Open Veins of Latin America: Five Centuries in the Pillage of a Continent* (originally published in Spanish in 1971), New York: Monthly Review Press.

Gardner, C. O. (1978) 'Population Improvement in Maize', in Walden (ed.) (1978) pp. 207–28.

George, Susan (1976) *How the Other Half Dies: The Real Reasons for World Hunger*, Harmondsworth: Penguin.

— (1990) *Ill Fares the Land*, Harmondsworth: Penguin.
Giddens, A. (1979) *Central Problems of Social Theory: Action, Structure and Contradiction in Social Analysis*, Basingstoke: Macmillan.
— (1990) *The Consequences of Modernity*, Cambridge: Polity Press.
— (1991) *Modernity and Self-Identity: Self and Society in the Late Modern Age*, Cambridge: Polity.
Glaeser, B. (ed.) (1987) *The Green Revolution Revisited: Critique and Alternatives*, London: Allen and Unwin.
Gleissman, S. R. (1986) 'Plant Interactions in Multiple Cropping Systems', in Francis (ed.) (1986) pp. 82–95.
— (ed.) (1990) *Agroecology: Researching the Ecological Basis for Sustainable Agriculture*, New York: Springer-Verlag.
— (1990) 'Integrating Trees into Agriculture: The Home Garden Agroecosystem as an Example of Agroforestry in the Tropics', in Gleissman (ed.) (1990) pp. 160–68.
Gómez-Pompa, A. (1978) 'An Old Answer to the Future', *Mazingara*, vol. 5, pp. 51–55.
Goodman, D., B. Sorj and J. Wilkinson (1987) *From Farming to Biotechnology: A Theory of Agro-industrial Development*, Oxford: Basil Blackwell.
Goodman, D. and J. Wilkinson (1990) 'Patterns of Research and Innovation in the Modern Agro-food System', in Lowe *et al.* (eds) (1990) pp. 127–46.
Gould, S. J. (1980) *The Panda's Thumb: More Reflections in Natural History*, New York: W. W. Norton.
Grabowski, R. (1988) 'The Theory of Induced Institutional Innovation: A Critique', *World Development*, vol. 16, no. 3, pp. 385–94.
— (1995) 'Induced Innovation: A Critical Perspective', in Koppel (ed.) (1995) pp. 73–92.
Graham, F. Jr. (1980) 'The Witch-hunt of Rachel Carson', *The Ecologist* vol. 10, no. 3, pp. 75–77.
Graham-Tomasi, T. (1991) 'Sustainability: Concepts and Implications for Agricultural Research Policy', In Pardey *et al.* (ed.) (1991) pp. 88–102.
GRAIN (Genetic Resources Action International) (1989) 'The IARCs and the Privatisation of Biotechnology', *Seedling*, vol. 6, nos. 3–4, July.
— (1991) 'Patenting Life Forms at the IARCs', *Disclosures*, no. 3, October.
— (1992a) 'Why Farmer-based Conservation and Improvement of Plant Genetic Resources?' in Cooper *et al.* (ed.) (1992) pp. 1–16.
— (1993a) 'A Decade in Review', *Seedling*, vol. 10, no. 1, pp. 3–12.
— (1993b) 'Cotton Has Been Patented', *Seedling*, vol. 10, no. 1, pp. 18.
— (1994a) 'Threats From the Test-tubes: Towards a Protocol on Biosafety', *GRAIN Biobriefing* no. 4, Part 4, June.
— (1994b) 'Intellectual Property Rights for Whom?' *GRAIN Biobriefing*, no. 4, Part 2, June.
Gray, R. B. and R. M. Merrill (1930) 'Corn-Borer Battle Enlists Many Kinds of Farm Machinery', in USDA (1930) pp. 175–78.
G. Alliance (1994) *Why Are Environmental Groups Concerned About Release of Genetically Modified Organisms Into the Environment*? a Green Alliance Briefing Document, October.
Greenpeace (1995) 'Greenpeace Hails Euro-Patent Bio-tech Decision: Decisive Move to Defend Nature', *Greenpeace Press Release*, 21 February.

Greenpeace Business (1996/7) 'Manufacturers and Retailers Reject Monsanto's Genetically Altered Soya Beans', *Greenpeace Business*, December/January, pp. 2–3.

Griliches, Z. (1958) 'The Demand for Fertilizer: An Economic Interpretation of a Technical Change', *Journal of Farm Economics*, vol. 40, pp. 591–606.

Habermas, J. (1968) *Toward a Rational Society: Student Protest, Science, and Politics*, Boston: Beacon Press.

Hadwiger, D. F. and W. P. Browne (eds) (1978) *The New Politics of Food*, Lexington, MA: Lexington Books.

— (1987) *Public Policy and Agricultural Technology: Adversity Despite Achievement*, Baisingstoke: Macmillan.

Hallauer, A. R. (1978) 'Potential of Exotic Germplasm for Maize Improvement', in Walden (ed.) (1978) pp. 229–47.

Hambridge, G. and E. Bressman (1936) 'Foreword and Summary', in USDA (1936) pp. 119–52.

Hamilton, C. (1939) 'The Social Effects of Recent Trends in the Mechanization of Agriculture', *Rural Sociology*, vol. 4, no. 1, pp. 1–24.

Hanusch, Horst (ed.) (1988) *Evolutionary Economics: Application of Schumpeter's Ideas*, Cambridge: Cambridge University Press.

Hardin, C. M. (1978) 'Agricultural Price Policy: The Political Role of Bureaucracy', in Hadwiger and Browne (eds) (1978) pp. 7–13.

Harding, S. (1986) *The Science Question in Feminism*, Ithica: Cornell University Press.

Harlan, H. V. and M. L. Martini (1936) 'Problems and Results in Barley Breeding', in USDA (1936) pp. 303–46.

Harlan, J. R. (1971). 'Agricultural Origins: Centers and Noncenters', *Science*, 174, 29 October, pp. 468–74.

— (1975) *Crops and Man*, Madison, Wisconsin: American Society of Agronomy and Crop Science Society of America.

— (1976) 'Genetic Resources in Wild Relatives of Crops', *Crop Science*, vol. 16, no. 3, pp. 329–33.

Harrar, J. G. (1954) 'A Pattern for Internatioonal Collaboration in Agriculture', *Advances in Agronomy*, vol. vi, pp. 95–119.

Harrington, L. W. (1992a) 'Measuring Sustainability: Issues and Options', in Hiemstra *et al.* (eds) (1992) pp. 3–16.

— (1992b) *Interpreting and Measuring Sustainability: Issues and Options*, Paper Presented at the NARS-CIMMYT-IRRI Workshop on Measuring Sustainability Through Farmer Monitoring, 6–9 May, Kathmandu, Nepal.

Harris, D. R. and G. C. Hillman (eds) (1989) *Foraging and Farming: The Evolution of Plant Exploration*, London: Unwin-Hyman.

Harvey, D. (1991) 'The CAP and Research and Development Policy', in Ritson and Harvey (eds) (1991) pp. 185–204.

Harwood, R. R. (1990) 'A History of Sustainable Agriculture', in Edwards *et al.* (eds) (1990) pp. 3–19.

Haskell, S. B. (1923) 'Agricultural Research in its Service to American Industry', *Journal of the American Society of Agronomy*, vol. 15, no. 12, pp. 473–81.

Hawkes, J. G. (1983) *The Diversity of Crop Plants*, Cambridge, Massachussets: Harvard University Press.

Hawthorn, G. (1991) *Plausible Worlds: Possibility and Understanding in History and the Social Sciences*, Cambridge: Cambridge University Press.

Hayami, Y. and K. Otsuka (1994) 'Beyond the Green Revolution: Agricultural Development Strategy into the New Century', in Anderson (ed.) (1994) pp. 15–42.

Hayami, Y. and V. Ruttan (1971) *Agricultural Development: An International Perspective*, Baltimore and London: Johns Hopkins University Press.

— (1985) *Agricultural Development: An International Perspective* (Revised and Expanded Edition), Baltimore and London: Johns Hopkins University Press.

Hayes, H. K. (1963) *A Professor's Story of Hybrid Corn*, Minneapolis, MN: Burgess Publishing Company.

Hayes, H. K. and R. J. Garber (1919) Synthetic Production of High-Protein Corn in Relation to Breeding, *Journal of the American Society of Agronomy*, vol. 11, no. 8, pp. 309–18.

Hearn, S. (1997) Communicating with Farmers, *Synopsis of a Paper Prepared for the Pesticides Forum*, February 5, PF/15.

Hecht, S. B. (1987) 'The Evolution of Agroecological Thought', in Altieri (1987a) pp. 1–20.

— (1989) 'Indigenous Soil Management in the Amazon Basin: Some Implications for Development', in Browder (ed.) (1989) pp. 166–81.

Heertje, A. (1988) 'Schumpeter and Technical Change', in Hanusch (ed.) (1988) pp. 71–89.

Henning, J. (1994) 'Economics of Organic Farming in Canada', in Lampkin and Padel (eds) (1994) pp. 143–60.

Hernandez X., Efraim (1985) 'Maize and Man in the Greater Southwest', *Economic Botany*, vol. 39, no. 4, pp. 416–30.

Hewitt de Alcantara, C. (1976) *Modernizing Mexican Agriculture: Socioeconomic Implications of Technological Change 1940–70*, Geneva: UNRISD.

Hicks, S. J. (1932) *A Theory of Wages*, London: Macmillan.

Hiemstra, W., C. Reintjes and E. van der Werf (eds) (1992) *Let Farmers Judge: Experiences in Assessing the Sustainability of Agriculture*, London: Intermediate Technology.

Higa, T. (1989) *Effective Microorganisms: A Biotechnology for Mankind*, Keynote Speech, Conference held at Khon Kaen University, Thailand, Tuesday October 17.

Hightower, J. (1973) *Hard Tomatoes, Hard Times*, Cambridge, MA: Schenkman Publishing.

Hindmarsh, R. (1991) 'The Flawed 'Sustainable' Promise of Genetic Engineering', *The Ecologist*, vol. 21, no. 5, pp. 196–205.

— (1992) Agricultural Biotechnologies: Ecosocial Concerns for a Sustainable Agriculture, in Lawrence *et al.* (eds) (1992) pp. 278–303.

Hobbelink, H. (1991) *Biotechnology and the Future of World Agriculture*, London: Zed Books.

Hodgson, G. M. (1988) *Economics and Institutions*, Cambridge: Polity Press.

— (1993) *Economics and Evolution: Bringing Life Back Into Economics*, Cambridge: Polity Press.

Holden, J., J. Peacock and T. Williams (1993) *Genes, Crops and the Environment*, Cambridge: Cambridge University Press.

Holderness, B. (1985) *British Agriculture Since 1945*, Manchester: Manchester University Press.

Howard, A. (1940) *An Agricultural Testament*, Oxford: Oxford University Press.

Howes, M. (1979) 'The Uses of ITK in Development', *IDS Bulletin*, vol. 10, no. 2, pp. 12–23.

Howes, M. and Chambers, R. (1979) 'Indigenous Technical Knowledge: Analysis, Implications and Issues', *IDS Bulletin*, vol. 10, no. 2, pp. 5–11.

Hull, F. H. (1945) 'Recurrent Selection for Specific Combining Ability in Corn', *Journal of American Society of Agronomy*, vol. 37, pp. 134–45.

ICDA (International Coallition for Development Action) (1994) 'Signature of the Uruguay Round', *ICDA Update*, no. 15, April-June, pp. 1–2.

— (1995) 'The Start of the World Trade Organisation', *ICDA Update*, no. 17, November 1994–February 1995, pp. 1–7.

IDS (Institute for Development Studies) (1979) 'Rural Development: Whose Knowledge Counts?' *IDS Bulletin*, vol. 10, no. 2.

IFOAM (International Federation of Organic Agriculture Movements) (1992) *Basic Standards of Organic Agriculture*, Tholey-Theley, Germany: IFOAM.

IRRI (International Rice Research Institute) (1962) *1961–62 IRRI Annual Review*, Los Banos, Philippines: IRRI.

— (1963) *Annual Report, 1963*, Los Banos, Philippines: IRRI.

— (1965) *The Mineral Nutrition of the Rice Plant: Proceedings of a Symposium at the International Rice Research Institute, February 1964*, Baltimore: John Hopkins for the International Rice Research Institute.

(1972) *Rice Breeding*, Los Banos, Philippines: IRRI.

Jackson, M. T. and B. V. Ford-Lloyd (1990) 'Plant Genetic Resources – A Perspective', in Jackson *et al.* (eds) (1990) pp. 1–17.

Jackson, M. T., B. V. Ford-Lloyd and M. L. Parry (eds) (1990) *Climatic Change and Plant Genetic Resources*, London: Belhaven.

Jaffé, W. and M. Rojas (1994) 'Maize Hybrids in Latin America: Issues and Options', *Biotechnology and Development Monitor*, no. 19, June, pp. 6–8.

James, C. (1998) 'Global Status and Distribution of Commercial Transgenic Crops in 1997', *Biotechnology and Development Monitor*, no. 35, June, pp. 9–12.

Janssens, M. J. J., I. F. Neumann and L. Froidevaux (1990) 'Low-Input Ideotypes', in Gleissman (ed.) (1990) pp. 130–45.

Jaura, R. (1996) '"No!" to Funds for Plant Genetic Conservation', *Third World Resurgence*, no. 72/73, Aug/Sep, pp. 13–14.

Jefferson, R. (1993) 'Agricultural Biotechnology: By Whom and For Whom?' *Biotechnology and Development Monitor*, no. 14, p. 24.

— (1994) Apomixis: A Social Revolution for Agriculture? *Biotechnology and Development Monitor*, no. 19, June, pp. 14–16.

Jenkins, M. T. (1978) 'Maize Breeding During the Development and Early Years of Hybrid Maize', in Walden (ed.) (1978) pp. 13–28.

— (1940) 'The Segregation of Genes Affecting Yield of Grain in Maize', *Journal of the American Society of Agronomy*, vol. 32, pp. 55–63.

— (1936) 'Corn Improvement', in USDA (1936) pp. 455–522.

Jennings, B. H. (1988) *Foundations of International Agricultural Research: Science and Politics in Mexican Agriculture*, Boulder and London: Westview Press.

Jennings, P. (1964) 'Plant Type as a Rice Breeding Objective', *Crop Science*, vol. 4, pp. 13–15.

Jiggins, J. (1986) 'Gender-related Impacts and the Work of the International Agricultural Research Centres', *CGIAR Study Paper* no. 17, Washington, D.C.: CGIAR.

JNC (Joint Nature Conservation Committee), English Nature and RSPB (Royal Society for the Protection of Birds) (1997) 'The Indirect Effect of Pesticides on Birds', Ongar: RSPB.

Johnson, A. W. (1972) 'Individuality and Experimentation in Traditional Agriculture', *Human Ecology*, vol. 1, no. 2, pp. 149–59.

Johnson, D. G. (1950) *Trade and Agriculture: A Study of Inconsistent Policies,* London: Chapman and Hall.

Juma, C. (1989) *The Gene Hunters: Biotechnology and the Scramble for Seeds,* African Centre for Technology Studies, Research Series, no. 1. London: Zed Press.

Junne, G. (1990) 'Takeover of Genentech–Lessons for Developing Countries?' *Biotechnology and Development Monitor,* no. 3, pp. 3–5.

— (1992) 'The Impact of Biotechnology on International Commodity Trade', in Da Silva *et al.* (eds) (1992) pp. 165–89.

Katz, M. and C. Shapiro (1985) 'Network Externalities, Competition and Compatability', *American Economic Review,* vol. 75, pp. 424–40.

— (1986) 'Technology Adoption in the Presence of Network Externalities', *Journal of Political Economy,* vol. 94, pp. 822–41.

Kauffman, S. A. (1988) 'The Evolution of Economic Webs', in Anderson *et al.* (eds) (1988) pp. 125–46.

— (1989) 'Cambrian Explosion and the Permian Quiescence: Implications of Rugged Fitness Landscapes', *Evolutionary Ecology,* vol. 3, pp 274–81.

— (1991) 'Antichaos and Adaptation', *Scientific American,* August, pp. 64–70.

Keller, E. F. (1985) *Reflections on Gender and Science,* New Haven: Yale University Press.

Kemp, R. (1993) 'An Economic Analysis of Cleaner Technology: Theory and Evidence', in Fischer and Schott (eds) (1993) pp. 79–113.

— (1997) *Environmental Policy and Technical Change: A Comparison of the Technological Impact of Policy Instruments,* Cheltenham: Edward Elgar.

Kenney, M. (1986) *Biotechnology: The University-Industrial Complex,* London: Yale University Press.

— (1995) 'University-Industry Relations in Biotechnology', in Fransman *et al.* (eds) (1995) pp. 302–10.

Kimmelman, B. (1983) 'The American Breeders' Association: Genetics and Eugenics in an Agricultural Context, 1903–13',. *Social Studies of Science,* vol. 13, pp. 163–204.

King, F. H. (1927) *Farmers of Forty Centuries, or Permanent Agriculture in China, Korea and Japan,* London: Jonathan Cape.

King, J. K. (1953) 'Rice Politics', *Foreign Affairs,* vol. 31, no. 3, pp. 453–60.

Kishor, N. M. (1992) *Pesticide Externalities, Comparative Advantage, and Commodity Trade: Cotton in Andhre Pradesh, India,* Policy Research Working Papers, Trade Policy, WPS 928, Country Economics Department, The World Bank, July, Washington, D.C.: World Bank.

Kislev, Y. and W. Peterson (1981) 'Induced Innovations and Farm Mechanization', *American Journal of Agricultural Economics,* vol. 63, pp. 562–65.

Kloppenburg, J. R. Jr. (1988) *First the Seed: The Political Economy of Plant Biotechnology, 1492–2000,* Cambridge: Cambridge University Press.

— (ed.) (1988) *Seeds and Sovereignty: The Use and Control of Plant Genetic Resources,* London: Duke University Press.

— (1993) 'Planetary Patriots or Sophisticated Scoundrels?' *Biotechnology and Development Monitor,* no. 16, p. 24.
Kloppenburg, J. R. Jr. and D. Kleinman (1988) 'Seeds of Controversy: National Property Versus Common Heritage', in Kloppenburg (ed.) (1988) pp. 173–203.
Knorr, D. (ed.) (1983) *Sustainable Food Systems,* Chichester: Ellis Horwood.
Knudson, M. and L. Hansen (1990) *Intellectual Property Rights and the Private Seed Industry,* USDA Economic Research Service, Agricultural Economics Report, no. 654. Washington, D.C.: USDA.
Komen, J. (1992) 'CGIAR Statement on Genetic Resources and Intellectual Property', *Biotechnology and Development Monitor,* no. 11, p. 20.
Koppel, B. M. (ed.) (1995) *Induced Innovation Theory and International Development: A Reassessment,* Baltimore: John Hopkins University Press.
— (1995a) 'Why a Reassessment?' in Koppel (ed.) (1995) pp. 3–21.
— (1995b) 'Induced Innovation Theory, Agricultural Research, and Asia's Green Revolution: A Reappraisal', in Koppel (ed.) (1995) pp. 56–72.
Koppel, B. M. and E. K. Oasa (1987) 'Induced Innovation Theory and Asia's Green Revolution: A Case Study of an Ideology of Neutrality', *Development and Change,* vol. 18, no. 1, pp. 29–67.
Kuhn, T. (1970) *The Structure of Scientific Revolutions* (2nd Enlarged Edition–First Published in 1962), Chicago: University of Chicago Press.
Kuznets, S. (1966) *Modern Economic Growth,* New Haven: Yale University Press.
Lacy, W. B. and L. Busch (1991) 'The Fourth Criterion: Social and Economic Impacts of Agricultural Biotechnology', in MacDonald (ed.) (1991) pp. 153–68.
Lakatos, I. (1970) 'Falsification and the Methodology of Scientific Research Programmes', in Lakatos and Musgrave (eds) (1970) pp. 91–195.
Lakatos, I. and A. E. Musgrave (eds) (1970) *Criticism and the Growth of Knowledge,* Cambridge: Cambridge University Press.
Lamola, L. M. (1995) 'The Role of Intellectual Property Rights in Modern Production Agriculture', in MacDonald (ed.) (1995) pp. 87–95.
Lampkin, N. H. (1985) 'The Profitability of Organic Farming Systems', *Paper Presented at the UCW Agricultural Society Conference,* Aberystwyth, November.
— (1990) *Organic Farming,* Ipswich: Farming Press.
— (ed.) (1990) *Collected Papers on Organic Farming,* 2nd Revised and Updated Edition), Centre for Organic Husbandry and Agroecology, University College of Wales, Aberystwyth.
— (1994a) 'Organic Farming: Sustainable Agriculture in Practice', in Lampkin and Padel (eds) (1994) pp. 3–9.
— (1994b) 'Researching Organic Farming Systems', in Lampkin and Padel (eds) (1994) pp. 27–43.
Lampkin, N. H. and S. Padel (eds) (1994) *The Economics of Organic Farming,* Wallingford, Oxon.: CAB International.
Lang, T. and H. Raven (1994) 'From Market to Hypermarket: Food Retailing in Britain', *The Ecologist,* vol. 24, no. 4, pp. 124–129.
Lappé, F. M. and J. Collins (1982) *Food First* (First Published in Great Britain in 1980), London: Abacus.
Lawrence, G., F. Vanclay and B. Furze (1992) *Agriculture, Environment and Society: Contemporary Issues for Australia,* Melbourne: Macmillan.
Leach, D. (1996) 'Clarifications in Roundup Ready Licensing', *Seed and Crops Industry,* vol. 47, no. 3, pp. 8 and 40.

LeBaron, H. (1991) 'Distribution and Seriousness of Herbicide Resistant Weed Infections', in Caseley *et al.* (eds) (1991) 27–43.

Lehmann, V. (1998) Patent on Seed Sterility Threatens Seed Saving, Biotechnology and Development Monitor, no. 35, June, pp. 6–8.

Lele, U., R. E. Christiansen and K. Kadiresan (1989) *Fertilizer Policy in Africa: Lessons From Development Programs and Adjustment Lending, 1970–87*, MADIA Discussion Paper 5, Washington D.C.: World Bank.

Lesser, W. (1990a) 'Sector Issues II: Seeds and Plants', in Siebeck (ed.) (1990) pp. 59–68.

— (1990b) 'An Overview of Intellectual Property Systems', in Siebeck (ed.) (1990) pp. 5–15.

— (1991) 'Needed Reforms in the Harmonization of U.S. Patent Law', in MacDonald (ed.) (1991) pp. 120–29.

Levin, S. (1991) 'An Ecological Perspective', in Davis (ed.) (1991) 45–58.

Levy, E. (1982) 'The Responsibility of the Scientific and Technological Enterprise in Technology Transfers', in Anderson *et al.* (eds) (1982) pp. 277–97.

Lewin, R. (1993) 'Genes From a Disappearing World', *New Scientist*, 29 May, pp. 25–29.

— (1993a) *Complexity: Life at the Edge of Chaos*, London: J. M. Dent.

Lewis, D. (1991) 'The Gene Hunters', *Geographical Magazine*, January, 36–9.

Liebman, M. (1987) 'Polyculture Cropping Systems', in Altieri (1987a) pp. 115–25.

Lindert, P. (1978) *Fertility and Scarcity in America*, Princeton: Princeton University Press.

Ling, C. Y. (1994) *US Interpretation Threatens Biodiversity Convention*, Posting from Third World Network to conference gen.biotech <gen.biotech@conf.igc.apc.org>, August 30.

Lipton, M. (1994) 'Agricultural Reserach Investment Themes and Issues: A View from Social-Science Research', in Anderson (ed.) (1994) pp. 601–616.

Lipton, M. with R. Longhurst (1989) *New Seeds and Poor People*, London: Unwin Hyman.

Lockeretz, W., G. Shearer and D. H. Kohl (1984) 'Organic Farming in the Corn Belt', *Science*, vol. 211. pp. 540–7.

Lockeretz, W. *et al.* (1975) *Organic and Conventional Crop Production in the Corn Belt*, Washington: Center for the Biology of Natural Systems, Washington University.

Loeb, J. (1912) *The Mechanistic Conception of Life: Biological Essays*, Chicago: University of Chicago Press.

Lohmann, L. (1991) 'Who Defends Biological Diversity? Conservation Strategies and the Case of Thailand', *The Ecologist*, vol. 21, no. 1, pp. 5–13.

Long, N. and A. Long (eds) (1992) *Battlefields of Knowledge: The Interlocking of Theory and Practice in Social Research and Development*, London: Routledge.

Long, N. and Villarreal, M. (1993) 'Exploring Development Interfaces: From the Transfer of Knowledge to the Transformation of Meaning', in Schuurman (ed.) (1993) pp. 140–68.

Lonnquist, J. H. (1961) *Progress from Recurrent Selection Procedures for the Improvement of Corn Populations*. University of Nebraska Agricultural Experiment Station Research Bulletin, no. 197, July.

Low, A. (1992) 'On-farm Research and Household Economics', in Hiemstra *et al.* (eds) (1992) pp. 49–61.

Low, A., C. Seubert and J. Waterworth (1991) *Extension of On-farm Research Findings: Issues from Experience in Southern Africa,* CIMMYT Economics Working Paper 91/03, Mexico, D.F.: CIMMYT.

Lowe, P., G. Cox, M. MacEwen, T. O'Riordan and M. Winter (1986) *Countryside Conflicts: The Politics of Farming, Forestry and Conservation,* Aldershot: Gower.

Lowe, P., T. Marsden and S. Whatmore (1990) 'Introduction: Technological Change and the Rural Environment', in Lowe *et al.* (eds) (1990).

— (eds) (1990) *Technological Change and the Rural Environment,* London: David Fulton.

Lowitt, R. (1979) 'Henry A. Wallace and the 1935 Purge in the Department of Agriculture', *Agricultural History,* vol. 53, no. 3, pp. 607–21.

Lumsden, R. D., R. García-E., J. A. Lewis and G. A. Frías-T. (1990) 'Reduction of Damping-Off Disease in Soils From Indigenous Mexican Agroecosystems', in Gleissman (ed.) (1990) pp. 83–103.

MacDonald, L. H. (ed.) (1981) *Agro-forestry in the African Humid Tropics,* Tokyo: United Nations University.

MacDonald, J. F. (ed.) (1991) *Agricultural Biotechnology at the Crossroads: Biological, Social and Institutional Concerns,* NABC Report 3, Ithaca, NY: National Agricultural Biotechnology Council.

— (ed.) (1995) *Genes For the Future: Discovery, Ownership, Access,* NABC Report 7. Ithaca, NY: National Agricultural Biotechnology Council.

MacDougall, C. L. and R. Hall (1994) *Intellectual Property Rights and the Biodiversity Convention: The Impact of GATT,* London: Friends of the Earth.

Machlup, F. and E. Penrose (1950) 'The Patent Controversy in the Nineteenth Century', *Journal of Economic History,* vol. 10, no. 1, pp. 1–29.

MacIntyre, A. A. (1987) 'Why Pesticides Received Extensive Use in America: A Political Economy of Agricultural Pest Management to 1970', *Natural Resources Journal,* vol. 27, no. 3, pp. 533–78.

Madden, J. P. (1987) 'Towards a New Covenant for Agricultural Academe', in Hadwiger and Browne (eds) (1987) pp. 93–108.

MAFF (Ministry of Agriculture, Fisheries and Food) (1992) *Agriculture in the UK, 1991,* London: HMSO.

— (1994) *Consultation Paper: Plant Variety Rights: Proposed Amendment of the Plant Varieties and Seeds Act 1964,* unpublished Mimeo, November 1994, Plant Varieties and Seeds Division, MAFF, Huntingdon, Cambridge.

— (1995) *Community Plant Variety Rights Update,* 31 March, Plant Varieties and Seeds Division, MAFF, Huntingdon, Cambridge.

— (1996) *Guide to National Listing of Varieties of Agricultural and Vegetable Crops in the UK,* London: MAFF.

Maga, J. (1983) 'Organically Grown Foods', in Knorr (ed.) (1983) pp. 305–51.

Magnusson, L. and J. Ottosson (eds) (1997) *Evolutionary Economics and Path Dependence,* Cheltenham: Edward Elgar.

Maitland, A. (1996) 'Against the Grain', *Financial Times,* Tuesday October 15.

Malthus, T. R. (1976a) *An Essay on the Principle of Population* (First Published 1798), Edited With an Introduction by Antony Flew, Harmondsworth: Penguin.

— (1976b) 'A Summary View of the Principle of Population', in Malthus 1976a.

Mangelsdorf, P. C. (1949) *Letter from Mangelsdorf to Weaver,* July 26, Rockefeller Foundation Archives, 1.1/323/3/18.

— (1951) 'Hybrid Corn', *Scientific American,* August, pp. 133–39.

— (1952) 'Foreword', in Wellhausen *et al.* (1952) pp. 5–6.

Marcus, A. I (1985) *Agricultural Science and the Quest for Legitimacy: Farmers, Agricultural Colleges, and Experiment Stations, 1870–1890*, Ames, Iowa State University Press.

— (1987) 'Constitutents and Constituencies: An Overview of the History of Public Agricultural Research Instituitions in America', in Hadwiger and Browne (eds) (1987) pp. 15–29.

Marcuse, H. (1964) *One-Dimensional Man*, London: Routledge.

Marglin, S. A. (1974) 'What Do Bosses Do? The Origins and Functions of Hierarchy in Capitalist Production', *Review of Radical Political Economics*, vol. 60, pp. 60–112.

— (1990) 'Losing Touch: the Cultural Conditions of Worker Accommodation and Resistance', in Marglin and Marglin (eds) (1990) pp. 217–82.

— (1992) 'Farmers, Seedsmen, and Scientists: Systems of Agriculture and Systems of Knowledge', unpublished Mimeo, May 1991, Revised March 1992.

Marglin, F. A. and S. A. Marglin (eds) (1990) *Dominating Knowledge: Development, Culture and Resistance*, Clarendon Press: Oxford.

Marrozi, J. (1997) 'Green Revolution Body Falls Into Red', *Financial Times*, Friday Jan. 17.

Marsden, T., P. Lowe and S. Whatmore (eds) (1992) *Labour and Locality: Uneven Development and the Rural Labour Process*, London: David Fulton.

Marsh, P. (1988) 'What Biotechnology Could Offer in the Future: Techniques That Promise Benefits May Be Harmful', *Financial Times*, 6 September.

Martinez-Alier, J. (1990) *Ecological Economics: Energy, Environment and Society*, Oxford: Blackwell.

Martínez Saldaña, T. (1991) 'Agricultura y Estado en Mexico: Siglo XX', in Teresa Rojas Rabiela (ed.) (1991) pp. 301–402.

Marx, K. (1959) *Capital. Volume I* (First Published in German in 1867), Moscow: Foreign Languages Publishing House.

— (1973) *Grundrisse: The Foundations of Political Economy* (Rough Draft) (First Published in German in 1939, Drafted in 1857–8), Translated With a Foreword By Martin Nicolaus, London: Allen Lane with New Left Review.

— u.d. *The Poverty of Philosophy* With Introduction by Frederick Engels, Drafted in 1846–7, London: Martin Lawrence.

Matheson, N., B. Rusmore, J. R. Sims, M. Spengler and E. L. Michelson (1991) *Cereal-legume Cropping Systems: Nine Farm Case Studies in the Dryland Northern Plains, Canadian Prairies and Intermountain Northwest*, Helena, Montana: Alternative Energy Resources Organization.

Maurya, D. M. (1989) 'The Innovative Approach of Indian Farmers', in Chambers *et al.* (eds) (1989) pp. 9–14.

Maurya, D. M., A. Bottrall and J. Farrington (1988) 'Improved Livelihoods, Genetic Diversity and Farmer Participation: A Strategy forRice Breeding in Rainfed Areas of India', *Experimental Agriculture*, vol. 24, pp. 311–20.

Maxwell, S. (1984) 'The Social Scientists in Farming Systems Research', *IDS Discussion Paper*, no. 199, November, Brighton: IDS.

McCalla, A. F. (1978) 'Politics of the Agricultural Research Establishment', in Hadwiger and Browne (eds) (1978) pp. 77–91.

— (1994) 'Ecoregional Basis for International Research Investment', in Anderson (ed.) (1994) pp. 96–105.

McGuahey, M. (1986) *Impact of Forestry Initiatives in the Sahel,* Washington DC: Chemonics International.

McGuire, S. (1996) 'Privatisation of Agricultural Research: A Case Study of Winter Wheat Breeding in the United Kingdom', unpublished M.Sc. Dissertation Submitted to the School of Development Studies of the University of East Anglia, September.

Meadows, D. H., D. L. Meadows, J. Randers and W. W. Behrens III (1972) *The Limits to Growth: A Report for the Club of Rome's Project on the Predicament of Mankind,* London: Earth Island.

Mearns, R. (1991) 'Environmental Implications of Structural Adjustment: Reflections on Scientific Method', *IDS Discussion Paper,* no. 284, February.

Mellon, M. (1991a) 'An Environmentalist Perspective', in Davis (ed.) (1991) pp. 61–78.

— (1991b) 'Biotechnology and the Environmental Vision', in MacDonald (ed.) (1991) pp. 65–70.

MELU (1977) 'Auswertung Drei-jahriger Erhebungen in Neun Biologisch-dynamisch Bewirtschafteten Betrieben', Baden-Wurttemberg. Stuttgart: Ministreium fur Ernahrung, Landwirtschaft und Umwelt.

Mends, C., T. L. Dobbs and J. D. Smolik (1989) *Economic Results of Alternative Farming Systems Trials at South Dakota State University's Northeast Research Station 1985–1988,* Research Report 89–3. Brookings, S. Dakota: Economics Department of South Dakota State University.

Mensch, G. (1979) *Stalemate in Technology: Innovation Overcomes the Depression,* Cambridge, MA: Ballinger.

Merchant, C. (1980) *The Death of Nature: Women, Ecology and the Scientific Revolution,* New York: Harper and Row.

Merrigan, K. A. (1995) 'Property is Nothing More Than Persuasion', in MacDonald (ed.) (1995) pp. 61–7.

Mies, M. and V. Shiva (1993) *Ecofeminism,* London: Zed Books.

Milton, F. B. (1989) 'Velvetbeans: An Alternative to Improve Small Farmers' Agriculture', *ILEIA Newsletter,* vol. 5, no. 2, pp. 8–9.

Mirowski, P. (1987) 'The Philosophical Bases of Institutional Economics', *Journal of Economic Issues,* vol. xxi, no. 3, pp. 1001–38.

— (1989) *More Heat Than Light: Economics as Social Physics, Physics as Nature's Economics,* Cambridge: Cambridge University Press.

— (1993) 'What Could Mathematical Rigour Mean?' *History of Economics Review,* vol. 20, pp. 41–60.

— (1994) 'What Are The Questions?' in Blackhouse (ed.) (1994) pp. 50–74.

Mitchell W., W. (1993) *Complexity: The Emerging Science at the Edge of Order and Chaos,* London: Viking.

Mollison, B. (1988) *Permaculture: A Designers' Manual,* Tyalgum, New South Wales: Tagari.

Mooney, P. R. (1983) 'The Law of the Seed', *Development Dialogue,* vol. 1–2, pp. 1–172.

— (1992) 'Towards a Folk Revolution', in Cooper *et al.* (eds) (1992) pp. 125–45.

Morales, H. L. (1984) 'Chinampas and Integrated Farms: Learning From the Rural Traditional Experience', in Di Castri *et al.* (eds) (1984) pp. 189–95.

Morse, L. (1996) 'From Petri Dish to Supper Plate', *Financial Times,* Tuesday October 15.

Mowery, D. and N. Rosenberg (1989) *Technology and the Pursuit of Economic Growth,* Cambridge: Cambridge University Press.

Mugabe, J. and E. Ouko (1994) 'Control Over Genetic Resources', *Biotechnology and Development Monitor,* no. 21, December, pp. 6–7.

Munson, A. (1993) 'Genetically Manipulated Organisms: International Implications and Policy-making', *International Affairs,* vol. 69, no. 3, pp. 497–517.

Murdoch, J. and J. Clark (1993) 'Sustainable Knowledge', *Paper Presented to the Working Group on the Social Construction of Agrarian Knowledge,* XVth European Congress of Rural Sociology, Wageningen, 2–6 August.

Murphy, M. (1992) *Organic Farming as a Business in Great Britain,* Cambridge: Agricultural Economics Unit, University of Cambridge.

Murphy, R. (1994) 'The Sociological Construction of a Science Without Nature', *Sociology,* vol. 4, no. 4, pp. 957–74.

Mutsaers, H. J. W. (1978) 'Mixed Cropping Experiments With Maize and Groundnuts', *Netherlands Journal of Agricultural Science,* vol. 26, pp. 344–53.

Myers, N. (1979) *The Sinking Ark: A New Look at the Problem of Disappearing Species,* Oxford: Pergamon Press.

NABC (National Agricultural Biotechnology Council) (1993) *Agricultural Biotechnology: A Public Conversation About Risk,* NABC Report 5, Ithaca New York: NABC.

— (1996) *Agricultural Biotechnology: Novel Products and New Partnerships,* NABC Report 8, Ithaca New York: NABC.

— (1985) 'Native Crop Diversity in Aridoamerica: Conservation of Regional Gene Pools', *Economic Botany,* vol. 39, 387–99.

Nabhan, G. P. (1985) 'Native Crop Diversity in Aridoamerica: Conservation of Regional Gene Pools', *Economic Botany,* vol. 39, 387–99.

Nana-Sinkam, S., M. Tankou and A. Haribou (1992) *Biotechnology Revolution: A Panacae or Myth to African Agriculture and Food Crisis,* Joint UNECA/FAO Monograph, Rome: UNECA/UNFAO

Nanda, M. (1991) 'Is Modern Science a Western, Patriarchal Myth? A Critique of the Populist Orthodoxy', *South Asia Bulletin,* vol. xi, Nos. 1&2, pp. 32–61.

Nations, J. D. (1988) 'The Lacandon Maya', in Denslow and Padoch (eds) (1988) pp. 86–88.

NBER (National Bureau for Economic Research) (1962) *The Rate and Direction of Inventive Activity: Economic and Social Factors,* Princeton: Princeton University Press.

Nelson, R. R. (1981) 'Research on Productivity Growth and Productivity Differences', *Journal of Economic Literature,* vol. 19, no. 3, pp. 1029–64.

— (1987) *Understanding Technical Change as an Evolutionary Process,* Amsterdam: Elsevier Science.

— (1990) 'Capitalism as an Engine of Progress', *Research Policy,* vol. 19, pp. 193–214.

Nelson, R. R. and S. G. Winter (1977) 'In Search of a Useful Theory of Innovation', *Research Policy,* vol. 6, pp. 36–76.

— (1982) *An Evolutionary Theory of Economic Change,* Cambridge, MA: Harvard University Press.

NIAB (National Institute of Agricultural Botany) (1996) *Cereal Variety Handbook: NIAB Recommended Lists of Cereals 1996,* Cambridge: NIAB.

Nicholas, R. B. (1991) 'Biotechnology at the Crossroads: Is Regulation the Gatekeeper?' in MacDonald (ed.) (1991) pp. 111–19.

Nicholson, J. (1978) 'Agricultural Scientists and Agricultural Research: The Case of Southern Corn Leaf Blight', in Hadwiger and Browne (eds) (1978) pp. 93–102.

Niebur, W. S. (1993) 'Maize', in OECD, pp. 113–21.

Nigh, R. B. (1976) *Evolutionary Ecology of Maya Agriculture in Highland Chiapas, Mexico*. PhD thesis, Stanford University, Ann Arbor, Michigan: University Microfilms International.

Nilsson, A. (1992) 'International Efforts to Prevent Bio-hazards', *Biotechnology and Development Monitor*, no. 10, March, pp. 16–18.

Noble, D. (1984) *Forces of Production*, Oxford: Oxford University Press.

Nolan, P. (1988) *The Political Economy of Collective Farms*, Cambridge: Polity Press.

Nordland, D. A., R. B. Chalfont and W. J. Lewis (1984) 'Arthropod Populations, Yield and Damage in Monocultures and Polycultures of Corn, Beans and Tomatoes', *Agro-Ecosystems*, vol. 11, pp. 353–67.

Norgaard, R. B. (1987) 'The Epistemological Basis of Agroecology', in Altieri (1987a) pp. 21–27.

— (1988a) 'The Rise of the Global Exchange Economy and the Loss of Biological Diversity', in Wilson (ed.) (1988) pp. 206–11.

— (1988b) 'Sustainable Development: A Co-Evolutionary View', *Futures*, vol. 29, no. 6, pp. 606–620.

— (1992) 'Coevolution of Economy, Society and Environment', in Ekins and Max-Neef (eds) (1992) pp. 76–88.

— (1994) *Development Betrayed: The End of Progress and a Coevolutionary Revisioning of the Future*, London: Routledge.

Nott, R. (1992) 'Patent Protection for Plants and Animals', *European Intellectual Property Review*, vol. 14, no. 3, March, pp. 79–86.

NRC (National Research Council) (1972) *Genetic Vulnerability of Major Crops*, Washington, D.C.: National Academy of Sciences.

— (1989) *Alternative Agriculture*, Washington DC: National Academy Press.

Oasa, E. K. (1981) *The International Rice Research Institute and the Green Revolution: A Case Study on the Politics of Agricultural Research*, PhD thesis, University of Hawaii, Ann Arbor, Michigan: University Microfilms International.

— (1987) 'The Political Economy of International Agricultural Research: A Review of the CGIAR's Response to Criticisms of the Green Revolution', in Glaeser (ed.) (1987) pp. 15–55.

OECD (1988) *Biotechnology and the Changing Role of Government*, Paris: OECD.

— (1992) *Biotechnology, Agriculture and Food*, Paris: OECD.

— (1993) *Traditional Crop Breeding Practices: An Historical Review to Serve as a Baseline For Assessing the Role of Modern Biotechnology*, Paris: OECD.

— (1996) *Science Technology Industry Review no. 19: Special Issue on Biotechnology*, Paris: OECD.

Oelhaf, R. C. (1978) *Organic Agriculture: Economic and Ecological Comparisons with Conventional Methods*, Chichester: J. Wiley.

Oldroyd, D. (1986) *The Arch of Knowledge: An Introductory Study of the History of the Philosophy and Methodology of Science*, London: Methuen and Co.

Oleson, A. and J. Voss (eds) (1979) *The Organization of Knowledge in Modern America, 1860–1920*, Baltimore and London: Johns Hopkins University Press.

Orsenigo, L. (1989) *The Emergence of Biotechnology*, London: Pinter.

OTA (US Congress Office of Technology Assessment) (1988) *New Developments in Biotechnology*, Special Report, OTA-BA-360, July, Washington, D.C.: US Government Printing Office.

— (1995) 'The Biotechniques in Agricultural Research', in Fransman *et al.* (eds) (1995) pp. 209–15.

Otero, G. (1995) ' The Coming Revolution of Biotechnology: A Critique of Buttel', in Fransman *et al.* (eds) (1995) pp. 46–61.

Outerbridge, T. (1987) 'The Disappearing Chianampas of Xochimilco', *The Ecologist*, vol. 17, no. 2, pp. 76–83.

Paarlberg, D. (1964) *American Farm Policy*, London: J. Wiley.

— (1978) 'A New Agenda For Agriculture', in Hadwiger and Browne (eds) (1978) pp. 135–40.

Padel, Susanne, and Nicholas Lampkin 1994. 'Farm-level Performance of Organic Farming Systems: An Overview', in Lampkin and Padel (eds) (1994) pp. 201–19.

Pallodino, P. (1994) 'Wizards and Devotees: On the Mendelian Theory of Inheritance and the Professionalization of Agricultural Science in Great Britain and the United States, 1880–1930', *History of Science*, vol. xxxii, pp. 409–44.

Pandey, B. and S. Chaturvedi (1993) 'India's Changing Seed Industry', *Biotechnology and Development Monitor*, no. 17, December, pp. 10–12.

Pardey, P., J. Roseboom and J. Anderson (eds) (1991) *Agricultural Research Policy: International Quantitative Perspectives*, Cambridge: Cambridge University Press.

Parr, J. F., R. I. Papendick and I. G. Youngberg (1983). Organic Farming in the United States: Principles and Perspectives. *Agro-Ecosystems*, vol. 8, pp. 183–201.

Parry, M. (1990) *Climate Change and World Agriculture.* London: Earthscan.

Pearce, D. and R. Tinch (1998) 'The True Price of Pesticides', in William Vorley and Dennis Keeney (eds) (1998) *Bugs in the System: Redesigning the Pesticide Industry for Sustainable Agriculture*, London: Earthscan, pp. 50–93.

Pearse, A. (1980) *Seeds of Plenty, Seeds of Want: Social and Economic Implications of the Green Revolution*, Oxford: Clarendon Press.

Penrose, E. (1951) *The Economics of the International Patent System*, Baltimore: Johns Hopkins University Press.

Pereira, W. (1991) 'Traditional Rice Growing in India', *The Ecologist*, vol. 21, no. 2, pp. 97–100.

Perez, C. (1983) 'Structural Change and Assimilation of New Technologies in the Economic and Social Systems', *Futures*, vol. 15, no. 5, pp. 357–75.

Perlas, N. (1993) 'Detoxifying the Green Revolution', *World Sustainable Agriculture Association Newsletter*, vol. 2, no. 11 (received by electronic mail–pagenumber unknown).

Perrin, R. M. (1977) 'Pest Management in Multiple Cropping Systems', *Agro-Ecosystems*, vol. 3, pp. 93–118.

Persley, G. J. (1990). *Beyond Mendel's Garden: Biotechnology in the Service of World Agriculture*, Wallingford, Oxon: CAB International for the World Bank.

Pieters, A. J. (1900) 'Seed Selling, Seed Growing, and Seed Testing', in USDA (1900), pp. 549–74.

Pimentel, D., G. Berardi and S. Fast (1983) 'Energy Efficiency of Farming Systems: Organic and Conventional Agriculture', *Agro-Ecosystems*, vol. 9, pp. 359–72.

Pimentel, D., W. Dazhong and M. Giampetro (1990) 'Technological Changes in Energy Use in US Agricultural Production', in Gleissman (ed.) (1990) pp. 305–21.

Pimentel, D. *et al.* (1993) *The Pesticide Question.*

Pinch, T. J. and W. E. Bijker (1987) *The Social Construction of Facts and Artefacts: Or How the Sociology of Science and the Sociology of Technology Might Benefit Each Other,* in Bijker *et al.* (eds) (1987) pp. 17–50.

Piñeiro, M. E. and E. J. Trigo (eds) (1983) *Technical Change and Social Conflict in Agriculture: Latin American Perspectives,* Boulder, Colorado: Westview.

Piñeiro, M. E., E. J. Trigo and R. Fiorentino (1979) 'Technical Change in Latin American Agriculture: A Conceptual Framework for its Interpretation', *Food Policy,* vol. 4, no. 3, pp. 169–77.

Piñeiro, M. E., R. Fiorentino, E. J. Trigo, A. Balcázar and A. Martínez (1983) 'Social Relations of Production, Conflict and Technical Change: The Case of Sugar Production in Colombia', in Piñeiro and Trigo (eds) (1983) pp. 47–69.

Pioneer (1997) Information from Pioneer website http://www.pioneer.com/CUSTOMER/RESEARCH/ROUND.HTML

Pistorius, R. (1992) 'Was the US' Refusal to sign the Biodiversity Convention Necessary?' *Biotechnology and Development Monitor,* no. 12, September, pp. 8–9.

Plant V. and Seeds (1989) 'Special Volume on UPOV', *Plant Varieties and Seeds,* vol. 2, no. 1, April.

Plotkin, M. (1988) 'The Outlook for New Agricultural and Industrial Products from the Tropics', in Wilson (ed.) (1988) pp. 106–16.

Plucknett, D. L. (1994) 'Sources of the Next Century's New Technology', in Anderson (1994) (ed.) pp. 343–73.

Plucknett, D. L. and N. J. H. Smith (1986) 'Sustaining Agricultural Yields: As Productivity Rises, Maintenance Research is Needed to Uphold the Gains', *BioScience,* vol. 36, no. 1, pp. 40–45.

— (1988) 'Plant Quarantine and the International Transfer of Germplasm', *CGIAR Study Paper* no. 25, Washington, D.C.: CGIAR.

Plucknett, D., N. Smith, J. Williams and N. Murth-Anishetty (1987) *Gene Banks and the World's Food,* Princeton: Princeton University Press.

Popper, K. R. (1970) 'Normal Science and Its Dangers', in Lakatos and Musgrave (eds) (1970) pp. 51–58.

Poschen, P. (1986) 'An Evaluation of the *Acacia albida*-based Agroforestry Practices in the Hararghe Highlands of Eastern Ethiopia, *Agroforestry Systems,* vol. 4, pp. 129–43.

Posey, D. L. (1995) *Indigenous Peoples and Traditional Resource Rights: A Basis for Equitable Relationships,* Paper Prepared for a Workshop on Traditional Resource Rights and Indigenous Peoples at the Green College Centre for Environmental Policy and Understanding, University of Oxford, June 28.

Pray, C. E., S. Ribeiro, R. Mueller and P. P. Rao (1991) 'Private Research and Public Benefit: The Private Seed Industry for Sorghum and Pearl Millet in India', *Research Policy,* vol. 20, pp. 315–24.

Prescott-Allen, R. and C. Prescott-Allen (1982) 'The Case for In Situ Conservation of Crop Genetic Resources', *Natural Resources,* vol. 23, pp. 15–20.

Pretty, J. and R. Chambers (1993) 'Towards a Learning Paradigm: New Professionalism and Institutions for Agriculture', *IDS Discussion Paper* no. 334, Dec.

Pretty, J. and R. Howes (1993) *Sustainable Agriculture in Britain: Recent Achievements and New Policy Challenges.* IIED Research Series, vol. 2, no. 1, London: IIED.

Pretty, J. (1998) *The Living Land: Agriculture, Food and Community Regeneration in Europe,* London: Earthscan.

Prigogine, I. and I. Stengers (1985) *Order Out of Chaos: Man's New Dialogue With Nature* (First Published in 1984), London: Fontana.

Primo B., C. Alberto (1990) 'Guidance From Economic Theory', in Siebeck (ed.) (1990) pp. 17–32.

Quaintance, H. (1904) 'The Influence of Farm Machinery on Production and Labor', *Publication of the American Economic Association*, Third Series, vol. 5, no. 4, Nov.

Qualset, C. O. (1991) 'Plant Biotechnology, Plant Breeding, Population Biology and Genetic Resources: Perspectives From a University Science', in MacDonald (ed.) (1991) pp. 81–90.

Raeburn, P. (1995) *The Last Harvest: The Genetic Gamble That Threatens To Destroy American Agriculture,* London: Simon and Schuster.

RAFI (Rural Advancement Foundation International) (1993a) 'Patents, Indigenous Peoples, and Human Genetic Diversity', *RAFI Communique*, May.

— (1993b) 'Amendments to the Plant Variety Protection Act of 1993', Internet conference "env.biotech", 4 October.

— (1993c) 'Indigenous Peoples Protest U.S. Secretary of Commerce Patent Claim on Guaymi Indian Cell Line', *RAFI Press Release*, October 25.

— (1994a) *Conserving Indigenous Knowledge: Integrating Two Systems of Innovation,* Study Commisioned by UNDP, New York: UNDP.

— (1994b) 'PVPA Update', *RAFI Press Release*, 23 August.

— (1997a) 'Bioserfdom: Technology, Intellectual Property and the Erosion of Farmers' Rights in the Industrialized World', *RAFI Communiqué*, March/April.

RAFI and GRAIN (1996) *Open Letter to the Ministerial Level Meeting on the Renewal of the CGIAR* (Held in Lucerne, 9–10 Feb).

Raghavan, C. (1990) *Recolonization: GATT, the Uruguay Round & the Third World,* London and Penang: Zed Books and Third World Network.

Raikes, P. (1988) *Modernising Hunger*. London: Catholic Institute for International Relations with James Currey.

Rasmussen, W. D. (1987) 'Public Experimentation and Innovation: An Effective Past but Uncertain Future', *American Journal of Agricultural Economics*, vol. 69, no. 5, pp. 890–98.

Ravnborg, H. M. (1992) *The CGIAR in Transition: Implications for the Poor, Sustainability and the National Research Systems,* ODI Agricultural Research and Administration (Research and Extension) Network Paper 31, June. London: ODI.

Redman, M. (1996) *Industrial Agriculture: Counting the Costs,* Paper Prepared for the Soil Association, September.

Reed, C. A. (ed.) (1977) *Origins of Agriculture,* The Hague: Mouton Publishers.

Reintjes, C., B. Haverkort and A. Waters-Bayer (1992) *Farming for the Future: An Introduction to Low-External-Input and Sustainable Agriculture,* London: Macmillan Press.

Reitz, L. P. (1968) 'Short Wheats Stand Tall', in USDA (1968), pp. 236–239.

Rhoades, R. (1989) *Evolution of Agricultural Research and Development Since 1950: Toward an Integrated Framework,* Sustainable Agriculture Programme, Gatekeeper Series SA12' London: IIED.

Ricardo, D. (1973) *The Principles of Political Economy and Taxation* (Reprint of the Third Edition, first published in 1821), London: J. M. Dent.

Richards, P. (1985) *Indigenous Agricultural Revolution: Ecology and Food Production in West Africa,* London: Hutchinson.

— (1986) *Coping With Hunger: Hazard and Experiment in an African Rice-Farming System,* London: Allen and Unwin.

— (1989) 'Agriculture as a Performance', in Chambers *et al.* (eds) (1989) pp. 39–43.

— (1994a) 'Local Knowledge Formation and Validation: The Case of Rice in Central Sierra Leone', in Scoones and Thompson (eds) (1994) pp. 165–70.

— (1994b) 'The Shaping of Biotechnology: Institutional Culture and Ideotypes', *Biotechnology and Development Monitor,* no. 18, March, p. 24.

Richey, F. D. (1922) 'The Experimental Basis for the Present Status of Corn Breeding', *Journal of the American Society of* Agronomy, vol. 14, Nos. 1&2, pp. 1–17.

— (1927) 'Corn Breeding in New Experiments', in USDA (1927), pp. 247–49.

— (1933) 'Corn Hybrids Result From Crossing Carefully Selected Parent Lines', in USDA (1933) , pp. 182–90.

Rindos, D. (1984). *The Origins of Agriculture: An Evolutionary Perspective,* London: Academic Press.

— (1989) 'Darwinism and its Role in the Explanation of Domestication', in Harris and Hillman (eds) (1989) pp. 27–41.

Rissler, J. and Mellon, M. (1993) *Perils Amidst the Promise: Ecological Risks of Transgenic Crops in a Global Market,* Cambridge, MA: Union of Concerned Scientists.

Ritson, C. and D. Harvey (eds) (1991) *The Common Agricultural Policy and the World Economy: Essays in Honour of John Ashton,* Wallingford, Oxon: CAB International.

Rizvi, S. J. H. and V. Rizvi (eds) (1992) *Allelopathy: Basic and Applied Aspects,* London: Chapman and Hall.

Rizvi, S. J. H., H. Haque, V. K. Singh and V. Rizvi (1992) 'A Discipline Called Allelopathy', in Rizvi and Rizvi (eds) (1992) pp. 1–10.

Roberts, L. (1992) 'How to Sample the World's Genetic Diversity', *Science,* vol. 257, August 28, pp. 1204–05.

Robinson, R. A. (1996) *Return to Resistance: Breeding Crops to Reduce Pesticide Resistance,* Davis, California: agAccess.

Rocheleau, D., K. Wachira, L. Malaret and B. M. Wanjohi (1989) 'Local Knowledge for Agroforestry and Native Plants', in Chambers *et al.* (eds) (1989) pp. 14–24.

Rockefeller, N. A. (1951) 'Widening the Boundaries of National Interest', *Foreign Affairs,* vol. 29, no. 4, pp. 523–38.

Rodale, R. (1990) 'Sustainability: An Opportunity for Leadership', in Edwards *et al.* (eds) (1990) pp. 77–86.

Rojas Rabiela, Teresa (ed.) (1991) *La Agricultura en Tierras Mexicanas Desde Sus Origenes Hasta Nuestros Dias,* Mexico, D.F.: Grijalbo.

Rome, A. W. (1982) 'American Farmers as Entrepreneurs, 1870–1900', *Agricultural History,* vol. 56, no. 1, pp. 37–49.

Romeiro, A. R. (1987) 'Alternative Developments in Brazil', in Glaeser (ed.) (1987) pp. 79–110.

Roobeek, A. J. M. (1995) 'Biotechnology: A Core Technology in a New Technoeconomic Paradigm', in Fransman *et al.* (eds) (1995) pp. 62–88.

Rosenberg, C. (1976). *No Other Gods: On Science and American Social Thought,* Baltimore: Johns Hopkins Univ Press (1976).

— (1977) 'Rationalization and Reality in the Shaping of American Agricultural Research, 1875–1914', *Social Studies of Science,* vol. 7, no. 4, pp. 401–22.

Rosenberg, N. (1969) 'The Design of Technical Change: Inducement Mechanisms and Focusing Devices', *Economic Development and Cultural Change,* vol. 18, pp. 1–24.
— (1982) *Inside the Black Box: Technology and Economics,* Cambridge: Cambridge University Press.
— (1990). 'Why Do Firms Do Their Basic Research (With their Own Money)?' *Research Policy*, vol. 19. pp. 165–74.
— (1994) *Exploring the Black Box: Technology, Economics, and History,* Cambridge: Cambridge University Press.
Rosenblum, G. (1994) *On the Way to Market: Roadblocks to Reducing Pesticide Use on Produce,* Washington DC: Public Voice for Food and Health Policy.
Rosensweig, M. L., J. S. Brown and T. L. Vincent (1987) 'Red Queens and ESS: The Coevolution of Evolutionary Rates', *Evolutionary Ecology*, vol. 1, pp. 59–94.
Ross, E. D. (1938) 'The Father of the Land-Grant College', *Agricultural History,* vol. 12, pp. 151–186.
Rossiter, M. W. (1975) *The Emergence of Agricultural Science: Justus Liebig and the Americans, 1840–80,* New Haven and London: Yale University Press.
— (1979) 'The Organization of the Agricultural Sciences', in Oleson and Voss (eds) (1979) pp. 211–48.
Rothschild, K. W. (1988) 'Discussion', in Hanusch (ed.) (1988) pp. 90–4.
Rouse, J. (1987) *Knowledge and Power: Toward a Political Philosophy of Science,* Ithaca: Cornell University Press.
Roy, S. (1990) *Agriculture and Technology in Developing Countries: India and Nigeria,* London: Sage.
Ruivenkamp, G. (1992) 'Can We Avert An Oil Crisis?' in Ahmed (ed.) (1992) pp. 166–88.
Ruttan, V. W. (1971) 'Technology and the Environment', *American Journal of Agricultural Economics*, vol. 53, pp. 707–17.
— (1978) 'Induced Institutional Innovation', in Binswanger *et al.* (eds) (1978) pp. 327–57.
— (1983) 'An Induced Innovation Interpretation of Technical Change in Agriculture in Developed Countries', in Piñeiro and Trigo (eds) (1983) pp. 2–24.
— (1988a) 'Generation and Diffusion of Agricultural Technology: Issues, Concepts and Analysis', in Ahmed and Ruttan (eds) (1988) pp. 69–127.
— (1988b) 'Cultural Endowments and Economic Development: What Can We Learn From Anthropology?' *Economic Development and Cultural Change,* vol. 36, Supplementary Issue, April, pp. S248.
Ruttan, V. W. and Y. Hayami (1984) 'Toward a Theory of Induced Institutional Innovation', *Journal of Development Studies*, vol. 20, pp. 203–23.
— (1995a) 'Induced Innovation Theory and Agricultural Development: A Personal View', in Koppel (ed.) (1995) pp. 22–36.
— (1995b) 'Induced Innovation Theory and Agricultural Development: A Reassessment', in Koppel (ed.) (1995) pp. 169–88.
Ruttan, V. W., H. P. Binswanger, Y. Hayami, W. W. Wade and A. Weber (1978) 'Factor Productivity and Growth: A Historical Interpretation', in Binswanger *et al.* (eds) (1978) pp. 44–87.
Ryan, B. and N. C. Gross (1943) 'The Diffusion of Hybrid Seed Corn in Two Iowa Communities', *Rural Sociology*, March, pp. 15–24.

Ryerson, K. A. (1933) 'History and Significance of the Foreign Plant Introduction Work of the United States Department of Agriculture', *Agricultural History*. vol. 7, no. 2, pp. 110–28.

Sagoff, M. (1988) *The Economy of the Earth*, Cambridge: Cambridge University Press.

Sahs, W. W., G. W. Lesoing and C. A. Francis (1992) 'Rotation and Manure Effects on Crop Yields and Soil Characteristics in Eastern Nebraska', *Agronomy Abstracts*, vol. 84, p. 155.

Salas, M. A. (1994) '"The Technicians Only Believe in Science and Cannot Read the Sky": The Cultural Dimension of the Knowledge Conflict in the Andes', in Scoones and Thompson (eds) (1994) pp. 57–69.

Salazar, R. (1992) 'Community Plant Genetic Resources Management: Experiences in Southeast Asia', in Cooper *et al.* (eds) (1992) pp. 17–29.

Salter, W. E. G. (1960) *Productivity and Technical Change*, Cambridge: Cambridge University Press.

Sánchez G., J. de Jesús and J. A. R. Corral (1997) 'Teosinte Distribution in Mexico', in Serratos *et al.* (eds) (1997) pp. 18–35.

Sanders, J. H. and V. W. Ruttan (1978) 'Biased Choice of Technology in Brazilian Agriculture', in Biswanger *et al.* (eds) (1978) pp. 276–96.

Sattaur, O. (1988) 'Native is Beautiful', *New Scientist*, vol. 118, no. 1615, pp. 54–7.

— (1989) 'The Shrinking Gene Pool', *New Scientist*, vol. 123, no. 1675. pp. 37–41.

Sauer, C. O. (1941a) *Letter from Sauer to Willits*, February 5, Rockefeller Foundation Archives, 1.2/323/10/63.

— (1941b) *Letter from Sauer to Willits*, March 12, Rockefeller Foundation Archives, 1.2/323/10/63.

— (1945) *Letter from Sauer to Willits*. Feb. 12, Rockefeller Foundation Archives, 1.1/200/391/4636.

— (1952) *Agricultural Origins and Dispersals*, New York: American Geographical Society.

— (1969) *Seeds, Spades, Hearths, and Herds* (2nd Edition), Cambridge, MA: MIT Press.

Schilthuis, W. (1994) *Biodynamic Agriculture*, Edinburgh: Floris.

Schneider, K. (1985) 'US Opposes a UN Plan to Collect, Store Genes Essential to Food Supply', *International Herald Tribune*, Friday November 29.

— (1986) 'Argentines Angry Over US Rabies Vaccine Test', *New York Times*, November 11.

Schultz, T. (1964) *Transforming Traditional Agriculture* (1983 edition), Chicago: Chicago University Press.

Schumpeter, J. (1939) *Business Cycles: A Theoretical, Historical and Statistical Analysis of the Capitalist Process*, New York: McGraw-Hill.

— (1951) *Preface of Japanese edition of Theorie Der Wirtschaftlichen Entwicklung*, as translated by I. Nakayama and S. Tobata, Tokyo: I. Shoten (1937) (Reprinted in R. V. Clemence (ed.) (1951) *Essays of J. A. Schumpeter*, Cambridge, MA: Addison-Wesley.

— (1976) *Capitalism, Socialism and Democracy*, (5th Edition with a New Introduction, First Published in Great Britain in 1943), London: George Allen and Unwin.

— (1983) *The Theory of Economic Development: An Inquiry Into Profits, Capital, Credit, Interest and the Business Cycle* (Original English version 1934), London: Transaction Books.

Schuurman, F. (ed.) (1993) *Beyond the Impasse: New Directions in Development Theory,* London: Zed Books.

Scobie, G. M. (1984) *Investment in Agricultural Research: Some Economic Principles,* CIMMYT Working Paper, August, Mexico, D.F.: CIMMYT..

Scoones, I. and J. Thompson (eds) (1994) *Beyond Farmer First: Rural People's Knowledge, Agricultural Research and Extension Practice,* London: Intermediate Technology Publications.

Scoones, I. and J. Thompson (1993) 'Challenging the Populist Perspective: Rural People's Knowledge, Agricultural Research and Extension Practice', *IDS Discussion Paper*, no. 332, December.

Scott-Ram, N. (1995) 'The Patentability of Human Genes', in BIA 1995, pp. 11–14.

SEASAN (Southeast Asia Sustainable Agriculture Network) (1992) *On Sustainable Agriculture for the Lowlands,* Bangkok: SEASAN.

Sederoff, R. and Meagher, L. (1995) 'Access to Intellectual Property in Biotechnology: Constraints on the Research Enterprise', in MacDonald (ed.) (1995) pp. 71–8.

Seed and C. Industry (1996) 'The Concern Over Monsanto's Roundup Ready Technology', *Seed and Crops Industry,* vol. 47, no. 2, pp. 4–5, and 40.

Sehgal, S. (1996) 'IPR Driven Restructuring of the Seed Industry', *Biotechnology and Development Monitor,* no. 29, December, pp17–21.

Seigler, D. (ed.) (1977) *Crop Resources,* London: Academic Press.

Seiler, A. (1998) '*Sui Generis* Systems: Obligations and Options for Developing Countries', Biotechnology and Development Monitor, no. 34, March, pp. 2–5.

Serratos, J. A., M. C. Willcox and F. Castillo (eds) (1997) *Gene Flow Among Maize Landraces, Improved Maize Varieties and Teosinte: Implications for Transgenic Maize,* Mexico, D.F.: CIMMYT.

Sharp, M. (1990) 'Technological Trajectories and Corporate Strategies in the Diffusion of Biotechnology', in Deiaco *et al.* (eds) (1990) pp. 93–114.

Shiva, V. (1990) *Staying Alive: Women, Ecology and Development,* London: Zed Press.

— (1994) 'The Need For *Sui Generis* Rights', *Seedling,* vol. 12, no. 1, March, pp. 11–15.

— (1996) 'South Safeguards Farmers' Rights on Agricultural Biodiversity', *Third World Resurgence,* no. 72/73, August/September, pp. 7–9.

Shiva, V. and R. Holla-Bhar (1993) 'Intellectual Piracy and the Neem Tree', *The Ecologist,* vol. 23, no. 6, November/December, pp. 223–7.

Siebeck, W. E. (ed.) (1990) *Strengthening Protection of Intellectual Property in Developing Countries: A Survey of the Literature.* World Bank Discussion Papers, no. 112, Washington D.C.: World Bank.

Siebeck, W. E. (1990) 'Introduction', in Siebeck (ed.) (1990) pp. 1–3.

Silva, Eugênio da Costa E. (1995) 'The Protection of Intellectual Property for Local and Indigenous Communities', *European Intellectual Property Review,* vol. 17, no. 11, pp. 546–9.

Simmonds, N. W. (1976) *Evolution of Crop Plants,* London: Longman.

— (1979) *Principles of Crop Improvement,* London: Longman.

Smith, A. (1976) *An Enquiry into the Nature and Causes of the Wealth of Nations, vol. 1* (1979 Reprint, 3rd Edition, First Published 1784, First Edition Published 1776, General Editors R. H. Campbell and A. S. Skinner, Textual Editor W. B. Todd), Oxford: Clarendon Press.

Smith, J. S. C. (1992) 'Plant Breeders' Rights in the USA; Changing Approaches and Appropriate Technologies in Support of Germplasm Enhancement', *Plant Varieties and Seeds*, vol. 5, no. 2, pp. 183–99.

Solleiro, J. L. (1995) 'Ownership of Biodiversity: A Developing Country's Perspective on an Open International Debate', in MacDonald (ed.) (1995) pp. 109–16.

Spitz, P. (1987) 'The Green Revolution Re-examined in India', in Glaeser (ed.) (1987) pp. 56–73.

Sprague, G. F. (1983) 'Heterosis in Maize: Theory and Practice', in Frankel (ed.) (1983) pp. 47–70.

Sprague, G. F. and M. T. Jenkins (1943) 'A Comparison of Synthetic Varieties, Multiple Crosses and Double Crosses in Corn', *Journal of the American Society of Agronomy*, vol. 35, pp. 137–47.

Stakman, E.C., Bradfield, R. and Mangelsdorf, P. (1967) *Campaigns Against Hunger,* Cambridge, MA: Belknap Press of Harvard University.

— (1951) *The World Food Problem, Agriculture, and the Rockefeller Foundation,* June 21, Rockefeller Foundation Archives, RG 3/915/3/23.

Stanhill, G. (1990) 'The Comparative Productivity of Organic Agriculture', *Agriculture, Ecosystems and the Environment,* vol. 30, pp. 1–26.

Steele, L. (1978) 'The Hybrid Corn Industry in the United States', in Walden (ed.) (1978) pp. 29–40.

Steinmann, R. (1983) *Der Biologischer Landbau – ein Betriebswirtschaftlicher Vergleich,* Schriftenreihe der FAT, 19. Forschungsanstalt fur Betriebswirtschaft und Landtechnik FAT, Tänikon, Switzerland.

Stewart, F. and J. James (eds) (1982) *The Economics of New Technology in Developing Countries,* London: Pinter.

Stout, B. (1990) *Handbook of Energy for World Agriculture,* Barking: Elsevier Science.

Stringfield, G. H. (1964) 'Objectives in Corn Improvement', *Advances in Agronomy*, vol. 16, pp. 101–37.

Sutherland, P. (1994) 'Seeds of Doubt: Assurance on Farmers' Privelege', *Times of India*, March 15.

Sydor, W. J. (1976) 'Genetic Feedback and the Evolution of Plant Resistance', *Agro-Ecosystems*, vol. 3, pp. 55–65.

Symes, D. and A. Jansen (1994) *Agricultural Restructuring and Rural Change in Europe,* Wageningen Studies in Sociology, no. 37, Wageningen; Wageningen Agricultural University.

Tate, W. B. (1994) 'The Development of the Organic Industry and Market: An International Perspective', in Lampkin and Padel (eds) (1994) pp. 11–25.

The Ecologist (1992) *Whose Common Future? Reclaiming the Commons,* London: Earthscan.

Thelwall, A. D. and T. M. Clucas (1992) *Innovation in Agro-biotechnology: Case Study on Plant Breeding Technology,* SAST Project no. 4, Luxembourg: Commission of the European Communities.

Thiele-Wittig, M.-H. (1992) 'Development of Technical Work at UPOV', *Plant Varieties and Seeds,* vol. 5, no. 3, pp. 129–42.

Third World Network (1990) *Return to the Good Earth: Damaging Effects of Modern Agriculture and the Case for Ecological Farming,* Penang: Third World Network.

Thirtle, C. G. and V. W. Ruttan (1987) *The Role of Demand and Supply in the Generation and Diffusion of Technical Change,* Chur. Switzerland: Harwood Academic.

Thurston, H. D. (1992) *Sustainable Practices for Plant Disease Management in Traditional Farming Systems,* Oxford: Westview Press.

Trenbath, B. R. (1974) 'Biomass Productivity in Mixtures', *Advances in Agronomy,* vol. 26, pp. 177–210.

Tribble, J. L. (1995) 'Gene Ownership Versus Access: Meeting the Needs', in MacDonald (ed.) (1995) pp. 97–103.

Trigo, E. and M. Piñeiro (1983) 'Foundations of Science and Technology Policy for Latin American Agriculture', in Piñeiro and Trigo (eds) (1983) pp. 165–73.

Tripp, R. (1989) 'Farmer Participation in Agricultural Research: New Directions or Old Problems?' *IDS Discussion Paper,* no. 256, February. Brighton: IDS

Tripp, R., P. Anandajayasekeram, D. Byerlee and L. W. Harrington (1991) 'FSR: Achievements, Deficiencies and Challenges for the 1990s', *Journal of Asian Farming Systems Associaton,* vol. 1, pp. 259–71.

US Congress (1909) *Report of the Country Life Commission,* Senate Document no. 705, 60th Congress, 2nd Session (1909).

UNCTC (United Nations Centre on Transnational Corporations) (1988) *Transnational Corporations in Biotechnology,* Doc. ST/CTC/61, New York: United Nations.

UNDP (1992) *Benefits of Diversity: An Incentive Towards Sustainable Agriculture,* Environment and Natural Resources Group, Programme Development and Support Division, Bureau for Programme Policy and Evaluation, New York: UNDP.

UPOV (1991) *International Convention for the Protection of New Varieties of Plants of December 2, 1961, as Revised at Geneva on November 10, 1972, on October 23, 1978, and on March 19, 1991,* Geneva: International Union for the Protection of New Varieties of Plants.

USDA (1900) *Yearbook of the USDA, 1899,* Washington, DC: USGPO.

— (1927) *Yearbook of Agriculture, 1926,* Washington, DC: USGPO.

— (1928) *Yearbook of Agriculture, 1927,* Washington, DC: USGPO.

— (1930) *Yearbook of Agriculture, 1930,* Washington, DC: USGPO.

— (1933) *Yearbook of Agriculture, 1933,* Washington, DC: USGPO.

— (1934) *Yearbook of Agriculture, 1934,* Washington, DC: USGPO.

— (1935) *Yearbook of Agriculture, 1935,* Washington, DC: USGPO.

— (1936) *Yearbook of Agriculture, 1936,* Washington, DC: USGPO.

— (1960) *Yearbook of Agriculture, 1960: Power to Produce,* Washington, DC: USGPO.

— (1968) *Yearbook of Agriculture 1968: Science for Better Living,* Washington, DC: USGPO.

— (Study Team on Organic Farming) (1980) *Report and Recommendations on Organic Farming,* Washington DC: USGPO.

van den Belt, H. and A. Rip (1987) 'The Nelson-Winter-Dosi Model and Synthetic Dye Chemistry', in Bijker *et al.* (eds) (1987) pp. 135–58.

van den Bosch, R. (1980) 'The Pesticide Mafia', *The Ecologist,* vol. 10, no. 3, pp. 78–82.

Vandermeer, J. (1989) *The Ecology of Intercropping,* Cambridge: Cambridge University Press.

Vandermeer, J. and B. Schultz (1990) 'Variability, Stability, and Risk in Intercropping: Some Theoretical Explorations', in Gleissman (ed.) (1990) pp. 205–29.

van der Ploeg, J.D. (1990) *Labour, Markets, and Agricultural Production,* Boulder, Colorado: Westview Press.

— (1992) 'The Reconstruction of Locality: Technology and Labour in Modern Agriculture', in Marsden *et al.* (eds) (1992) pp. 19–43.

van Marrewijk, Gisbert A. M. (1995) 'Application of *In Vitro* Techniques in Plant Breeding', in Fransman *et al.* (eds) (1995) pp. 115–36.

van Wijk, J. (1989) 'Plant Varieties Patentable in Mexico', *Biotechnology and Development Monitor,* no. 9, December, p. 20.

— (1990) 'Patents and the GATT', *Biotechnology and Development Monitor,* no. 3, June, p. 24.

— (1993) 'Farm Seed Saving in Europe Under Pressure', *Biotechnology and Development Monitor,* no. 17, December, pp. 13–14.

— (1994) 'Hybrids, Bred for Superior Yield or for Control?' *Biotechnology and Development Monitor,* no. 19, June, pp. 3–5.

van Wijk, J., J. I. Cohen and J. Komen (1993) *Intellectual Property Rights for Agricultural Biotechnology: Options and Implications for Developing Countries,* ISNAR Research Report, no. 3, The Hague: International Service for National Agricultural Research.

Vaughan, D. and Te-Tzu Chang (1992) 'In Situ Conservation of Rice Genetic Resources', *Economic Botany,* vol. 46, no. 4, pp. 368–83.

Vavilov, N. I. (1951) 'The Origin, Variation, Immunity and Breeding of Cultivated Plants', *Chronica Botanica,* vol. 13, no. 1, pp. 1–364.

Vellvé, R. (1992) *Saving the Seed: Genetic Diversity and European Agriculture,* London: Earthscan.

Vine, A. and Bateman, D. (1981) *Organic Farming Systems in England and Wales: Practice, Performance and Implications,* Aberystwyth: Department of Agricultural Economics, UCW Aberystwyth.

Walden, D. B. (ed.) (1978) *Maize Breeding and Genetics,* Chichester: John Wiley and Son.

Walden, H. T. (1966) *Native Inheritance: The Story of Corn in America,* New York and London: Harper Row.

Wallace, H. A. (1919) 'Corn Show Evolution', *Wallace's Farmer,* vol. 44, no. 51, pp. 2509–2524.

— (1934) 'Foreword', in USDA (1934) p. III.

Wallace, H. A. and E. A. Bressman (1949) *Corn and Corn Growing* (5th Edition), London: Chapman and Hall.

Wallace, H. A. and Brown, W. L. (1956) *Corn and its Early Fathers,* Michigan: Michigan State University Press.

Wallerstein, I. (1991) *Unthinking Social Science: The Limits of Nineteenth Century Paradigms,* Cambridge: Polity.

Ward, N. (1993) 'The Agricultural Treadmill and the Rural Environment in the Post-productivist Era', *Sociologia Ruralis,* vol. xxxiii, no. 3/4, pp. 348–64.

Ward, N., J. Clark, P. Lowe and S. Seymour (1993) 'Water Pollution From Agricultural Pesticides', *Research Report,* Newcastle-Upon-Tyne: University of Newcastle-Upon-Tyne, Department of Agricultural Economics and Food Marketing (Centre for Rural Economy).

Watkins, K. (1992) *Fixing the Rules: North-South Issues in International Trade and the GATT Uruguay Round,* London: Catholic Institute for International Relations.

Watts, M. (1990) 'Peasants Under Contract: Agro-Food Complexes in the Third World', in Bernstein *et al.* (eds) (1990) pp. 149–62.
— (1994) 'Living Under Contract: The Social Impacts of Contract Farming in West Africa', *The Ecologist,* vol. 24, no. 4, pp. 130–34.
WCED (World Commission on Environment and Development) (1987) *Food 2000: A Report to the World Commission of Environment and Development,* London: Zed Books.
Weaver, W. (1950) *Inter-Office Correspondence with Chester I. Barnard,* September 21, Rockefeller Foundation Archives.
Webber, H. and E. Bessey (1900) 'Progress of Plant Breeding in the United States', in USDA (1900) pp. 465–90.
Wellhausen, E. J. (1950) 'Rockefeller Foundation Collaboration in Agricultural Research in Mexico', *Agronomy Journal,* vol. 42, no. 4, pp. 167–75.
— (1978) 'Recent Developments in Maize Breeding in the Tropics', in Walden (ed.) (1978) pp. 59–84.
Wellhausen, E. J., L. M. Roberts and E. Hernandez X., with P. C. Mangelsdorf (1952) *Races of Maize: Their Origin, Characteristics and Distribution,* Cambridge: The Bussey Institution of Harvard University.
West, R. C. (1979) *Carl Sauer's Fieldwork in Latin America,* Ann Arbor, Michigan: University Microfilms International (for Syracuse University Dept. Of Geography).
— (ed.) (1982) *Andean Reflections: Letters From Carl O. Sauer While on a South American Trip Under a Grant From the Rockefeller Foundation, 1942,* Westview Press Dellplain Latin American Studies no. 11, Boulder: Westview.
Widdowson, R. W. (1987) *Towards Holistic Agriculture: A Scientific Approach,* Oxford: Pergamon.
Wiebe, G. A. and J. D. Hayes (1960) 'The Role of Genetics in the Use of Agricultural Chemicals', *Agronomy Journal,* vol. 52, no. 12, pp. 685–6.
Wildavsky, A. (1991) 'Public Policy', in Davis (ed.) (1991) pp. 80–101.
Wilken, G. C. (1987) *Good Farmers: Traditional Agricultural Resource Management in Mexico and Central America,* Berkeley: University of California Press.
Wilkes, H. G. (1977a) 'Native Crops and Wild Food Plants', *The Ecologist,* vol. 7, no. 8, pp. 312–17.
— (1977b) 'The Origin of Corn–Studies of the Last Hundred Years', in Seigler (ed.) (1977) pp. 211–23.
— (1977c) 'Hybridization of Maize and Teosinte in Mexico and Guatemala and the Improvement of Maize', *Economic Botany,* vol. 31, pp. 254–93.
— (1992) *Strategies for Sustaining Crop Germplasm Preservation, Enhancement, and Use.* Issues in Agriculture, no. 5. Washington, DC: CGIAR.
— (1997) 'Teosinte in Mexico: Personal Retrospective and Assessment', in Serratos *et al.* (eds) (1997) pp. 10–17.
Williams, J. T. (1988) 'Identifying and Protecting the Origins of Our Food Plants', in Wilson (ed.) (1988) pp. 240–7.
Wilson, E. O. (1978) *On Human Nature,* Cambridge, MA: Harvard University Press.
— (ed.) (1988) *Biodiversity,* Washington D.C.: National Academy Press.
Willits, J. (1942a) *Letter from Willits to Sauer,* 25 June, Rockefeller Foundation Archives, 1.1/393/4633.
— (1942b) *Letter from Willits to Stewart,* 9 March, Rockefeller Foundation Archives, 1.1/391/4632.

— (1942c) *Letter from Willits to Staff*, 2 April, Rockefeller Foundation Archives, 1.1/391/4632.

Winkelmann, D. L. (1991) 'Comments From Management', in CIMMYT (1991) pp. 2–7.

— (1994) 'A View of Quintessential International Agricultural Research', in Anderson (ed.) (1994) pp. 85–95.

Winner, L. (1977) *Autonomous Technology: Technics Out of Control as a Theme in Political Thought*, London: MIT Press.

WIPO (World Intellectual Property Organisation) (ed.) (1987) *Symposium on the Protection of Biotechnological Inventions*, Ithaca, New York, June 4 and 5, Geneva: World Intellectual Property Organisation.

Wirth, T. (1994) 'Testimony of the Honorable Timothy E. Wirth, Counselor, Department of State Senate Foreign Relations Committee, United States Senate. April 12 1994', Posting to Conference <gen.biotech@conf.igc.apc.org>, April 28.

Wolfe, M. S. (1981) 'The Use of Spring Barley Cultivar Mixtures as a Technique for the Control of Powdery Mildew', *Proceedings of the 1981 British Crop Protection Conference*, vol. 1, pp. 233–239.

— (1985) 'The Current Status and Prospects for Multiline Cultivars and Varieties Mixtures for Disease Resistance', *Annual Review of Phytopathology*, vol. 23, pp. 251–73.

Wookey, B. (1987) *Rushall: The Story of an Organic Farm*, Oxford: Blackwell.

Wright, S. (1933) 'The Roles of Mutation, Inbreeding, Crossbreeding, and Selection in Evolution', in Jones (ed.) (1933) pp. 356–66.

WSAA (World Sustainable Agriculture Association) (1993/4) 'Nature Farming Rice Crop Succeeds Despite Record Cold Summer: Story is Front-Page News in Japan', *WSAA Newsletter*, vol. 3, no. 12, Dec. 1993, Jan/Feb. 1994, p. 1.

WTO (World Trade Organisation) (1996) 'Environment and TRIPS. Committee on Trade and Environment, WT/CTE/W/8', *Trade and Environment*, 11 March (first published as a Restricted Document, 8 June 1995), Geneva: WTO.

Yoxen, E. (1983) *The Gene Business: Who Should Control Biotechnology?* London: Pan Books.

Zohary, D. (1970) 'Centers of Diversity and Centers of Origin', in Frankel and Bennett (eds) (1970) pp. 33–42.

Zuckerman, B. M., M. B. Dicklow, G. C. Coles, R. García E. and N. Marban-Mendoza (1989) 'Suppression of Plant Parasitic Nematodes in the Chinampa Agricultural Soils', *Journal of Chemical Ecology*, vol. 15, pp. 1947–55.

Index